When Things Grow Many

When Things Grow Many

Complexity, Universality and Emergence in Nature

L. S. Schulman

OXFORD
UNIVERSITY PRESS

OXFORD
UNIVERSITY PRESS

Great Clarendon Street, Oxford, OX2 6DP,
United Kingdom

Oxford University Press is a department of the University of Oxford.
It furthers the University's objective of excellence in research, scholarship,
and education by publishing worldwide. Oxford is a registered trade mark of
Oxford University Press in the UK and in certain other countries

Impression: 2

Published in the United States of America by Oxford University Press
198 Madison Avenue, New York, NY 10016, United States of America

British Library Cataloguing in Publication Data

Data available

Library of Congress Control Number: 2021936303

ISBN 978-0-19-886188-1 (hbk.)

DOI: 10.1093/oso/9780198861881.001.0001

Printed and bound by
CPI Group (UK) Ltd, Croydon, CR0 4YY

Dedicated to
my wife, Claire

Preface

This is a book about *statistical mechanics*, but not so much about statistical *physics*. The distinction is in the applications, not the methods. The central idea is that when dealing with objects that have a large number of constituents qualitatively new phenomena emerge. This is a message one can take from physics and apply to a myriad of topics. Much, but not all, of the material is based on courses given at Clarkson University and the Georgia Institute of Technology. The presentation is aimed at advanced undergraduates and graduate students, including—in fact, *especially*—those not directly involved in traditional areas of physics. The general theme is that statistical mechanics is useful in many areas of science, both those where the subject developed, such as chemistry, as well as more recent applications in biological, financial, linguistic, social and other fields.

The organization of the book is chaotic—perhaps as befits the subject matter. At one point I thought it best to sort things out along topics, biology here, sociology in its own chapter, and so forth. But as I wrote and thought, the areas became so intertwined that I abandoned this plan. There are applications of power laws to linguistics, sociology, economics, and more. Ditto for maximum entropy methods; ditto for self-organized criticality and so on. So in the end, I jump around a bit, sometimes focusing on the application, sometimes on the method. This means that if you're interested in economics—and although you might care about firefly synchronization, it's not your priority—you'll have to hunt around a bit. I have tried to make the index compensate for the lack of clear boundaries.

The first eight chapters are an introduction to the methods of statistical physics, although many examples from Nature are provided along the way (in Appendix B more formal results in statistical physics are given). The problems dealt with are simplified, but the methods are those used in richer situations. And richer situations follow; quite a few individual applications are given in the final chapters although they are only a small fraction of those treated in the literature. (Appendix A presents the notation we use.)

As to the mathematical level, I have been free in the use of equations, although in many cases footnotes or appendices are provided with some of the necessary background. The goal is to make this readable to as large an audience as possible. The footnotes, which are plentiful, serve three purposes. They fill in material that may be unfamiliar to the reader, they cover advanced material whose details often can be ignored, and finally there are ordinary footnotes, guides to citations or extra material whose insertion in the main text would be a distraction.

And now for the mea culpa: First, I confess to a certain amount of myopia. I did not read every paper on every subject. I didn't even read a significant fraction of them. So in many cases I'm out of date, in many cases I report what I'm familiar with, omitting significant other work. I apologize to those I slight. There are also value judgements: The data are inadequate for the conclusion, or some article misses the point, and so forth. I think I'm right, but maybe not. So take those "conclusions" with a grain of salt. It also must be said that this book has an emphasis different from books on a particular subject matter. I'm not going to teach anyone ecology, linguistics, finance or any specific field. Rather this *should* teach approaches to a problem. (Witness, the number of times I express ignorance. And ignorance can take two forms: my own or that of the human race.) There's also an apology due over the selection of MATLAB™ for the sample programs. I once knew Fortran and a smattering of other computer languages, but in the last few years I've focused on MATLAB™. Yes, I know that (for example) Python is (1) free and (2) superior in some respects (Airy functions as of 2015). True, some books give generic programs, but I'll have to leave such translations to readers. There is also the issue of classification. Does linguistics (Zipf's law) belong in the social sciences? Does an electronic application of the Kuramoto model belong in biological sciences? Maybe yes, maybe no, but please don't worry about it.

And finally it's likely that some instructors will develop new topics that I don't treat. Tell me about it![1] There might be a second edition. This also goes for interesting applications that don't appear in the present volume. I'm sure there are topics that are omitted through my own ignorance.

---

[1]My email address is schulman (at) clarkson.edu and variations are given on both of my (out of date) Clarkson websites. I can also be reached at schulman137 (at) gmail.com.

Acknowledgments

I am grateful to people at all institutions where this course has been taught and especially to the students who participated in the presentation of these ideas. Those students who contributed were Rachel E. Barker, Jonathan Brassard, Pablo Bravo, Isabel Dengos, Lawson Glasby, Nick Gravish, Conner J. Herndon, Shane J. Jacobeen, Matthew W. Jensen-Bukovinsky, Shengkai Li, Michael Lovall, Megan G. Matthews, Christopher R. McBryde, James (Cordy) McCord, James McInerny, Karan P. Mehta, Collin Reff, Karan Shah, James Taylor, Jacob Troupe, and Mackenna Wood. In addition many colleagues have contributed to my own understanding as well as posing questions to which I often had no answer. Those who have directly influenced this work are John Beggs, Bruce Boghosian, Marcos da Luz, Bernard Gaveau, Dan Goldman, Nicholas Gotelli, John Harte, Ady Mann, Eva Mihóková, David Storch, Lucas Wetzel, and Marina Wosniack

Contents

1

Introduction

You can often solve a one-body problem, $r(t)$, the trajectory of a particle as a function of time. You can sometimes solve a two-body problem, and every once in a while you can find the paths of three particles.[1] But what happens if you have 10^{23} of them?

It turns out that

More is simpler.

That doesn't mean you can solve the 10^{23}-body problem; it does mean you can make statements about the collective behavior of a large number of particles, statements that are informative, even though they lack detail.

And once you've learned to do this for gases and physical systems, you can often do it for other systems having a large number of constituents, for example neurons in the brain, or starlings in flight, or stars in galaxies, or people buying houses in a neighborhood, or stock market prices, or epidemics, or evolution .... The list goes on.

There is a second phenomenon that will also interest us. This is epitomized in the title of an article by Phil Anderson [9],[2]

More is different.

Anderson was opposing the (then) accepted wisdom, for example the idea that sociology is simply applied psychology or solid state physics an application of elementary particle physics. Here the point is complexity and emergence. At the end of his article he quotes a (probably fictitious) exchange between Scott Fitzgerald and Ernest Hemingway: SF: The rich are different from us. EH: Yes, they have more money. Leaving aside positive and negative nuances in this exchange, it is nevertheless clear that in the United States annual salaries of $\$10^6$ and $\$10^4$ yield qualitatively different experiences.

The first part of this book will be devoted to how some of these ideas can be applied to gases, social systems, (pseudo-)epidemiology and ferromagnetism. In these

---

[1]Apropos of *solving* there's an amusing history of *not* solving. I've heard the following sequence from various sources, including the names of the people who showed what can't be solved. Here's one version: "In eighteenth-century Newtonian mechanics, the three-body problem was insoluble. With the birth of general relativity around 1910 and quantum electrodynamics in 1930, the two- and one-body problems became insoluble. And within modern quantum field theory, the problem of zero bodies (vacuum) is insoluble. So, if we are out after exact solutions, no bodies at all is already too many!" This quote is ascribed to G. E. Brown and appears in [149].

[2]P. W. Anderson, American, Nobel Prize, condensed matter physics, Bell Labs, Princeton, 1923–2020. Held strong opinions and told you about them. In this article it's clear that Anderson was annoyed by his perception of the arrogance of particle physicists.

When Things Grow Many. Lawrence S. Schulman, Oxford University Press.
© Lawrence S. Schulman (2022). DOI: 10.1093/oso/9780198861881.003.0001

examples considerable simplification has been made in order to develop notions used in more realistic applications.

1.1   Building

The plan is to build gradually to situations where the collective behavior becomes more and more complicated. First we will consider situations where you simply have a lot of participants, but they seldom interact with each other. If there are N participants we will see that fluctuations—deviations from the simplest predictions—go like $\sqrt{N}$. The next level of complexity has the participants/particles/constituents interacting with one another. Now you can get "phase transitions," sudden changes in behavior as an external parameter is varied. Finally you can get critical points, situations where the details of the interaction really do seem to disappear and many systems—systems that have different microscopic interactions—all behave in the same way. The archetype of this behavior was noted by Guggenheim many years ago.[3] He found that when certain properties of fluids are plotted in a particular way, they all fall on a common curve. See Fig. 1.1. To explain that figure requires a bit of background on critical phenomena. A gas and a liquid may seem different, but there is a temperature and a pressure where the two concepts merge and one describes the system as a "fluid." For steam and water if you heat H_2O to $647\,K$ (about $705°F$) and keep it at a pressure of 218 atmospheres, the difference between liquid and gas disappears. Below that temperature there is a phase transition, namely a temperature and pressure at which the two phases, liquid and gas, can coexist. When they do, there are two different allowed densities for the substance.

What Guggenheim did was look at the corresponding critical points for eight different substances, Ne, A, Kr, etc.[4] For each of them there is a critical temperature and density at which liquid and gas are identical; call them T_c and ρ_c. He then looked at temperatures below T_c (where there *was* a phase transition) and found the densities of the liquid and of the gas at that particular temperature. The plot shows the scaled quantities, T/T_c and ρ/ρ_c. For each temperature (below T_c) there are two values of ρ/ρ_c and this is what is plotted.

The absolutely remarkable feature of this graph is that for all eight substances and near the critical point the points fall on the same curve.[5] A more subtle feature concerns the curve itself: it turns out that $\Delta\rho \sim \text{const} \cdot (\Delta T)^{1/3}$ (for $\Delta T \geq 0$). That exponent $(1/3)$ remained a mystery for more than 20 years and we will not derive it, but we will see how it *could* come about.[6]

It also happens that at the critical point the fluctuations are larger. For a simple model of a ferromagnet it's not difficult to show that the relative fluctuations go like

---

[3]Guggenheim considered his graph an illustration of the "law of corresponding states." This is still some way from recognizing renormalization theory. For a history of the subject see Kadanoff [121]. See also a history of earlier years in [120].

[4]Ne is neon, A argon, Kr krypton, Xe xenon, N_2 molecular nitrogen, O_2 molecular oxygen, CO carbon monoxide, and CH_4 methane. At room temperature and 1 atmosphere all these are gases.

[5]They also are pretty close away from the critical point, but that's another matter.

[6]It's not exactly 1/3, but it's close. See Footnote 2, Chapter 8.

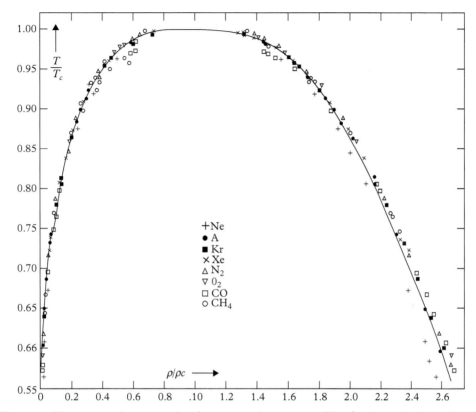

Fig. 1.1 Illustrating the principle of corresponding states. The fluids studied are Ne, A, Kr, Xe, N_2, O_2, CO, and CH_4.[4] At each temperature at which there is a first-order phase transition, two densities are plotted: that of the liquid and that of the gas. Both density and temperature are normalized by their critical values, that is, for each element or compound the actual values are divided by the pair of values at the critical point (where there is no longer a distinction between liquid and gas). This figure is adapted from [92].

one over the *fourth* root of N, much larger than the usual (one over the) square root. (This will be derived in Sec. 5.3.)

The fact that many critical point exponents (like the $1/3$ just noted) are the same for a variety of substances, substances for which the detailed interactions can be different, is a manifestation of *universality*.

At this point the other mysteries enter. First, a lot of systems seem poised at criticality. One of the characteristics of criticality is the appearance of *fractal*[7] dimensions in the description. (For example, near the fluid transition mentioned above, the surfaces of constant density have fractal structure.) Many systems exhibit this behavior (e.g., the coast of England is of fractal dimension). Nevertheless, *criticality is non-generic*, meaning it requires special values of the governing parameters to occur. Why

---

[7]See Appendix C for a tutorial on fractals.

then is it so common? One answer is self-organized criticality (SOC). The idea is that a system, on its own, will adjust its parameters to achieve criticality. (More on this later.)

Closely related to fractality and criticality is the occurrence of *power laws*. This will also be discussed more fully later, but let me give one example that I have recently encountered in my own research. The question is, how rare is a rare gene? Naively you would expect that the distribution of genes is "normally distributed," that is, it is proportional to $\exp(-\text{const} \cdot (x - x_0)^2)$ (x is the probability of finding a particular gene and there is a central value x_0); after all that's how height is distributed and height is governed by genes.[8] But it turns out that the dropoff is *much* slower. (For large deviations a power law is slower than an exponential, and smaller yet than the exponential of a square.) I was studying evolution in the "Tangled Nature" model[9] and found that the probability distribution for genomes was not normal, but instead drops off like $\Pr(x) \sim x^{-\alpha}$, with α less than 2.[10] This is a significant issue since it could contribute to rapid evolution (including antibiotic resistance), so I was eager to see if this property held in Nature, not only in the Tangled Nature model. I soon learned that indeed it did—and that questions of this sort had been investigated once there was sufficient information available on natural DNA. See [77]. So rather than a discovery about Nature I had discovered another way in which the Tangled Nature model gave natural results; and, as remarked earlier, the long-tailed distribution gave us something more to worry about with respect to antibiotic resistance—the presence of rare genes means that antibiotics will wipe out *most* bacteria, but there might be a few that resist the onslaught. And these few will rapidly multiply in the absence of competition.

Two other words often make their appearance, when "things grow many." They are *complexity* and *emergence*. Complexity is an ill-defined notion that seems to be common to the kind of systems we'll be looking at: ecology, cell dynamics, neuron behavior, financial markets, …. Complexity is a bit like obscenity (to paraphrase US Supreme Court Justice Potter Stewart): you know it when you see it, but it can be difficult or impossible to come up with a clear definition.[11] Many of the systems where statistical mechanics gives useful insights are of this sort. One of the puzzles of complexity is the side-by-side features of robustness and fragility. To quote Carlson and Doyle [37],

> The [Boeing-]777 is robust to large-scale atmospheric disturbances, variations in cargo loads and fuels, turbulent boundary layers, and inhomogeneities and aging of materials, but could be catastrophically disabled by microscopic alterations in a handful of very large-scale integrated chips or by software failures.

This is another aspect of "more": More may induce fragility: robust with what's anticipated, fragile with respect to what's not solvable—even if anticipated.

---

[8]Height also depends on nutrition but presumably the genetic part alone is normally distributed.

[9]The Tangled Nature model is due to H. J. Jensen and collaborators. See [97; 197] and Sec. 10.3.

[10]The variable x is here the (normalized) number of appearances of a particular genome. Sometimes the power in the "power law" is such that $\int_1^\infty dx\, x^{-\alpha}$ does not converge ($\alpha \le 1$), implying that there must be some large-x cutoff.

[11]See Appendix H for some of my own ideas on this subject.

A related notion, implicit in Anderson's article [9], is *emergence*. This is typified by the flight of a flock of starlings: you can give simple rules governing the behavior of a bird, making reference only to a few of the bird's neighbors, and from those rules *emerge* well-defined collective behavior, in this case flocking. Presumably this is what happens with starlings, minimizing the amount of instructions that need to be genetically encoded as instinct.[12]

The second part of this book was originally planned around phenomena: linguistic, stock market bubbles, and so on. However, as ideas began to merge it seems that $1/f$ noise and Zipf's law (to take only two examples) have more in common (or *might* have more in common) than do the underlying fields. For this reason I have taken a more quantitative approach, looking for commonalities in mathematical properties rather than in underlying phenomena. It's not my choice, but Nature's.

As a result, if you're, say, interested in power laws for genes you go to the "power law" chapter first and genes would be one of the applications. Similarly for someone who cares about segregation, the Schelling model might be relevant and is found in the section on phase transitions in social phenomena—it's close to an Ising model after all. And some material might appear in more than one place; for example, in our percolation examples I talk a bit about "self-organized criticality" (SOC), but later in the book discuss it more fully. In this connection, you may find the index useful.

---

[12]The "presumably" at the beginning of this sentence covers all its assertions. Simple rules can account for flocking, yielding a sufficient condition but not a necessary one. Moreover, with more information about animal learning, the issue of instinct is no longer so clear.

2
Ideal gas

Applications discussed in this chapter: Ideal gas, Independent degrees of freedom

---

In a way this is the simplest example. You have a gas, but you consider each particle's motion on its own. As usual there's a bit of a fairy tale here. On the one hand, you assume there's enough interaction so that the gas particles reach equilibrium at some temperature, T. On the other hand, we pretend that for our calculation they don't talk to each other (i.e., interact with each other). What we're looking to obtain is the well-known "ideal gas law,"

$$PV = Nk_BT, \tag{2.1}$$

where P is the pressure in the gas, T the temperature, V the volume, k_B Boltzmann's constant ($\sim 1.38 \times 10^{-23}$ J/K) and N the number of atoms.[1]

We look at the kinematics of this ideal gas. Our aim is to find the pressure the atoms exert on a wall. Let the "wall" be the y-z plane, so that momentum transfer with that wall will only take place if the particles have positive velocity in the x direction. And we only need to consider a small area of this wall, call it A, so that we could imagine on a larger scale the wall is curved, but locally it looks flat. For pressure we divide by A anyway, so as long as we don't go down to atomic size areas, we're fine.

The force on this area arises from all the impacts. We focus on a short time interval, Δt, which we'll eventually divide by, so it too will disappear. Now assume that the particles have mass m. If a particle is reflected off the wall and had velocity in the x-direction, v_x, then the change in momentum is $\Delta p = 2mv_x$. Velocities in the other directions are irrelevant. The contribution to the momentum change during the interval Δt of this one impact is thus $\Delta p = 2mv_x$.

Not all particles will hit the area A of the wall—only those whose x-velocity is such they are within a distance $v_x\Delta t$ of the wall. To handle this we introduce a number density (# per volume) of atoms, namely $n(v_x)dv_x$ is the number of particles per volume that have velocity between v_x and $v_x + dv_x$. The force exerted by those with this velocity within this interval will be $\frac{\Delta p}{\Delta t}$, or

---

[1] Some readers may be more familiar with the form $PV = nRT$ (Pressure × Volume = [number of moles] × [Gas Constant] × Temperature), which is the same as Eq. (2.1), with $nR = Nk_B$. In this equation # moles = $n = N/N_A$, with N the total number of particles, N_A Avogadro's number $\sim 6.022 \times 10^{23}$ (the number of atoms in a mole), and $R = k_B N_A \sim 8.314$ J/(mol-K). This language is more common in chemistry.

When Things Grow Many. Lawrence S. Schulman, Oxford University Press.
© Lawrence S. Schulman (2022). DOI: 10.1093/oso/9780198861881.003.0002

$$\frac{\text{combined momentum transfer for velocity between } v_x \text{ and } v_x + dv_x}{\Delta t}$$

$$= \frac{\text{momentum change of 1 impact} \times \text{number per volume of relevant velocities} \times \text{volume}}{\Delta t}$$

$$= \frac{2mv_x \times n(v_x) \, dv_x \times (Av_x \Delta t)}{\Delta t}. \tag{2.2}$$

("Relevant" means the interval $[v_x, v_x + dv_x)$.) To find the total force we integrate over v_x, so that the pressure, force per unit area, is

$$P = F/A = \int_0^\infty 2mv_x \, n(v_x) v_x \, dv_x. \tag{2.3}$$

(Note that the integral goes over positive v_x only.) One can define an average velocity that looks very much like the integral in Eq. (2.3). In particular the average of *the square of the velocity in the positive x-direction* is

$$\langle v_x^2 \rangle = \frac{V}{N/2} \int_0^\infty n(v_x) v_x^2 dv_x, \tag{2.4}$$

where the "2" (in $N/2$) comes because only half the particles are being counted, those moving in the direction of the wall. It has also been assumed that the distribution function $n(v_x)$ is the same for negative v_x as for positive values. Putting Eqs. (2.3) and (2.4) together yields

$$P = 2m \frac{N/2}{V} \langle v_x^2 \rangle. \tag{2.5}$$

Of course the x direction was only adopted for convenience. What really interests us is the total velocity, $v^2 = v_x^2 + v_y^2 + v_z^2$ (where v with no subscript is the magnitude of the vector v). Assuming isotropy—all directions are equivalent—we have $\langle v_x^2 \rangle = \frac{1}{3} \langle v^2 \rangle$ so that we get

$$P = \frac{1}{3} m \frac{N}{V} \langle v^2 \rangle. \tag{2.6}$$

The quantity $m \langle v^2 \rangle$ is twice the average kinetic energy of the particles, and since the gas is "ideal" it's the only energy they have. Letting $\langle \epsilon \rangle$ be the average energy per particle $(= \frac{1}{2} m v^2)$ we have

$$P = \frac{2}{3} \frac{N}{V} \langle \epsilon \rangle. \tag{2.7}$$

This equation is of the same form as Eq. (2.1) *if* the average energy and the temperature are identified, up to proportionality. In particular for our kinetic derivation to reproduce the ideal gas equation, we must set

$$\langle \epsilon \rangle = \frac{3}{2} k_B T. \tag{2.8}$$

But this is perfect. The relation Eq. (2.8) is what is known as *equipartition*,[2] namely that each degree of freedom in the system has, on the average, energy $\frac{1}{2}k_B T$. There are three kinematic degrees of freedom for each ideal gas particle, hence the 3 in Eq. (2.8). The result of all these manipulations is Eq. (2.1), $PV = Nk_B T$.

2.1   Fluctuations of the ideal gas

Although on the average $PV = Nk_B T$, individual strikes of the walls are effectively random. So sometimes the pressure will be higher, sometimes lower. To estimate the fluctuations I'll use a random walk picture. As a by-product I'll present a cartoon proof of the most salient property, the $\sqrt{N}$ behavior of (uncorrelated) fluctuations.

Suppose the box of our example has a partition down its middle, so there is volume $V/2$ on each side. *On the average* there will be $N/2$ particles on each side, so that we still have $P\frac{V}{2} = k_B T\frac{N}{2}$; the pressure will be the same as for the full volume. However, if we want to know the *fluctuations* in pressure, we'll need to know the fluctuations in N. We immediately[3] have $\Delta P/P = \Delta N/N$, so we are left with determining ΔN.

Since the particles don't talk to each other the choice of which side of the partition a given particle goes to is random, independent of all the others. Let each particle (molecule, atom, whatever) be labeled j and let X_j be its choice, $X_j = +1$ when it's on one side of the partition, $X_j = -1$ when it's on the other. Now the number on the plus side is $\sum_j \mathcal{T}(X_j > 0)$, where $\mathcal{T}$ is a *truth function*, and takes the value 1 when its argument is true, 0 otherwise. Therefore[4]

$$\Delta N = \left(\frac{N + \Delta N}{2}\right) - \left(\frac{N - \Delta N}{2}\right) = \sum_j \mathcal{T}(X_j > 0) - \sum_j \mathcal{T}(X_j < 0) = \sum_j X_j. \qquad (2.9)$$

Since for each X_j there is a 50% chance of being on one side or the other it's clear that the *expectation value* of ΔN is zero, that is,

$$\langle \Delta N \rangle = \sum_j \langle X_j \rangle = 0. \qquad (2.10)$$

---

[2]The equipartition principle is one reason that classical mechanics had to be replaced by quantum mechanics. The electromagnetic field inside a cavity has an infinity of degrees of freedom and should therefore have infinite energy. It does not, and it was Max Planck who was led to the idea that by quantizing that field you could get a dropoff, a weakening of equipartition for high-energy excitations. The relation of light quanta—photons—to their energy is $E = h\nu$, with ν the frequency of the radiation and h Planck's constant, in honor of Planck. The idea of quantizing the energy of photons had later been used by Einstein (in connection with the photoelectric effect) and was (strangely) the reason the Swedes gave him the Nobel Prize.

[3]Well, maybe it's not so immediate. Suppose there are $\frac{N+\Delta N}{2}$ on one side and $\frac{N-\Delta N}{2}$ on the other. Then $(P \pm \Delta P)\frac{V}{2} = k_B T\left(\frac{N \pm \Delta N}{2}\right)$. Now subtract the minus equation from the plus one, yielding $V\Delta P = k_B T\Delta N$. To lowest order we still have $PV = k_B TN$, so that (again to lowest order) dividing by the latter equation we get $\Delta P/P = \Delta N/N$. By the way, for this calculation N or ΔN can be odd, leading to half-integer "number of particles." This is not a problem and could be dealt with by insisting that N be even—but for simplicity I've allowed the half-integer values.

[4]See Footnote 3 if you're worried about the 2's.

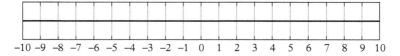

Fig. 2.1 One-dimensional walk. After N steps the average position is zero.

Notice that I have used the fact that the expectation value ($\langle \cdot \rangle$) of a sum is the sum of the expectation values.[5] The zero I have just obtained is the average over many experiments. Clearly in any one experiment you are unlikely to have things break exactly half-half. So you expect a range of values, different in each experiment, but which average to zero. To find this range you look at the square of ΔN, that is,

$$\langle (\Delta N)^2 \rangle = \left\langle \left[\sum_j X_j \right]\left[\sum_{j'} X_{j'} \right] \right\rangle = \sum_j \langle X_j^2 \rangle + \sum_{j \neq j'} \langle X_j X_{j'} \rangle = \sum_{j=1}^{N} 1 + 0 = N. \quad (2.11)$$

The first equality (in Eq. (2.11)) is the definition; the second the distributive law; the third (with a bigger equal sign than the others) needs more elaborate justification. First consider an individual $\langle X_j^2 \rangle$. Since X_j^2 is always 1, its average is also 1. Therefore every term in $\sum_j \langle X_j^2 \rangle$ gives a 1 and there are N such terms. Next we consider $\langle X_j X_{j'} \rangle$ for $j \neq j'$. Since the random variables X_j and $X_{j'}$ are independent[5] of one another, $\langle X_j X_{j'} \rangle = \langle X_j \rangle \langle X_{j'} \rangle = 0 \cdot 0 = 0$. Every such term is zero.

For convenience let me define $\delta N \equiv \sqrt{\langle (\Delta N)^2 \rangle}$. (Note that δN and ΔN are conceptually different quantities. ΔN is a random variable, while δN is a number.) Then we've found

$$\delta N = \sqrt{N}. \quad (2.12)$$

What this tells you is not to worry about the pressure fluctuations in your bicycle tire. Suppose your pressure measurements involve 10^{20} particles (less than a thousandth of a mole), then $\delta N / N$ is about 10^{-10}, which is the same as your (relative) pressure variation.

The general message is that fluctuation of independent actors (agents, atoms, ...) goes like the square root of their number.

This feature has been established by Eq. (2.11), but you still may feel uncomfortable that something that averages to zero (ΔN, the random variable) nevertheless has a square whose average grows. Let me give you a cartoon demonstration of this fact. Let $S_N \equiv \sum_{j=1}^{N} X_j$, which will be the position of a random walker (on a straight line) after n steps to the left or right. (Note that X_j still takes values ± 1 with equal probability.) It's clear from Fig. 2.1 that *on the average* our walker gets nowhere: $\langle S_N \rangle = 0$. But let's look at the square of S_N (which was computed in Eq. (2.11)). We suppose that, due to fluctuations, at step n the position of the walker is m.

The square of S_N is depicted in Fig. 2.2—it's a square. Whatever the N value, we've assumed that the walker has gotten to m, so the square of S_N can simply be represented by a square of side m. Now two things can happen. The next step can be

[5]See Appendix D, an introduction to probability theory.

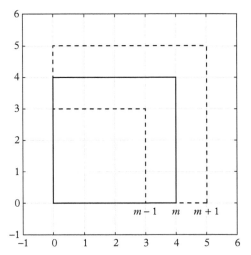

Fig. 2.2 Two-dimensional view of a one-dimensional walk. Although the average of the position doesn't change, the average of its *square* does.

to the right or to the left. If X_{N+1} (the next step) is +1, then S_N becomes the larger square,[6] as shown by the dashed line(s) of length $m + 1$ in Fig. 2.2. If the next step is to the left, you would get the smaller square shown in Fig. 2.2, with line(s) of length $m - 1$. The change in *the expected value* of S_N^2 is thus

$$\langle \Delta S_N^2 \rangle = \frac{1}{2} \cdot \left[(m + 1)^2 - m^2 \right] + \frac{1}{2} \cdot \left[(m - 1)^2 - m^2 \right] = 1, \tag{2.13}$$

so although $\langle S_N \rangle$ does not increase, $\langle S_N^2 \rangle$ *does*, averaging an increase of 1 on each time step.

---

[6]In Fig. 2.2 I take $m > 0$ but the actual equation, Eq. (2.13), is correct even if $m \leq 0$.

3

Rubber bands

Applications discussed in this chapter: Phase transitions, Mean field theory, Corrections to mean field theory

---

The game below has actually been played. It is a simple example of a phase transition.

3.1 The game

Each person in a large group is given a rubber band. It can be put on the left wrist or the right wrist—initially an arbitrary choice. Next, all persons are assembled and are broken into groups of three. (For simplicity assume the total number of people is a multiple of three.) Here is the rule on changes: If everyone in a given group (of three) has the rubber band on the left wrist, do nothing. Ditto for the right wrist. But if in some group it's two to one, go with the majority. The person in the minority moves the rubber band to the other wrist. The group breaks up and new groups are formed. This continues until ... until there's nothing to do. In fact, rather quickly everyone will have the rubber band on the same wrist. Which wrist will it be? We'll see.

3.2 Analysis

Let the total number of participants be N. Let p_L be the probability that there is a rubber band on an individual's left hand; in other words, $p_L = \frac{\#\text{ on left wrist}}{N}$. Similarly for p_R (R is right); they obviously sum to 1. What is the probability that a particular group has all left-wristers? All three members need to have the rubber band on the left, so it's just p_L^3. Similarly for three on the right. What about two lefts and one right? If you would say which of the three has it on the right, then the probability would be $p_L^2 p_R$, but you don't say. So there are three ways this could happen. Thus the probability of such a group is $3p_L^2 p_R$. Similarly for two rights.

Next look at transition probabilities. Designate the new value (after an encounter) with a prime. Thus

$$p_L' = p_L^3 + 3p_L^2 p_R \quad \text{and} \quad p_R' = p_R^3 + 3p_R^2 p_L. \tag{3.1}$$

In words, the new probability that a given person's rubber band is on the left hand is the sum of the probabilities that all three in the group already had it on the left hand

When Things Grow Many. Lawrence S. Schulman, Oxford University Press.
© Lawrence S. Schulman (2022). DOI: 10.1093/oso/9780198861881.003.0003

plus the probabilities that only one of them had it on the right hand. (Note that the new values again sum to 1.)

Now we simplify. Drop the subscript and let $p \equiv p_L = 1 - p_R$. The iteration equation becomes

$$p' = p^3 + 3p^2(1-p) = p^2(3-2p). \qquad (3.2)$$

To analyze this we first seek fixed points, that is, values of p, call them $\{\bar{p}\}$, such that $p' = p$. These are interesting because they must include the values for which there are no longer any changes—a situation I claimed was always the outcome of this "game." The fixed point requirement implies

$$\bar{p} = \bar{p}^2(3-2\bar{p}). \qquad (3.3)$$

Obviously one such solution is $\bar{p} = 0$. If you want other solutions you can divide Eq. (3.3) by $\bar{p}$ and obtain a quadratic equation that is easy to solve. For the equation divided by $\bar{p}$ you get $\bar{p} = \frac{3}{4} \pm \frac{1}{4}$, which is to say that the other solutions are 1/2 and 1 (which will later become obvious).

Does this tell you what Eq. (3.2) implies? Not quite, but it's almost the whole story. To see that, let's examine a finer property of the fixed points—their stability. That is, if I start near some $\bar{p}$, will I go toward it or away? This is very similar to putting a ball at the top of a hill or in a valley. In a valley, if you displace the ball, it will roll back to where it was; it's stable. On the top of a hill, it will roll away; it's unstable. (Conservation of energy is irrelevant in the present situation. In the valley excess energy is absorbed by friction; on a hill energy is drawn from the potential energy of being on a hill. For the stochastic process, which is what p undergoes, there is no conservation law.) To study stability quantitatively, let $\delta p = p - \bar{p}$. Then from Eq. (3.2) we have

$$\delta p' \equiv p' - \bar{p} = (\bar{p} + \delta p)^2 \left(3 - 2(\bar{p} + \delta p) \right) - \bar{p}. \qquad (3.4)$$

This is less of a mess than it looks, since terms not involving δp go away—which is because $\bar{p}$ is a fixed point. Expanding, subtracting and writing the result as a power series in δp gives

$$\delta p' = (6\bar{p} - 6\bar{p}^2)\delta p + (3 - 6\bar{p})(\delta p)^2 - 2(\delta p)^3. \qquad (3.5)$$

The first fixed point that we examine is $\bar{p} = 1/2$. For this Eq. (3.5) becomes $\delta p' = (3/2)\delta p$, with higher powers of δp neglected. Thus δp *increases* in magnitude on each time step. This implies instability. For the other fixed points, the term multiplying δp vanishes and we need to look to the next higher power. For $\bar{p} = 0$ the deviation vanishes quite rapidly, like $3(\delta p)^2$. Similarly at the other end you get the same dependence, except that because you're approaching from below it's $-3(\delta p)^2$. Since it's meaningless to go outside the interval $[0,1]$ this completes the analysis.[1]

---

[1] If p starts in $[0,1]$ it stays there, and the iteration Eq. (3.2) converges to one of the fixed points. Outside the interval $[0,1]$, p is unphysical. Nevertheless, the iteration converges for $1 \leq p \leq \frac{1+\sqrt{5}}{2} \equiv p_c$, the golden ratio. Moreover, for $p \in (1, p_c)$ the iteration converges to 1 or 0. Call the locale of the switches a_n. (The switch goes from 1 to 0 or from 0 to 1. Exactly at the switch you get the unstable attractor, 1/2.) Then $\frac{a_{n+1} - a_n}{a_n - a_{n-1}} = \frac{1}{6}$, which is reminiscent of the Feigenbaum ratio (but not the same).

Our earlier questions are now answered. If more than half of the people initially put the rubber band on the left wrist, the final state will be *all* left. Similarly for right. Half-and-half: unlikely. This is the conclusion of the mean field theory we have used until now, and, according to it, the left–right issue is settled.

That's the simplest theory. There are two reasons this might not work. The first is mathematical: I've ignored fluctuations. What happens if initially there is a majority of left-handed people but they mostly form uniform groups with all three persons left-wristers, while the few not in such groups are forced to switch. At an early stage this could change the balance. The second is social. What if people prefer to form groups involving their friends. So the mixing is not truly random. You may have persistent small groups of non-conformists for a relatively long time. (This is why epidemiology can be difficult, to say nothing of politics.)

3.3 Simulation

You can simulate the entire rubber band game or, more simply, carry out our analysis-based iteration. I've done the latter using MATLAB™. The equal sign in program-ming generally has a meaning different from that in the usual mathematics. It means set the object on the left equal to the quantity on the right.[2] The following program shows the results of iterating the equation $p' = p^2(3-2p)$ many times. Lines (or portions of lines) preceded by a symbol "%" are comments.

```
function rubber_band
% Rubber band is on left or right hand.
% Groups of 3 meet and majority wins.
% The process continues.
dp=.001;
p0=0:dp:1;         % These are the initial conditions for the iterations.
                   % The matlab expression means that p0 is the set
                   % {0, .001, .002, ..., .999, 1}.
np=length(p0);     % Number of elements in the set p0.
toler=1e-6;        % Tolerance (the desired degree of precision).
conv_time=zeros(1,np);pf=conv_time;   % Predefining saves time in matlab.
            % conv_time stands for convergence-time; pf is p-final
for k=1:np, % a loop, ended by (the second) "end"
    er=1;      % Error (will iterate until er < toler)
    p=p0(k);   % The initial condition for this iteration.
            % k-th element of the set p0.
    niter=0;   % Start the convergence count (number of iterations).
```

A similar situation obtains for $p < 0$. There are other peculiarities, such as the fact that $1 + \frac{-3+\sqrt{17}}{4} \approx$ 1.2807764 is invariant; in other words, that $\bar{p} + \delta p$ (with $\delta p = \frac{-3+\sqrt{17}}{4}$) might be a fixed point of the derivative equation although not a solution to the original.

[2]There can be another meaning to the equal sign as well. A statement such as a=b will yield a logical true (1) or false (0), depending on whether $a = b$ or not. (In MATLAB™ there would be a double equal sign, a == b.)

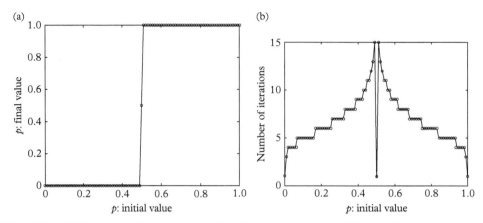

Fig. 3.1 (a) The mean field prediction: if initially p is less than $1/2$ the ultimate value is 0; if it's more than $1/2$, the value is one. If it's exactly $1/2$, the mean field prediction is $1/2$, despite the fact that it's unstable. But there is structure in how the system approaches this limit. (b) The number of iterations needed to reach within 10^{-6} of $\bar{p}$. Note the increase in value as $1/2$ is approached.

```
while er>toler,        % Another loop, conditional on er>toler,
                       % where "er" is error and "toler" is tolerance.
    pnew=p*p*(3-2*p);  % The basic iteration.
    er=abs(p-pnew);    % Change on this time step.
                       % "abs" is absolute value.
                       % When er is less than toler, the iteration stops.
    p=pnew;            % p becomes the new value, next in the iteration.
    niter=niter+1;     % Counts the number of iterations.
end                    % End of the "while" loop.
conv_time(k)=niter;    % Keeps track of how many time steps to
                       % convergence, for the given initial condition.
pf(k)=p;               % Final value of the iteration.
end
figure(100), plot(p0,pf,':o')      % plots final value vs. initial value
                       % figure number is arbitrary
figure(101), plot(p0,conv_time,'o:')% plots convergence time
```

In Fig. 3.1, I show the results.

Note that because $1/2$ was one of my initial conditions this unstable fixed point manages to make an appearance in Fig. 3.1a. Otherwise, they all end up going to 0 or 1. What's amusing is the increase in the number of iterations needed when the initial point is close to $1/2$.

3.4 Independent folk

To see yet another phenomenon imagine that there are some independent people who, when not confronted by friends, will spontaneously choose a wrist for their rubber band, maybe changing it, maybe not. Let the fraction of such independents be q, with $0 \le q \le 1$. For ease of calculation we change the scenario slightly. Randomly pick an individual. Suppose that individual is female. If she's an independent, then she chooses a wrist arbitrarily (with equal probability), related neither to what everyone else is doing nor to which wrist the rubber band is currently on. If she's not independent, then she meets two other (randomly chosen) individuals and goes with the majority rule, as before. The fundamental iteration, given by Eq. (3.2), is replaced by

$$p' = q \cdot \frac{1}{2} + (1-q) \cdot p^2(3-2p).$$
(3.6)

If you want to match time scales with the previous notion of breaking everyone into groups, you can do the random selection (leading to Eq. (3.6)) N times for each convening of the groups.

Exercise 2 Remember that $p \equiv p_L$. So in addition to Eq. (3.6) there will be a corresponding equation for p'_R. Check that $p'_L + p'_R = 1$.

As before, the first thing we look for is a fixed point, $\bar{p}$, satisfying

$$\bar{p} = \frac{q}{2} + (1-q)\bar{p}^2(3-2\bar{p}).$$
(3.7)

Presumably as $q \to 0$ we should get our usual rubber band solution, while for $q \to 1$ we would expect to lose the transition and find $\bar{p} \to 1/2$. We'll see below that this is *not* exactly true, but first let's examine the predictions of our probability equations. It turns out that Eq. (3.7) predicts that there's a magic value of q, call it q_c, such that, for $q > q_c$, $\bar{p}$ is already $1/2$.

There are a couple of ways to see this. One is to implement Eq. (3.6) starting with some initial p value and iterating. Beginning with p values near both 1 and 0 one gets Fig. 3.2. This makes it clear that there *is* a phase transition, so that $\bar{p}$ does not go gradually to one-half.

An analytic way to see this is again to look for fixed points, but now to let $r \equiv p - 1/2$ since we know that $1/2$ has a special role here. Consider the function

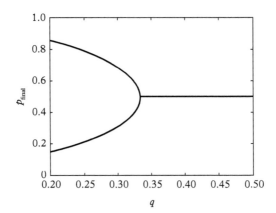

Fig. 3.2 Iterating Eq. (3.6) for 2000 time steps to obtain a limiting value of p (for various q). For each q two iterations were performed, one with initial value near 0, the other near 1.

$$f \equiv p - p' = p - \frac{1}{2}q - (1-q)p^2(3-2p) = r\left[\frac{3q}{2} - 2qr^2 + 2r^2 - \frac{1}{2}\right], \quad (3.8)$$

where I have made the substitution $r = p - 1/2$ in the final equality (and have done a bit of algebra). For a fixed point, f should be zero, which says that $r = 0$ (i.e., $p = 1/2$) is always a fixed point, although it may not be stable. To find the other fixed points we have only a quadratic to solve, which yields

$$r = p - \frac{1}{2} = \pm\sqrt{\frac{1-3q}{4(1-q)}}. \quad (3.9)$$

This clearly gives the result of Fig. 3.2 with a second-order (continuous) transition at $q = 1/3$. For $q > 1/3$ this gives imaginary, hence unphysical, fixed points, leaving $p = 1/2$ as the only candidate.

Exercise 3 Check stability of the root, $\bar{p} = \frac{1}{2}$. Hint: See Sec. 3.2. For $q \geq 1/3$ the fixed point $1/2$ had better be stable—it's the only one. For $q < 1/3$ that fixed point should become unstable. (The real values given in Eq. (3.9) become stable.)

3.4.1 What actually happens

As is often the case, the lowest approximation, a kind of mean field theory, isn't quite right. First, it's obvious that for large q you couldn't have exactly half and half for the distribution of rubber bands. That's because for large q each individual is independent and you expect $\sqrt{N}$ fluctuations (with N the number of individuals playing). When normalized, by dividing by N, this is our usual $1/\sqrt{N}$ deviation. So you'd need to have large N to see any kind of transition. For corresponding "games" involving atoms in a gas the numbers are enormous (typically 10^{23} or more) so this effect can be ignored.

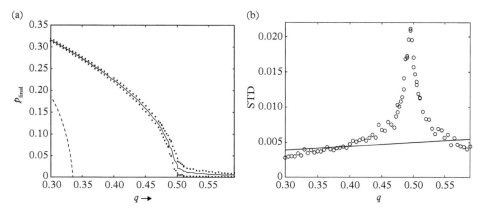

Fig. 3.3 For each q value groups of 20,000 were sampled 40,000 times. q values are shown on the horizontal axis. (a) The vertical axis shows the mean p values (as a solid line), the standard deviation of the sample (as dots above and below the line) and the prediction of Eq. (3.9) as the dashed line. (b) The vertical axis shows those same standard deviations (STD) as well as (the solid line) $\sqrt{qN}/N$.

The simulations I'm about to exhibit didn't use anything like those numbers, so that my interpretation is marred by these fluctuations.

But there's more to the deviations than just adding random fluctuations. What numerical evidence I've gathered suggests that there is a phase transition, but it occurs at about $q = 1/2$. Why not $1/3$? I presume it's those same fluctuations. The reason I say this is in Fig. 3.3. The solid line in Fig. 3.3a is a plot of the limiting p values for groups of $N = 20,000$, for each q value (with many samplings of each group member). In the same plot are dots indicating the standard deviation of those numbers based on those 40,000 repeats. To the left, a dashed line, is the prediction of Eq. (3.9). Fig. 3.3b compares those deviations with $\sqrt{qN}/N$. Note that deviations peak around q-values near $1/2$. Note also that the actual p values to the left of the (supposed) transition are always larger than predicted by Eq. (3.9). My speculation is that the freedom of the "independent folk" drives fluctuations in the others, shifting the phase transition to larger q values.

Exercise 4   Here's something that I don't know how to do. Maybe someone will solve this, and, who knows, maybe it has some practical value. The peak in Fig. 3.3 is not quite at $1/2$, and is closer to 0.495. Why? I've given a reason to move away from $1/3$, but that's just conversation. Can a precise value for the phase transition be found?

3.5   How often does the prediction go wrong?

We return to the "no independents" form of the game.

A quantity of practical interest is the odds of not ending in the state favored by the initial conditions. This is the fluctuation problem mentioned at the end of Sec. 3.2.

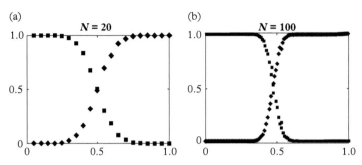

Fig. 3.4 The initial fraction of left-wrist persons is shown as the horizontal axis. On the vertical axis is the fraction with a particular outcome, all left (diamond) or all right (square). (a) The outcomes for 20 persons; (b) the outcomes for 100. Thus for the 20 person game, suppose you start with 40% (i.e., eight persons) with the rubber band on the left wrist. In the 1000 random trials (reflected in these output graphs) starting with these initial conditions, about 70% of the time all rubber bands ended on the right—but in the other cases, due to the random nature of the trials, things turned out the other way. Note that as the number of people increases ($N = 100$ vs. $N = 20$) the relative uncertainty is reduced.

Thus, it may happen that initially the number of L's exceeds the number of R's but that the system nevertheless ends "all R's." This will also depend on N, the number of participants. Note that we only deal with randomness. The tendency to form groups with friends or other human tendencies are not dealt with in this treatment.

Can you simulate this quantity? This would involve writing a program that actually "plays the game," rather than just looking at averages as done in the program in Sec. 3.3. Fig. 3.4 shows the results of such a program.

Exercise 5 Write a program to play the game (in MATLAB™ or in your favorite programming language). Check my results above.

4
Percolitis

Applications discussed in this chapter: Directed percolation, Mean field theory, Astrophysics, Neurology, Self–organized criticality (SOC), Epidemiology

---

First, the name: this model is secretly directed *percolation* (about which, more later) and "itis" is a word-ending that indicates a disease. The model was introduced in an article on galaxies [204] to explain some of the techniques used. In general, a Markov process can be rephrased as directed percolation.

But now for the epidemiology of this "disease."

4.1   An epidemic model

Oz is a city of N people and has little contact with the outside world. One of its male citizens returns from a rare trip with a nearly symptomless disease called percolitis (the only manifestation of the "disease" is a preternatural cheerfulness and a desire for coffee). After 24 hours he is[1] completely cured but he may have transmitted the disease to some of the people he came in contact with during his period of infection. We wish to study the progress of this disease in Oz.

Here are the salient properties: Percolitis lasts for 24 hours, from midnight to midnight. Each person with the disease has probability p of transmitting it to anyone with whom there is any contact. This includes self-transmission. There is no immunity. In Oz everyone comes in contact with everyone else every day.[2] See Fig. 4.1.

We first formulate a rough picture of how the disease progresses, or does not progress. Measure time, t, in days. Let $s(t)$ be the expected fraction of sick people on day-t. Since all people are equivalent (after the first day, anyway), $s(t)$ is also the probability that any particular individual is sick. We want to predict the expected number of sick people on day-$(t+1)$.

---

[1] Although the disease lasts 24 hours, at this stage we allow self-transmission, so it would be more accurate to say he is *usually* cured (if the transmission probability is small).

[2] The essential simplifying assumption here is that transmission to all possible individuals is equally likely. Neglecting immunity is adopted for simplicity alone. For application to both galaxies and neurons "immunity" is not assumed. On the contrary, the refractory period—during which a neuron does not fire or star formation cannot occur—is an essential part of those models.

When Things Grow Many. Lawrence S. Schulman, Oxford University Press.
© Lawrence S. Schulman (2022). DOI: 10.1093/oso/9780198861881.003.0004

Percolitis

A disease of the central nervous system with renal complications.

Etiology

Searches for specific viral or bacterial causes for this disease have so far not proved fruitful and those organic pathologies that exist (such as renal dysfunction) can be traced to behavioral sources (predilection for excess coffee consumption). Similarly, no vector for transmission has been identified. Some investigators have speculated that the disease may be entirely psychosomatic.

Symptoms and signs

Onset as well as termination is invariably at midnight. The following morning there is a mild cephalalgia for the first hour followed by preternatural joy. By midday coffee consumption mounts (hence the name of the disease) that strangely does not interfere with sleep.

Treatment

No specific therapy is known. Coffee substitutes are not well received; for example, in some victims (usually male) this may elicit the response, "Real men don't drink decaf."

Transmission and contagion

There is extensive empirical knowledge of transmission characteristics of percolitis despite the complete lack of biological information. A diseased individual can transmit the disease at any time before about 8 p.m. and irrespective of the degree of contact the probability, p, of transmission is the same.

Let α be a particular person in Oz. The probability that α is sick on day-$(t+1)$ is deduced as follows:

$$
\begin{aligned}
s(t+1) &\equiv \text{Probability of being sick on day-}(t+1) \\
&= 1 - \text{Probability of being healthy on day-}(t+1) \\
&= 1 - \text{Probability that no sick person on day-}t\text{ transmitted the disease to } \alpha \\
&= 1 - (1-p)^{\#\text{sick people on day-}t} = 1 - (1-p)^{Ns(t)} .
\end{aligned}
\tag{4.1}
$$

It is convenient to rescale p, writing $r \equiv pN$. This doesn't change anything. It's just that if p is on the order of unity, then *everyone* gets sick. So to see interesting phenomena

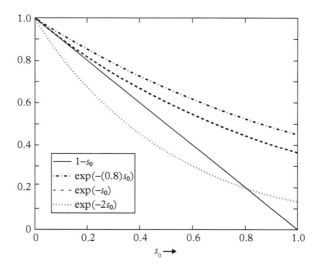

Fig. 4.1 Both $1 - s_0$ and $\exp(-rs_0)$, plotted against s_0, for various values of r.

we'll need much smaller values; moreover, as we shall see, just *how* much smaller depends on N. Eq. (4.1) becomes

$$s(t+1) = 1 - \left(1 - \frac{r}{N}\right)^{Ns(t)}. \qquad (4.2)$$

One last mathematical fact will bring the equation to its final form. That fact is

$$\lim_{M\to\infty} \left(1 - \frac{x}{M}\right)^{M} = \exp(-x). \qquad (4.3)$$

This is proved in a footnote.[3] Letting $M = Ns(t)$ and $x = rs(t)$ shows[4] that Eq. (4.2) becomes

$$s(t+1) = 1 - \exp(-rs(t)). \qquad (4.4)$$

This is a *mapping*, similar to the logistic equation studied in ecology and in physics. To find its properties we first seek its steady states, that is, s_0 such that

$$s_0 = 1 - \exp(-rs_0). \qquad (4.5)$$

An intuitive way to solve such a transcendental equation is first to rearrange, writing it as $1 - s_0 = \exp(-rs_0)$, and then plotting both the left-hand side and right-hand side of this equation against s_0. Fig. 4.1 illustrates this.

---

[3]The (natural) logarithm of the right-hand side is $-x$. The logarithm of the left-hand side is $M \log(1 - x/M)$. Using a Taylor expansion for the logarithm, this is $-M \cdot [x/M + (x/M)^2/2 + \cdots]$. Only the first term of the Taylor expansion survives the limit $M \to \infty$, giving $-x$, as claimed.

[4]The approximation becomes strained as $s \to 0$, but since we're letting $N \to \infty$, so long as $s > 0$, it's acceptable. However, for finite-N applications one must be careful.

By virtue of the meaning of s, s_0 must lie in the interval $[0, 1]$. As you can see, $s_0 = 0$ is always a solution. However, depending on r, the two curves may again meet for $0 < s_0 < 1$. The condition that guarantees a second meeting is that the slope of the exponential at $s_0 = 0$ must be more negative than the slope of the straight line $1 - s_0$. Examining the derivative of the exponent, this means $r > 1$.

So there is a critical value, $r_c = 1$, such that for $r > r_c$ there is a strictly positive solution to Eq. (4.4) (which is also less than 1 for $r < N$). For $r \le r_c$, the only solution is 0.

The special role played by $r = 1$ is nearly obvious. $r = 1 \Leftrightarrow p = 1/N$. With N people in town, a sick person is expected to infect pN people (assuming no double infection, which is acceptable for small s). The disease will thus sustain itself only if $pN > 1$.

To decide whether a solution will occur in reality we must also examine its *stability*. In other words, if you start near that solution, does the time development of the system approach it or does it run away? As we shall see, the answer is: for $r \le r_c$, $s_0 = 0$ is the stable solution; for $r > r_c$, the second solution, where the lines cross for $s_0 > 0$, is the stable solution.

4.2   Discussion

Let's pause for a moment. What has been illustrated so far? First, the model shouldn't be construed as being realistic. If you wanted to study a real epidemic, you would not have everyone equally likely to become infected. One of the difficulties in epidemiology is figuring just how much contact people have, with related questions about disease transmission.[5] Also, some people get sicker than others and stay sicker longer. So a lot of simplification has taken place. What is left?

For epidemics we have obtained a rough criterion for disease persistence or the absence thereof. We also have predictions for time dependence (as we shall see in greater detail below). This information is a platform on which to build refinements. One such refinement is immunity, which will be discussed in an exercise and in Sec. 4.6.1.

Beyond this, however, we have obtained a bona fide phase transition. In one parameter regime the system behaves one way; in another regime it is totally different. One expresses this using an *order parameter*. For a liquid–gas transition the order

---

[5]Isabel Dengos, a (former) Clarkson student, reported on her experience in doing *real* epidemiology. She studied the incidence of influenza (flu) based on data from June 2007. One feature that is important, but plays no role in my treatment of percolitis, is knowing how many people were affected. One can't always distinguish the flu from a bad cold, and many cases of the flu are not reported to medical authorities. So her method was indirect, namely finding out from Google how many people in each region of the United States searched for information on flu symptoms. If all regions are equally knowledgeable regarding online searches then this should yield proportional samples. She also mentioned a kind of mean field theory in which N individuals were divided into three groups: Susceptible (S), Infected (I) and Recovered (R). The "SIR" model relates these quantities as follows: $\frac{dS}{dt} = -\beta \frac{SI}{N}$, $\frac{dI}{dt} = \beta \frac{SI}{N} - \gamma I$, $\frac{dR}{dt} = \gamma I$, and finally $N = S + I + R$. Thus β and γ are rate constants. For more on this and related models see [107].

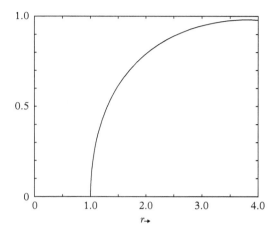

Fig. 4.2 Limiting values of percolitis density. s_0 versus r over a range of r. For each r, this value was obtained from Eq. (4.4) by starting at $1/2$ and iterating until the change in the s value on an iteration was less than 10^{-5}. This is a phase diagram for the percolitis transition. For $r \leq 1$ the curve coincides with the x-axis.

parameter is the difference in density between the two phases. Thus in Fig. 1.1 this is the length of a horizontal line between the two sides of the curve.[6] For our epidemic problem it is $s_0(r)$, where I've included the r-dependence of s_0 for emphasis. As a function of r, the derivative of s_0 is discontinuous. Another way to say this is that it is impossible to write $s_0(r)$ as a single convergent power series over a range of values on both sides of $r = 1$. Suppose you could write such an expansion around $r = 0.8$. The coefficients are easy to evaluate: they must all be zero, since $s_0(r)$ in a neighborhood of this value is identically zero. But if all coefficients are zero and the power series were good for all r, then, for $r > 1$, s_0 would have to be zero also—which it isn't. This implies a breakdown[7] in *analyticity*, which for many people is the ultimate criterion for the existence of a phase transition.

In Fig. 4.2 I show the function $s_0(r)$. As indicated, it is obtained by iteration. One could also deduce this curve through solving the equation $s_0 = 1 - e^{-rs_0}$, although such a solution could produce unstable values as well.

---

[6] Note that in Fig. 1.1, as $T \to T_c$ from below, the density difference goes to zero. This is analogous to our system, for which $r > 1$ is like $T < T_c$. For $r < 1$ our order parameter is zero, while in the liquid–gas case for $T > T_c$ the order parameter is not defined. As we will see in Sec. 5, in ferromagnetism the order parameter is the magnetization, which, in the absence of an external field, is zero above the critical temperature.

[7] There are functions that have a power series but are not analytic. For example the function $f(x)$ defined by $f(x) = 0$ for $x \leq 0$, $f(x) = \exp(-1/x)$ for $x > 0$. (Exercise: Find the power series.) Analyticity, however, is more stringent, namely the requirement that a function have derivatives in all directions in the complex plane. This in turn implies that there is a power series. The demonstration in the text is sufficient to show that $s_0(r_0)$ is *not* analytic.

Exercise 6 Among some scientists the hallmark of a phase transition is a breakdown in analyticity.[8] This stems from the seminal work of Onsager [171] and Lee and Yang [137; 138] showing that there is a breakdown in analyticity at a phase transition. (I remark that these ideas fail completely when it comes to understanding the metastable state.) Now there are two functions, $s_0(r)$, the physical probability of finding someone with percolitis, and something I'll call $\bar{s}_0(r)$, the solution(s) of the transcendental equation $s = 1 - \exp(-rs)$. For example, letting $\bar{s}_0 = x + iy$, there's a complex root that to a good approximation has $y = \frac{19\pi}{8r}$ and $x = 1 - y/\tan(ry)$ (which gets better as $r \to 1$). For $s_0(r)$ I've already demonstrated that it can't be analytic. What about $\bar{s}_0(r)$?

4.3 Behavior of the order parameter near the critical point

We study the solution to $s_0 = 1 - e^{-rs_0}$, focusing on r near 1.

For $r \le 1$, there isn't much to study: $s_0 = 0$, period.

Next consider $r > 1$, but not by much, so that $0 < \epsilon \equiv (r-1) \ll 1$. In this case, from our graphical solution we know that s_0 is also close to zero.[9] Expand Eq. (4.5) in both ϵ and s_0. After a bit of algebra, this yields

$$s_0 = s_0 + \epsilon s_0 - \frac{(1+\epsilon)^2 s_0{}^2}{2} + \cdots, \tag{4.6}$$

where the dots refer to higher powers of s_0. First, subtract the extra s_0 on both sides of the equation. Then, since we seek a non-zero value of s_0, divide by the s_0's on both sides of the equation, leading to

$$s_0 = \frac{2\epsilon}{(1+\epsilon)^2} + \cdots = 2\epsilon + O(\epsilon^2) \approx 2(r-1). \tag{4.7}$$

This is the behavior of the order parameter, $s_0(r)$, near $r = 1$. Consistent with Eq. (4.7) you can examine the slope of $s_0(r)$ just to the right of $r = 1$ in Fig. 4.2 and ascertain that indeed it is 2.

For r large there is no simple analytic expression for $s_0(r)$. But from the graphical solution it is clear that $s_0(r) \to 1$ as $r \to \infty$. Thus for $p = O(1)$ (so $r = O(N)$), $s_0(r) \sim 1$, as noted earlier.

---

[8]Being *analytic* is a powerful condition on a function. (If you are first encountering this concept, I'd suggest skipping this exercise.) The definition sounds simple: let D be a domain (an open, arcwise-connected set) of the complex plane and $z \in D$. Then $f(z)$ is analytic if its derivative, $df/dz = \lim_{h \to 0}(f(z+h) - f(z))/h$, exists for h coming from any direction in D. This seemingly mild restriction is enough to guarantee that there are infinitely many derivatives at z and that there exists a power series for f (as well as many other properties).

[9]That $s_0 \to 0$ as $\epsilon \to 0$ does not imply that they go to zero at the same rate (i.e., $0 < \lim |s/\epsilon| < \infty$). In our present example, in fact, $s_0 = O(\epsilon)$, but in other cases you could have other powers, such as $O(\epsilon^{1/2})$, $O(\epsilon^2)$, $O(\epsilon^{1/3})$. Therefore when you expand in ϵ you have to be careful about dropping s_0 terms, since a priori you don't know how they scale with ϵ.

4.4 Approach to equilibrium

We next examine stability. Let Oz start with $Ns(0)$ sick individuals,[10] where $s(0)$ is near the value $s_0(r)$. Stability depends on whether $\delta(t) \equiv s(t) - s_0(r)$ grows or shrinks (in absolute value) in time. Recalling that $s_0(r) = 1 - e^{-rs_0(r)}$, we have

$$\delta(t+1) \equiv s(t+1) - s_0(r) = 1 - e^{-rs(t)} - \left(1 - e^{-rs_0(r)}\right)$$
$$= e^{-rs_0(r)}\left(1 - e^{-r\delta(t)}\right) = (1 - s_0(r))\left(1 - e^{-r\delta(t)}\right). \quad (4.8)$$

Since $\delta(t)$ is assumed small, we expand in a power series:

$$\delta(t+1) = (1 - s_0(r))\left[r\delta(t) - \frac{1}{2}\left(r\delta(t)\right)^2 + \cdots\right]. \quad (4.9)$$

Now split the problem into three cases, $r < 1$, $r > 1$, and $r = 1$.
The easy case is $r < 1$. For this value, $s_0(r) = 0$ and Eq. (4.9) becomes

$$\delta(t+1) = \left[r\delta(t) - \frac{1}{2}\left(r\delta(t)\right)^2 + \cdots\right] \sim r\delta(t). \quad (4.10)$$

This implies that

$$\left|\frac{\delta(t+1)}{\delta(t)}\right| = r < 1, \quad (4.11)$$

so $\delta(t) \to 0$. Zero is thus a stable fixed point of the equation. Moreover, it's the only fixed point for r between zero and one, so the disease eventually peters out.

The case $r > 1$ is algebraically more complicated but conceptually the same. First notice that from Eq. (4.10) it follows that the root 0 is *unstable* (since now in Eq. (4.11) r is greater than 1). So we better find that the other solution is stable.

Going back to Eq. (4.9), the factor multiplying $\delta(t)$ is $r(1 - s_0(r))$, and it is this quantity whose magnitude we need to show is less than one. Of course r is greater than one, but $(1 - s_0(r))$ is less, and we need to see how they compete. But for $r = 1 + \epsilon$ and small ϵ, Eq. (4.7) provides the answer. Thus $r(1-s_0(r)) \approx (1+\epsilon)(1-2\epsilon) = 1-\epsilon+O(\epsilon^2)$. So indeed the positive root is stable. Showing that $re^{-rs_0(r)} < 1$ for larger r is messier, but in any case the product is less than unity. (In Fig. 4.3 I show the product $r(1 - s_0(r))$ as a function of r, based on the $s_0(r)$ values in Fig. 4.2. Clearly this product continues to decrease away from $r = 1$, and the *only* regime where it approaches 1 is near the critical point, $r = 1$.[11])

---

[10]This is not the initial condition in our first version of the fable, in which a single individual returns from a trip. Now either many people have arrived or we are starting our clock at a point where the disease is already present at the given level: $s(t_{start})N$ people are sick (which is close to $s_0(r)N$). (There is a slight abuse of notation in the previous sentence: s in one case is a function of time, while s_0—note the subscript—is a function of r.)

[11]A way to do the analytic demonstration for $r > 1$ is to consider the function $f \equiv r(1-s_0)$, nominally a function of r, as a function of s_0, with $0 < s_0 \leq 1$. From Eq. (4.5) we can write r as a function of

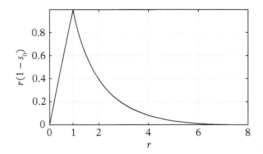

Fig. 4.3 $r(1 - s_0(r))$ as a function of r.

Note that in both $r \neq 1$ cases the convergence is exponential. That is, the ratio of the δ's is strictly less than 1, so that each successive time step lowers δ by a fixed factor.

The interesting case is $r = 1$. The root $s_0(r) = 0$ is a double root and Eq. (4.12) becomes

$$\delta(t + 1) = \delta(t) - \frac{1}{2}\delta(t)^2 + \cdots .\tag{4.12}$$

If we were to ignore the quadratic term we'd find $\delta(t + 1) = \delta(t)$, which means no change, so we must look further. Note though that we already have evidence that this converges to zero: we are subtracting the strictly positive $\delta(t)^2/2$ from the $\delta(t)$ term,[12] so $\delta(t+1)$ is surely smaller than $\delta(t)$ (there is convergence of the series in Eq. (4.9) to worry about, but let's not ...). To help guess the solution to this iteration and to learn the precise way in which this converges, we turn t into a continuous time variable, so that Eq. (4.12) becomes

$$\frac{d\delta(t)}{dt} = -\frac{1}{2}\delta(t)^2 .\tag{4.13}$$

The solution is $\delta(t) = 2/(t + t_0)$, a power law, which is *much* slower than an exponential.[13] To check how this works out for the discrete iteration we need to see whether

$$\delta(t + 1) = \frac{2}{t + t_0 + 1} \overset{?}{=} \delta(t) - \delta(t)^2/2 = \frac{2}{t + t_0} - \frac{1}{2}\left(\frac{2}{t + t_0}\right)^2 .\tag{4.14}$$

s_0, namely $r = -\log(1 - s_0)/s_0$, from which it follows that f, now written as a function of s_0, satisfies $f(s_0) = (1 - 1/s_0)\log(1 - s_0)$. We want to show this to be less than one, but we already know it is one at $s_0 = 0$, so all we need to show is that its derivative is negative (we also know that $r(1 - s_0)$ cannot turn negative). The derivative is $df/ds_0 = [\log(1 - s_0) + s_0]/s_0^2$. But $s_0^2 > 0$, so we only need to establish that $\phi(s_0) \equiv \log(1 - s_0) + s_0 < 0$. At $s = 0$, $\phi = 0$, so again all we need to show is that $d\phi/ds < 0$ for $s > 0$. But $d\phi/ds = -s/(1 - s)$, which is clearly negative for $s > 0$.

[12] Leaving $\delta(t + 1)$ to be strictly positive.

[13] See Sec. 4.7.

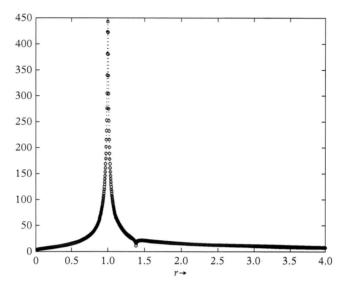

Fig. 4.4 Number of iterations of Eq. (4.4), starting from $1/2$, for the change in a single step to be less than 10^{-5}.

Exercise 7 Check this. Estimate the error in Eq. (4.14) for large times.

This power law behavior is typical of *critical phenomena*. If one were to assign a dynamics to the van der Waals gas[14] I expect that there would be special values of temperature and pressure for which the system relaxes very slowly. As an indication of the slowness of convergence, in the program that generated Fig. 4.2 I kept track of the number of iterations needed to arrive at an s value that changed less than 10^{-5} on the next iteration (and called this $s_0(r)$). In all cases I began from $s(0) = 1/2$. The number of iterations as a function of r is shown in Fig. 4.4. Note the increase near $r = 1$.

Exercise 8 Explain the small drop in the number of iterations slightly to the right of $r = 1$. (Nothing profound to be learned from this exercise—just a check to see whether you're paying attention.)

To gain perspective, go back to the $r < 1$ case. Recall that, to lowest order in δ (which is sufficient), we had $\delta(t) = r^t \delta(0)$. Since $r < 1$, I can write $r = \exp(-1/\tau)$ with $\tau > 0$. But then $\delta(t) = e^{-t/\tau} \delta(0)$, which is exponential decay. As noted earlier the same is true for $r > 1$. The quantity τ is a characteristic time for equilibration (to $s_0(r)$, whether it is strictly positive or not). We'll see later that this is essentially a *correlation length*, but let's not get ahead of ourselves.

---

[14]See Appendix E.

How does τ behave as a function of r? Just solve: $\tau = -1/\log r \approx 1/\epsilon$, where in the approximate equality I take $r = 1-\epsilon$ and ϵ small. This implies that $\tau \to \infty$ as $r \to r_c = 1$; the characteristic time diverges.

Exercise 9 Show that, as r approaches 1 from above, the analogously defined τ also diverges (and in the same way).

But we have also seen that in fact there *is* convergence when r is exactly one. What's going on? Having $\tau = \infty$ doesn't mean there is no convergence. Rather, it's telling us that there is now a different, and much slower, mode of convergence. Instead of an exponential, we've gone over to a polynomial—that is, one-over-some-power-of-t, in our case, t^1. This behavior is typical of critical points and, as we can see, is present in the percolitis problem.

What you can't see in percolitis is the divergence of spatial characteristic distances—there is *no spatial coordinate*. When a gas such as CO_2 is very close to its critical point, this otherwise transparent substance turns milky white. It's not absorbing light so much as bending it a lot. This is known as opalescence, a phenomenon that occurs in the gemstone opal, for similar reasons, as I'll now explain. Away from a critical point the characteristic distance for density fluctuations in CO_2 is on the order of the distance between molecules, a few nanometers at most. However, near the critical point large collective motions take place, and the density fluctuations can be as large as several hundred nanometers, on the scale of the wavelength of visible light. When a wave (such as light) encounters objects on the scale of its wavelength it is much more strongly scattered than usual (a resonance phenomenon), as a result of which it does not propagate freely through the substance: it is scattered a lot, as I said earlier. Critical opalescence is thus an indication of a tremendous increase in the characteristic length of density fluctuations. The gemstone opal is made of small spheres of transparent hard silica embedded in softer silica. Typically these spheres are a few hundred nanometers in size. This has two effects. First there is enhanced scattering and second the regularity of their placement leads to diffraction effects (the latter is not present in the phase transition). But whatever the cause, one obtains patterns that are pleasing to the eye.

Exercise 10 Immunity. For an epidemic, immunity should change the nature of our solution. Also, immunity is important in other applications of a percolitis-like formalism (as discussed in Sec. 4.9). In brain modeling, the Oz citizens become neurons and "infection" corresponds to the firing of the neuron. Neurons in general exhibit a refractory period during which they do not re-excite. (An important difference though is that neurons can fire both excitatory signals and inhibitory signals.) Similarly in the star formation process (the application to galactic morphology) a region where star formation has recently taken place is unlikely to allow a second episode of star formation for a considerable period—the gas is too hot.

(a)   (b)

Fig. 4.5 (a) Critical opalescence in carbon dioxide. The CO_2 is kept very close to its critical point and scatters light of many wavelengths, leading to a milky white appearance. The small balls at the top and bottom of the container are for additional measurements [81]. Courtesy of T. Golfinopoulis, MIT. (b) Opal. By Daniel Mekis - Own work, CC BY-SA 3.0, https://commons.wikimedia.org/w/index.php?curid=27929123 (accessed September 2020).

We therefore develop a theoretical description in which there is d-day immunity to percolitis. First consider 1 day. Show that the mean field expression is

$$1 - s(t+1) = \exp(-rs(t)) + s(t) - s(t)\exp(-rs(t)). \tag{4.15}$$

Hint: As in Sec. 4.5, being sick, transmitting disease and similar variables can be considered "yes" or "no," "on" or "off." These are modeled as Boolean random variables and take (only) the values 0 and 1. For two such variables, A and B, the or operator (symbolized by a vertical line, "|") satisfies $A|B = A + B - AB$.
Find the critical value r_c such that the disease fades away for $r < r_c$.
   One can extend Eq. (4.15) to d-day immunity. See Sec. 4.6.1.

Exercise 11   Consider another kind of immunity, one that operates through the transmission probability. Instead of a constant probability r (more precisely, r/N), one could take a variable rate that depends on the sickness level on the previous time step, for example $r[1 - s(t)]$. This model could apply where the transmission depends on some disease resource that is depleted when the illness is widespread. In the galaxy model (discussed in Sec. 4.9) this is a natural assignment, since star formation (what corresponds to being sick with percolitis) depends on having enough cool gas around, something that is less likely if there was a recent episode of star formation. Take the model then to have the iteration scheme

$$s(t+1) = 1 - e^{-rs(t)[1-s(t)]}. \tag{4.14}$$

(*r* is a constant.) You can easily check that the threshold, $r = 1$, is unchanged (why?). But this scheme has another threshold, a value of *r* for which the system *does not settle down*. It oscillates between two levels indefinitely (very much like the logistic map[15]). Can you find that next threshold analytically or numerically? (Answer: It's about 8.02659.[16]) As for the logistic map (see Appendix F) there is also a transition to chaos at *r* in the upper teens (later bifurcations: 14.361, 16.61, 17.15, 17.27, with chaos at about 17.3). There is evidence that such oscillations exist in galaxies.

4.5   Discreteness and fluctuations*

In this section we calculate second moments. This will introduce new formalism. Everything we've done until now treats the *expected* value of the number of sick people. If you're an epidemiologist you'd like to know the worst case scenario, or, if you're an optimistic epidemiologist, what's the lightest level of infection you're likely to see. Also, ignoring fluctuations overestimates the time it takes for the disease to expire for $r \lesssim 1$. Here's why: we've established that $s \to 0$. So, whatever N is, $s(t)N$ finally becomes a small number. As such, its fluctuations may become comparable to the remaining sick population and the disease will go from, say, five sick people to zero sick people, just because of a fluctuation. But once it does that the game is over: no more percolitis. So there's an asymmetry. Fluctuations that increase *s* have little permanent effect, but those that decrease *s* can cause a sudden halt to the process. There is thus a tendency to shorten the survival of the disease due to fluctuations — what you could call quantization effects.

Similar considerations arise in ecology, when you're trying to estimate the survival of a species. In that case, another simplification of ours will *under*estimate the lifetime of the species. This is because the system is not homogeneous (in percolitis homogeneity is the assumption that everyone contacts everyone every day). What has been found is that, because of irregularities in habitat and terrain, some animals will survive in out-of-the-way places.

Exercise 12   As an example of the effect of fluctuations and a finite size effect, consider the issue of extinction for $r > 1$. For $N \to \infty$ the number of people sick follows Eq. (4.4). However, for any finite system it can happen—and *will* happen with finite (even if small) probability—that *no one* gets sick on a given time step.

---

[15]$x(t+1) = \lambda x(t)(1 - x(t))$ is the physics form of the logistic map (see Appendix F). The curves $x(1-x)$ and $1 - \exp(-rs(1-s))$ have the same properties: for $0 \le (x \text{ or } s) \le 1$ there is a 1–1 mapping $x = X(s,r)$ and $\lambda = \Lambda(s,r)$ (one bump in the middle, with a monotonic decrease on both sides). Note that this does not mean the mapping can be written down as a simple expression.

[16]If you're really feeling ambitious you can check the usual criterion for the first bifurcation. Define $F(s) = 1 - \exp(-rs(1-s))$. Then at the first bifurcation you should have $F'(s) = -1$, where the prime means $\partial/\partial s$.

If r is close to 1 ($= r_{\text{critical}}$), this can be important and terminate the "disease." Do numerical simulations of the finite-N process (not using Eq. (4.4)) to determine the actual lifetimes as a function of N. I suggest using $r = 1.1$ and 1.5 and using rather small values of N.

To treat fluctuations we introduce a convenient formalism. Define collections of *random variables*. A random variable is a quantity that depends on the outcome of a trial. The number of spots showing on top after the toss of a die is such a variable. (See Appendix D for more details.) The value, heads or tails, of a coin flip is another example. In our case our first collection of random variables corresponds to whether or not the disease is transmitted. It was for this possibility that we defined a probability p. Now we say, *if* Mr. 44 was sick on day 17, does he or does he not infect Ms. 37, so that she will be sick on day 18? You can define the random variable whether or not Mr. 44 was sick. This random variable takes the value 1 (transmission) with probability p, and takes the value 0 (no transmission) with probability $1 - p$. Similarly, for all pairs of people and for all times we define

$$A(j,k,t) = \begin{cases} 1, \text{ if } j \to k \text{ transmission would be successful on day-}t \\ \quad \text{(only relevant if } j \text{ is sick on day-}t, \text{ making } k \text{ sick on day-}(t+1)), \\ 0, \text{ otherwise.} \end{cases}$$

$$(4.17)$$

People are reduced to numbers ($j, k = 1, \cdots, N$) and time is measured in days. By assumption $\langle A(j,k,t) \rangle = p$ for all j, k and t,[17] and all are independent. The A's are Boolean random variables, taking only the values 0 and 1.

A word about "random": If you had deep understanding of the percolitis transmission agent and of the genome of each individual you might be able to predict whether or not transmission would take place. Similarly if you had really good control in coin flipping you might be sure of the outcome, so an idealized random variable would no longer be a good description. That's life. Randomness is a model with which you can (often) get excellent predictions, but for classical phenomena it does not represent the deepest description. (On the other hand, many physicists consider quantum mechanics to be fundamentally random, although Einstein did not fall into that "many.")

We next define a second random variable — one that depends on the A's and on the initial conditions. This is the medical state of each individual, and at each time it is given by

$$S(n,t) = \begin{cases} 1, \text{ if person \#}n \text{ is sick on day-}t, \\ 0, \text{ otherwise.} \end{cases}$$

$$(4.18)$$

Now put everything together. The "equation of motion" for S is

$$S(n,t+1) = 1 - \prod_{\ell} [1 - A(\ell,n,t)S(\ell,t)].$$

$$(4.19)$$

---

[17]Explicitly: $\langle A(j,k,t) \rangle \equiv (+1) \cdot \Pr[A(j,k,t) = 1] + (0) \cdot \Pr[A(j,k,t) = 0] = p$.

The logic is as follows: In order for n to be sick on day-$(t+1)$, someone who was sick on day-t had to be the source of infection. This means that there is at least one ℓ for which both $S(\ell, t) = 1$ and $A(\ell, n, t) = 1$, meaning that the sick ℓ successfully transmits to n. In that case at least one term in the product is zero so that one minus the product is one.

This logic is made easy to follow by our having introduced the Boolean variables A and S: they are perfect for dealing with yes/no arguments. (Cf. Eq. (4.1).)

We next define $s^{(n)}(t) \equiv \langle S(n, t) \rangle$, with "$n$" the "name" of some individual. The brackets are the ensemble average, the average over the universe in which the A's take value 1 with probability r/N and so on. We next deduce the behavior of $s^{(n)}(t)$.

In the city of Oz all persons are equivalent (ignoring the fact that one individual brought the disease) so that it doesn't matter which individual (n) is used in this definition. So we drop the "n" in the notation. Thus we simply call the expectation $s(t)$. Taking the expectation of Eq. (4.19)

$$s(t+1) = 1 - \left\langle \prod_\ell (1 - A(\ell, n, t) S(\ell, t)) \right\rangle = 1 - \prod_\ell (1 - ps(t)) = 1 - (1 - ps(t))^N . \quad (4.20)$$

The A's are correlated neither with the S's nor with each other. It is also true that to a good approximation $S(\ell, t)$ is not correlated with $S(\ell', t)$, when $\ell \neq \ell'$. (But we don't prove this here.) Writing $r \equiv Np$ and using Eq. (4.3), this becomes

$$s(t+1) = 1 - \exp(-rs(t)). \quad (4.21)$$

This is precisely the equation we had before. It describes the development of the expectation of S, and we didn't need all this formalism to arrive at its equation of motion.

Where the formalism does matter is in calculating fluctuations. We give an outline of the calculation.

1. Define

$$\tilde{s}(t) = \frac{1}{N} \sum_n S(n, t) \quad (4.22)$$

   (note the tilde on the (lower case) letter "s"). The *expectation* of this quantity, the ensemble average $\langle \tilde{s}(t) \rangle$, is what I have previously called $s(t)$. The difference is that the new $\tilde{s}(t)$ is a random variable rather than an average. As we just saw

$$s(t+1) \equiv \langle \tilde{s}(t+1) \rangle = 1 - \prod_\ell \left(1 - p\langle S(\ell, t) \rangle \right) \approx 1 - \exp(-rs(t)). \quad (4.23)$$

2. More notation: A bar over a quantity indicates an expectation value and means the same thing as a pair of brackets. Thus $\bar{u}$ is the same as $\langle u \rangle$. This alternative notation is occasionally used for ease of writing or reading. (See Appendix A.)

3. Define $\Delta s(t) \equiv \sqrt{\langle \tilde{s}(t)^2 \rangle - \langle \tilde{s}(t) \rangle^2}$, which measures the fluctuations in $\tilde{s}$. $\Delta s(t)$ is a number, rather than a random variable. Note the difference in the location of the 2 involved in squaring $\tilde{s}$. As a first remark, it is true that

$$[\Delta s(t)]^2 \equiv \overline{\tilde{s}(t)^2} - \overline{\tilde{s}(t)}^2 = \overline{[1 - \tilde{s}(t)]^2} - \overline{[1 - \tilde{s}(t)]}^2, \tag{4.24}$$

which is useful, since evaluating $1-\tilde{s}$ is easier than evaluating $\tilde{s}$. Thus $\langle 1 - \tilde{s}(t+1) \rangle = \langle (1 - p\tilde{s}(t))^N \rangle \approx \exp(-rs(t))$.

4. Looking at the square of $(1-\tilde{s})$ leads to a double sum and the $(n = m)$ vs. $(n \neq m)$ distinction is essential,

$$\langle (1 - \tilde{s}(t+1))^2 \rangle = \frac{1}{N^2} \sum_m \sum_n \left\langle \prod_\ell \prod_j [1 - S(\ell,t)A(\ell,n,t)][1 - S(j,t)A(j,m,t)] \right\rangle. \tag{4.25}$$

For $n = m$ the two products are *identical*. Since they are either zero or one, the square of the product is again the product itself. Therefore each of these "diagonal" terms gives $(1 - ps)^N$. For $n \neq m$, the two products are *independent* (or very nearly so, but let's not worry about that) and the expectation of the products is $(1 - p\langle \tilde{s} \rangle)^{2N} = (1 - ps)^{2N}$. Note that there are N "diagonal terms" and $N^2 - N$ other terms. In the case where you square before averaging, all N^2 terms are "off-diagonal," that is, they give $(1 - ps)^{2N}$.

5. Combining the terms and doing a certain amount of algebra yields

$$[\Delta s(t+1)]^2 \equiv \langle [1 - \tilde{s}(t+1)]^2 \rangle - [\langle 1 - \tilde{s}(t+1) \rangle]^2 = \frac{1}{N}\left[(1 - ps)^N - (1 - ps)^{2N}\right]. \tag{4.26}$$

For large N, and bearing in mind that $e^{-rs} = 1 - s$, this can be written

$$\Delta s(t+1) = \sqrt{\frac{s(1-s)}{N}}. \tag{4.27}$$

By now the $\sqrt{N}$ is no surprise, but what I wanted to emphasize is that, when $s \to 0$, the decline in Δs is proportional to s itself. When s drops exponentially, so do the fluctuations. If s drops like a power law, the fluctuations also survive longer. Finally, when s goes to a non-zero limit (for $r > 1$), there is a residual, persistent level of fluctuation. Note too that in this case the number multiplying $1/\sqrt{N}$ is O(1), so that (as usual) total fluctuations are $\sqrt{N}$.

Exercise 13 For $r > 1$ and out of the critical region, estimate the necessarily finite time before extinction of the disease. The idea is that, with N people, it will eventually happen that—despite its being unlikely—all attempts at transmission fail. See Footnote 30 (this chapter) for a discussion of extinction.

4.6   Self-organized criticality (SOC): Applications to galaxies and mean field theory

To get water to its critical state you need to get the pressure and temperature just right: the pressure should be 218 atmospheres and the temperature 374°C (approximately). When it is, or is near, critical you get power laws, fractal structure and other properties that should be as non-generic[18] as the critical point itself. For example, in the percolitis iteration you don't get power law lifetimes unless the parameter r is exactly 1.

Why then are there so many signs of criticality all over the place? Although some of these examples are disputed there are fractals in coast lines, power laws in the sizes of cities, in music, in $1/f$ noise and in many time series.

A general explanation was proposed in Bak et al. [19]. It is called self-organized criticality (SOC) and is a mechanism by which a system automatically becomes critical. The original example was called a sand pile.[19] I won't go into details about sand piles, but will give two examples. The first comes from the world of galactic morphology. The second is a mean field version. In the first you *almost* get SOC; in the second the SOC works but fails to account for fluctuations. A third example, freed of these defects, will not be taken up in this chapter, but will be mentioned. It will also be the subject of computer modeling in Chapter 12.

But be careful. I have recently noted that you can get power laws even if you're not at criticality. Feedback can get you close enough to see power laws, while at the same time being away from the critical point. SOC does occur, but be aware that power laws can be a sign of being near criticality, not necessarily at it. The problem is that the data are always finite, allowing for multiple interpretation.

4.6.1   SOC in galaxies

The application to galactic morphology will be discussed at length, so I won't go into details here. (See Secs. 4.9 and 11.3.) Suffice to say that the shape of spiral galaxies can be looked upon as a directed percolation problem.

A phenomenon that cannot be ignored in galaxies is hot gas. After an episode of star formation the region in which the stars are formed is no longer suitable for an explosion that gives rise to new stars. In effect there is immunity. The time scales are such that this immunity lasts longer than the time intervals for star formation, which means that, in simulations of galaxies, this effect must be taken into account. We did this in a complicated way, keeping track of the amount of hot gas that would follow star formation [206]. Nevertheless, we'll see that this can be modeled with percolitis.

But, first, a global look. Let p be the probability that a site in the galaxy where star formation has taken place gives rise to star formation in an adjacent region on

---

[18]Meaning, requiring special values of the control parameters: temperature and pressure in this case.

[19]In Bak et al. [19] the system modifies itself so as to approach criticality. The model used is called a sand pile, but actually it is a cellular automaton. The picture—in terms of sand—is roughly as follows: As sand drops from a small source it creates a pointed pile that grows ever sharper until it suddenly collapses, causing an avalanche. The timing of this catastrophe (in the model) was such that the number of pieces of sand varied wildly, showing the signs of criticality. Amusingly, actual sand does not seem to have the indicated features although sticky rice does. See Chap. 7 Sec. 6.

the next time step. In our model most sites had six nearest neighbors, so you'd expect to ignite $6p$ of them (less duplicate ignitions). Moreover if you took a value of p that was less than $1/6$ you'd expect the process to die out.

It turns out that neither expectation is correct. The principal shift is because we are dealing with an actual spatial model, and not a mean field model. But it's also true that a site does not always have exactly six neighbors, and there is a changing of neighbors because of differential rotation in the galaxy. Moreover, because of the heating of the gas, some of a site's neighbors had star formation in the recent past, so only $5p$ or $4p$ might be the probability of star formation in one of the neighbors. Effectively, the value of p has been cut down. (Of course when p gets close to $p_{\text{critical}} \approx 1/6$ the likelihood that one of a site's neighbors was recently excited also goes to zero.) In any case, the *effective* value of p has been driven closer to its critical point. This is not quite full (self-organized) criticality, but it can be close.

The other expectation—that everything dies for $p < p_{\text{critical}} \approx 1/6$—is also false for a technical reason. We include a probability for *spontaneous* star formation (i.e., *not* induced by a nearby supernova). We do that because that's what Nature does. So suppose you began with a galaxy rich in cold gas but with no star formation. With small but non-zero probability an episode of star formation would occur and suddenly the whole galaxy would light up. Then, just as suddenly (well, we're talking millions of years), all life[20] would be extinguished. This, by the way, turns out to explain the peculiarities of dwarf galaxies, which (in our static picture) show a wide variety of brightness: they are being caught at different phases of their evolution [76].

The feature of partial SOC can be demonstrated by percolation with immunity. For this we have to formulate the general theory of long-term immunity. Our usual logic is that to be healthy on day-$(t + 1)$ you need to have all attempts to make you sick fail, and the probability for that is (approximately) $\exp(-sr)$. But now there are other ways to avoid sickness. For d-day immunity, you could have been sick on day-t or day-$(t-1)$ ... or day-$(t-d+1)$. (Note that d is a number of days, not a differential.) For two Boolean variables, say A_1 and A_2, the "or" relation (with symbol "$|$") becomes

---

[20]Speaking of life, I'll mention the history of the galaxy work. It began with Conway's *Game of Life*. Martin Gardner, writing in the Mathematical Games column in *Scientific American*, featured John Conway's creation in several issues. (You can read about this at https://en.wikipedia.org/wiki/Conway%27s_Game_of_Life, accessed September 2020.) Shortly after that Seiden had a sabbatical at the Technion (Haifa, Israel), where I was a faculty member, and we explored properties, including randomness, of this cellular automaton—we even wrote a learned paper on the subject [201]. Seiden, whose regular place was at IBM (Yorktown Heights, New York, United States), invited me to visit. In those days IBM indulged in a certain amount of pure science. One of the people there was Humberto Gerola, an astrophysicist, who proposed that Conway's *Game of life,* plus randomness, was the way star formation in galaxies might work. It turned out that even simple simulations (ignoring immunity) already gave realistic images of galaxies. See [204] or [206].

But we were far from the only ones enamored by the Conway game and inspired by Gardner. The rules were simple: Every square on a monochrome checker board was either alive or dead, 1 or 0. On the next time step those living squares that had two or three live neighbors (out of eight) stayed alive. Dead squares having exactly three living neighbors came to life. In all other situations the given square was dead on the next generation. The calculation of each square's future was simultaneous. A fantastic variety of patterns emerged from this simple rule, and many were discovered in fits of playfulness. The name "life" reflected John von Neumann's creation of a self-reproducing "organism" and Conway's simplification of those rules.

$A_1|A_2 = A_1 + A_2 - A_1 A_2$.[21] Let A_1 be 1 if no one of the day-t sick people transmitted the disease to a given individual and let A_2 be 1 if that individual was sick any time in the last d days (in both cases the Boolean variable is zero otherwise). Then the mean field expectation immediately gives[22]

$$1 - s = e^{-rs} + ds - e^{-rs}ds. \tag{4.28}$$

Expanding $\exp(-rs) \approx 1 - rs + r^2 s^2/2 + \cdots$ and keeping only terms to order s^2 this implies

$$s = \frac{r-1}{r(d+r/2)}. \tag{4.29}$$

Recall that, without immunity, s was $2(r-1)$, so that immunity has cut down s considerably (remember, $r > 1$). Focusing on r one could say that one has started with some r that may be far from 1, but its effect (as measured by its distance from 1) has been reduced by a factor $r(2d+r)$.[23]

4.6.2   SOC in a mean field theory

This example [68] is a simple variation of percolitis. It is artificial and a chemistry metaphor is probably better,[24] but I will stick to the language of epidemiology. Now, however, instead of immunity for a day, we'll make it a lifetime. All die at the same time and all are born in the same season (so we're not talking about humans). Percolitis is spread by contact (as before), with all having contact with all. This time though the disease has some negative consequences. The next generation is created by pairs.[25] The negative consequence is that if *both* parents had percolitis there is a small lowering of fertility. On the other hand, if neither parent had percolitis or if only one did, then the average number of offspring is two.

   In qualitative terms it's clear what will happen. The transmission probability (which stays constant) may initially be sufficient to have the percolitis probability well above zero. In that case there will be many pairs with both partners stricken, so

---

[21]To see that $A|B = A + B - AB$ just consider the four possible cases, letting each of A and B take the values 0 and 1. (These A's are different from those considered in Sec. 4.5.)

[22]In general for d Boolean variables $A_1|A_2|\cdots|A_d = \sum A_k - \sum A_k A_{k'} + \sum A_k A_{k'} A_{k''} - \cdots$. The sums run over all possible single variables, pairs, triplets, etc., until you run out of new combinations. For our situation it is not necessary to go to this level of generality. A single Boolean variable can be 1 if *at any time* in the last d day(s) the individual was sick, zero otherwise.

[23]The case $d = 0$ is acceptable because the formula for s is actually $s \approx 2(r-1)/r^2$; it's just that near $r = 1$ the denominator can be replaced by 1.

[24]The chemistry metaphor is a plausible reaction. There is a total of N molecules, consisting of three sorts: S, I and H (sick, inert and healthy for percolitis, but a disease would not be the metaphor now). You have S→H, a spontaneous reaction with a certain rate (1/day in the usual percolitis, which this is not). Then you have H going to S *in the presence of an* S (H+S →2S) at another rate and finally 2S→ S+I. An alternative is 3S→2S+I, which surely makes more sense as chemistry. See Chapter 12, where there is a slight change of notation (R (recovered) in place of H (healthy)). The case of three "parents" is considered—the three is why the chemistry metaphor is better.

[25]Another requirement is that the pair be randomly selected. Since this is a mean field model, there's no harm in having the parameter α take a smaller value to allow for the male/female distinction. No such requirement exists for the chemistry model.

there will be reduced fertility and "N" will decrease, until the transmission probability becomes critical for that population. (Recall that $r = pN$, so that for constant p decreasing N, means decreasing the r.) On the other hand, the *rate* of decrease will itself decrease, since, with few getting percolitis, pairs of sick individuals will become increasingly unlikely.

In quantitative terms two changes are necessary. First, N becomes a function of time, $N(t)$. Second, it is no longer best to scale the transmission probability with N, so that p is simply a small number, designated $1/M$ with $M \gg 1$. Define $\nu(t) \equiv N(t)/M$ and assume $\nu(0) \gg 1$. The time evolution is then

$$s(t+1) \equiv \frac{S(t+1)}{N(t+1)} = 1 - \left(1 - \frac{1}{M}\right)^{S(t)} \to 1 - e^{-s(t)\,\nu(t)}, \tag{4.30}$$

with $S(t)$ the number sick at time-t. This equation is derived exactly as in Eq. (4.4). For $N(t)$ we need an additional equation. It is

$$N(t+1) = N(t)\left(1 - \alpha\left(\frac{S(t)}{N(t)}\right)^2\right), \tag{4.31}$$

with α a parameter that gives the degree to which fertility is reduced. In words, the number of pairs of percolitis sufferers is proportional to the square of $S(t)$. Divide by M to get

$$\nu(t+1) = \nu(t)\left(1 - \alpha s(t)^2\right), \tag{4.32}$$

which is just Eq. (4.31) divided by M.

We first look for fixed points. The fixed points of Eq. (4.30), call them $s_0(r)$, are given by solutions of $s_0(r) = 1 - \exp(-\nu s_0(r))$. They are already known and are

$$s = 0 \text{ for all } \nu, \text{unstable for } \nu > 1, \tag{4.33}$$

$$s > 0 \text{ for } \nu > 1. \tag{4.34}$$

But the solution must also be a fixed point of Eq. (4.32). These are

$$s = 0 \text{ allows all } \nu, \tag{4.35}$$

$$s > 0 \text{ implies } \nu = 0. \tag{4.36}$$

We conclude that the only stable fixed points are $(\bar{\nu}, \bar{s}) = (\nu, 0)$ for $\nu \leq 1$. More to the point, starting from $\nu > 1$ we must arrive at $(1, 0)$, and the system stops there. But having ν approach 1 from above leads to the critical point, because $\nu = 1$ implies $N = M$ and the disease dies out. It has reached its critical point.

It follows that, even though for pure percolitis having the transmission probability equal to $1/N$ is not generic, in this model the system spontaneously arrives at the critical point. This is self-organized criticality.

Now the point $(\nu, s) = (1, 0)$ had better be stable, and we can also check the rate of approach to equilibrium. Taking Eqs. (4.30) and (4.32) to lowest order and setting $\nu = 1 + \delta$ (so both s and δ are small) gives

$$s' = s + \delta s - \frac{1}{2}s^2 \quad \text{and} \quad \delta' = \delta - \alpha s^2, \tag{4.37}$$

where no prime means the quantity is evaluated at time-t and primed variables are evaluated at time-$(t + 1)$. After sufficient time, the changes in s and δ will be small and Eq. (4.37) can be replaced by

$$\frac{ds}{dt} = \delta s - \frac{1}{2}s^2 \quad \text{and} \quad \frac{d\delta}{dt} = -\alpha s^2. \tag{4.38}$$

These equations are satisfied by setting $s = \frac{\bar{s}}{t+t_0}$ and $\delta = \frac{\bar{\delta}}{t+t_0}$ provided

$$-\bar{s} = \bar{\delta}\bar{s} - \frac{1}{2}\bar{s}^2 \quad \text{and} \quad -\bar{\delta} = -\alpha\bar{s}^2. \tag{4.39}$$

$\bar{\delta}$ is eliminated and since s is assumed positive we can divide by $\bar{s}$. This yields

$$\bar{s} = \frac{1}{4\alpha}\left[1 \pm \sqrt{1 - 16\alpha^2}\right]. \tag{4.40}$$

The conclusion is that the self-organized critical behavior has the same time dependence as the non-generic critical behavior, provided $\alpha < 1/4$. In Chapter 12 we discuss the case of $3S \rightarrow 2S+!$, which, even when fluctuations are included, exhibits SOC. For the transition matrix see [69]; Chapter 12 discusses the stochastic process.

4.7   The truth about percolitis

Back to non-SOC.

As remarked in Footnote 4, as $r \rightarrow 1$, the limits $s_0(r) \rightarrow 0$ and $N \rightarrow \infty$ are not independent. In particular we used the identity $e^{-x} = \lim_{M \rightarrow \infty}\left(1 - \frac{x}{M}\right)^M$ (see Footnote 3 for a proof), where what does the job of M is the product $Ns(t)$ (for a particular r and ultimately $s(t) \rightarrow s_0(r)$). In Fig. 4.6 I show how $\left(1 - \frac{x}{M}\right)^M$ approaches its limit.

This implies that the critical properties we studied in connection with percolitis were properties of the mapping, Eq. (4.4), rather than of the stochastic process defined by percolitis. Near criticality, the mapping thus takes on a life of its own, but if one is curious about percolitis itself it is better to use the transition matrix $R(k, j)$, the probability that a state with j sick people goes over to a state with k sick people. This will be used (in Footnote 30) to look at extinction rates for high r values, but now it is used near criticality. Since it is exact, it can be used in either context.

First we justify the formula given there:

$$\text{Prob}(j \text{ sick lead to } k \text{ sick on the next time step})$$
$$\equiv R(k, j) = C_k^N q^{j(N-k)}[1 - q^j]^k, \tag{4.41}$$

with C_k^N the combinatorial coefficient, $C_k^N = N\text{-choose-}k = \binom{N}{k} = \frac{N!}{k!(N-k)!}$ and $q = 1 - r/N = 1$ minus the transmission probability. Here is the justification: You have j

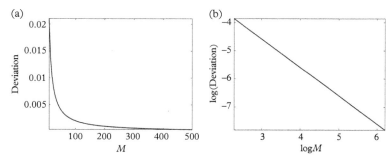

Fig. 4.6 Quality of the limit. The value of x used is $1.1/N \sim 0.028$ with $N = 400$. In the text I explain why I use this particular x value. Figure (a) is the difference of e^{-x} from $\left(1 - \frac{x}{M}\right)^{M}$ and figure (b) is the logarithm of the difference. Note, especially from figure (b) which has slope very close to -1, that the difference goes to zero like $1/M$.

sick, it doesn't matter which j. Assume the k *to-be-sick* individuals are specified (they could be 1 through k). Each of them is sick, which, being 1 minus the probability of being healthy, means that the probability of getting sick is 1 minus $(1 - p)^{j} = q^{j}$, the latter the probability that all attempts and transmission fail. Raising $(1 - q^{j})$ to the kth power (once for each sick individual) gives the factor $[1 - q^{j}]^{k}$, which appears on the right of Eq. (4.41). All the other individuals must be healthy—that is, all attempts at transmission fail. For one individual this would be q^{j}; for $N - k$ of them you simply take that power. Finally those k individuals could be any k from among the N available, leading to the combinatorial coefficient.

For the extinction problem it can be useful to add a small probability of spontaneous development of the sickness, so as to make R irreducible, but here we take the pure process. Our population (N) values will be large enough so that aspect is negligible.[26]

For our purposes there are two principal differences between true percolitis and Eq. (4.4) (the basic mean field iteration). First, the true probability distribution goes to zero exponentially for $r = 1$. In fact, it *always* goes to zero, but it can be exponentially small in N (what we've been calling extinction) and the coefficient also depends on r. At $r = 1$ the value of R's *second* eigenvalue dominates, the first always being 1 (a property of stochastic matrices). Even for N values that one would think are large, this second eigenvalue is well separated from 1 as well as from the eigenvalue that's smaller than it. We have in mind an eigenvalue expansion (a.k.a. spectral expansion) of R:

$$R = \sum_{\ell} \lambda_\ell |\ell\rangle\langle\ell| , \qquad (4.42)$$

with ℓ labeling the eigenvalues, say in decreasing value, and the projection $|\ell\rangle\langle\ell|$ being the product of the right and left eigenvectors of R.[27] (In general you need to say

---

[26] As remarked, our matrix R is reducible. It therefore could in principle have additional eigenvalues that are 1 (consider the identity matrix). But for our system we know that there is no other permanent state, so that possibility doesn't arise.

[27] Here is more detail for those unfamiliar with bra-ket notation. Eq. (4.42) can be written more fully as $R(x, y) = \sum_{\ell} \lambda_\ell \langle x|\ell\rangle\langle\ell|y\rangle$, where $\langle x|\ell\rangle$ is a right eigenvector, (i.e., $R(u, v)\langle v|\ell\rangle = \lambda_\ell \langle u|\ell\rangle$ and $\langle\ell|y\rangle$

more—such as details concerned with Jordan forms, complex λ and degeneracy—but here this is not necessary.) We can call the eigenvalue 1, λ_0, so that it would be the behavior of λ_1 that would dominate—provided λ_2 is not too close to λ_1. In practice this is the case.[28] The state $|0\rangle$ is the stationary state, but in this case it is trivial: nobody is sick, which if you'd written in terms of numbers of people would give $\langle n|0\rangle = \delta_{n0}$, the latter being the Kronecker delta.[29] Thus the probability of any non-zero number being sick is dominated by R^t acting on a state with a small number, n, of sick people, which is to say $\lambda_1^t \langle n|1\rangle$, an exponential (where the numbers on the left and right are different labels, one for the state ("1") and one for the number of people ("n"), something like the different arguments appearing in $\langle x|\psi\rangle$ in quantum mechanics).

Remark: It's remarkable that even in the exact theory, $r = 1$ retains a special property. It's either at that value (1), or close to it, that the second eigenvaiue (i.e., λ_1, the first being 1) begins to be a power of the size of the city (N). This is also true of λ_2, λ_3 and probably others (I haven't checked). Thus $\lambda_1 \sim 2.1518 N^{0.4772}$, $\lambda_2 \sim 0.5242 N^{0.4965}$ and $\lambda_3 \sim 0.2670 N^{0.5032}$, to excellent approximations. (For $r < 1$ this continues, but for $r > 1$ the power law is no longer a good fit.) Note, by the way, that the exponent of N is increasing faster than linearly, meaning that nothing catches the exponential decay—in other words, one does not have a situation like that described in Eq. (4.43) below.

Given the time dependence governed by R, one can compare the long time limit with that of Eq. (4.4). For the mapping $s(t+1) = 1 - \exp(-rs(t))$ (which is Eq. (4.4)) we already know its limiting values (shown in Fig. 4.2). To compare this with true percolitis we should look at R^T on some initial state (anything with a reasonable number of sick people, so it doesn't vanish identically) for a large value of T. As for the single number $s_0(r)$ that we obtained, we can define a limiting value for the distribution. For N large enough and T not super-large the mean value of the number of sick people takes the place of $s_0(r)$. In Fig. 4.7 I show a comparison of the two results for $N = 100$. Note that the results of using R are in excellent agreement with $s_0(r)$, but *only* for $r \gtrsim 1.28$. For lower values there is significant divergence. This point of divergence decreases with increasing N, so that, for example, for $N = 200$ and 400 they diverge at 1.16 and 1.10, respectively. (This explains the 1.1 used in Fig. 4.6.)

---

is a left eigenvector $(\langle \ell|u\rangle R(u,v) = \lambda_\ell \langle \ell|v\rangle)$. Usually we use the notation "p_ℓ" for right eigenvectors and "A_ℓ" for left eigenvectors, but here we try to connect to those who find the quantum mechanical notation more transparent. Note however that left and right eigenvectors can differ from one another. Also, $\langle x|\ell\rangle\langle \ell|y\rangle = p_\ell(x)A_\ell(y)$ (either expression) is a projection operator.

[28] At $r = 1$ I list the first few eigenvalues for various N values.

N	λ_1	λ_2	λ_3
100	0.9912	0.8890	0.7689
200	0.9982	0.9275	0.8495
400	0.9999	0.9404	0.8990

The dominance of $|1\rangle$ over $|2\rangle$ can be measured by $-1/\log(\lambda_2/\lambda_1)$, which for the three N values listed here is 9.19 (100), 13.61 (200) and 16.30 (400). Although, with increasing N, the state $|2\rangle$ does survive longer (and make the decay non-exponential for the indicated time periods), compared to the value of T that I use (10,000) it is negligible. My expectation is that for much, much larger N, times longer than 10,000 may be needed, but eventually the decay would be exponential.

[29] The Kronecker delta, δ_{jk}, is one for $j = k$, zero otherwise.

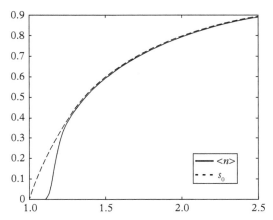

Fig. 4.7 Comparison, for $N = 100$, of the predicted asymptotic values of the transition matrix (designated $\langle n \rangle$ and $s_0(r)$). The dashed line is the result of using the relation $s(r) = 1 - e^{-rs(r)}$; the solid line is the exact result.

So the values of $s_0(r)$ obtained earlier are incorrect, except that the larger the N, the better the value of $s_0(r)$ for a given r.

What about SOC? This was studied in Sec. 4.6. In principle you should write down the appropriate transition matrix and look at the spectrum near 1. Since there is always a gap you would always get an exponential, but there would be signs of a power law if the spectrum looked like it was going to an appropriate limit. Thus you might find that the spectrum decreases in fixed amounts from 1, that is, $\lambda_n = 1 - n\Delta\lambda$ with $\Delta\lambda = 1/N$. This gives a power law in the $N \to \infty$ limit:

$$f(t) = \sum \lambda_\ell^t \to \int_0^1 dx\,(1-x)^t = \int_0^1 du\,u^t = \left.\frac{u^{t+1}}{t+1}\right|_0^1 = \frac{1}{1+t}, \tag{4.43}$$

a power law.

It turns out that it's easier to write down a master equation by allowing Δt, the interval between transitions, to go to zero. In that limit the mean field SOC studied in Sec. 4.6 does *not* lead to a power law. But if you could imagine three parents for each child, you *would* get SOC. If imagining three parents does not conform to your image of Nature, change metaphors. The chemical metaphor is that there are three species and the total number remains constant. You can think of the species as S = sick, H = healthy and I = inert. Thus the model of Sec. 4.6 would have a rate for S $\to$ H and a rate for S + H $\to$ 2S. In addition there would be a reduction in the susceptible population. This would be another rate constant for 2S $\to$ S + I—so you never reduce the total population, but some members become "inert," never to be sick again. Using the chemical picture you would have a reaction 3S$\to$2S+I (instead of 2S$\to$ S+I), which *suppresses* fluctuations and allows SOC.

The proof of the assertions of the last paragraph is quite technical and will not be given here. See [69]. In that article R is "recovered" and plays the role of H, meaning "healthy" here. The idea is to look at the spectrum of the time evolution matrix. As remarked earlier, there is always exponential decay, but in the 3S$\to$2S+I case the

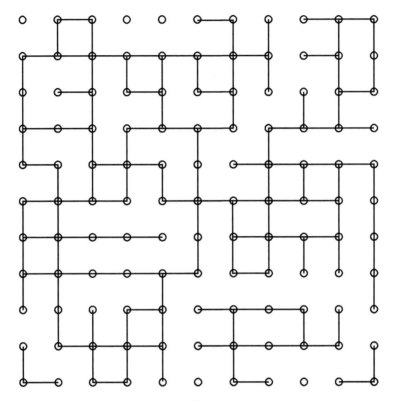

Fig. 4.8 A wire grid with approximately 60% of its links intact.

closest eigenvalue to one is at a distance $1/\sqrt{N}$ from one and in the limit (of $N \to \infty$) gives a power law decay. See Chapter 12.

4.8   Abstract percolation

A simple model of percolation is a wire mesh with a fraction $(1-p)$ of the node-to-node wires cut (so a fraction p survives). See Fig. 4.8. One asks: Does this mesh conduct electricity from the left to the right? Equivalently, can one find a path along uncut wires from one side to the other?

Obviously, when the cutting is random, sometimes it will and sometimes it won't. If p is close to one, usually it will; close to zero, usually it won't. The remarkable thing though is that, for large grids, there's a critical value, p_c, such that there is almost always conduction if $p > p_c$ and almost never conduction if $p < p_c$. How accurate the "almost always" is depends on the grid size. An example is shown in Fig. 4.9.

The percolitis epidemic can be thought of in this way, but as a variant called *directed percolation*. Imagine a horizontal line with N nodes, each node representing a resident of Oz. Now draw a copy of that line and place it above the first. This is a second copy of the town. Next, with probability p, from each node on the first line draw a line with an arrow to a node on the second line, above it. (So on the average

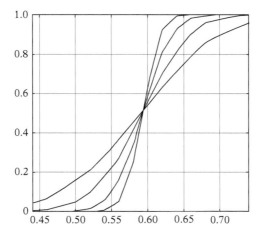

Fig. 4.9 Percolation probability as a function of p (bond occupation probability) on an N-by-N square lattice. Values: $N = 10, 20, 40, 80$ and the probability, p, taken from 0.44 to 0.74 in intervals of 0.02. The probability of having a path from left to right (in a diagram such as Fig. 4.8) is a function of p. Several values of N are shown, from small (the most gentle dependence) to large, where the connectivity is much steeper. The number of repetitions is 6400, hence the deviations. It is known that the threshold for percolation is $p \approx 0.59$.

there will be Np such lines.) Think of the lower line as day-1 and the line above it as day-2. A line from (say) lower node #31 to upper node #17 is interpreted as "*if #31 is sick on day 1, #17 will be sick on day 2.*" So the filling in of such a line is *exactly* the same as having the random variable $A(31, 17, 1)$ (defined in Eq. (4.17)) take the value 1.

Continue with day-3, day-4, and so on. Now imagine you start from a sick node at the bottom of the rows. If you can continue the line upward indefinitely, then, by our first interpretation, this means the epidemic persists. But in that case you can draw a line from one end (the bottom) to the other ("infinity" at the "top").[30] Hence

---

[30] A technical point is that, since $N < \infty$, eventually the "disease" (or the line) *will* stop, no matter how large p is, so long as it is less than 1. The time for this to happen grows exponentially with N, which can be seen as follows. If the typical number of sick people is n (which is proportional to N when the epidemic settles down, so $n = s_0 N$, with $0 < s_0 < 1$), then the probability that *every* attempt to infect on the next step fails for *every* individual is $(1 - p)^{nN}$, or, writing $1 - p = 1 - r/N$, it is $\exp(-s_0 r N)$. Since N is assumed large, we usually ignore this exponentially small eventuality. More precise estimates of the likely time to extinction can be obtained using the transition matrix. (See Appendix D.5.) One looks at the second largest eigenvalue of the master equation. Let $R(k, j)$ be the probability of going from exactly j sick to exactly k sick. Then $R(k, j) = C_k^N q^{j(N-k)}[1-q^j]^k$, with C_k^N the combinatorial coefficient, N-choose-k and $q = 1 - r/N = 1$ minus the transmission probability. (See Sec. 4.7.) This is a stochastic matrix with largest eigenvalue one. It is reducible and its *second* largest eigenvalue (call it λ_1) is associated with the long-lived metastable state (for $r > 1$). The lifetime is then $\tau = -1/\log(\lambda_1)$. Numerically, $\log \tau$ appears to have the dependence $r s_0 N$, but τ is roughly 50% smaller for r in the neighborhood of 1.5. It would seem that the system finds a better decay route than the simple one of my estimate, but I haven't investigated its nature. Alternatively one can make R *irreducible* by including a tiny probability for *spontaneous* development of percolitis—call this p_{spon}. So long as p_{spon} is small compared with $(1 - \lambda_1)$ (which is exponentially small in N) this will not affect conclusions about the system drawn from the modified R.

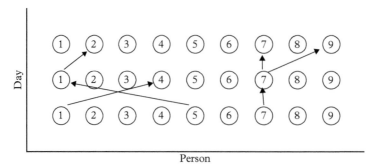

Fig. 4.10 Percolitis as directed percolation. On the first day, persons #1, #5 and #7 are sick. #1 infects #4 (the next day), #5 infects #1 and #7 infects herself. The following day #1 infects #2, #4 infects no one, and #7 infects both herself and #9.

this epidemic has a percolation interpretation. See Fig. 4.10. ("Indefinitely" in this case doesn't mean indefinitely, since there's always the chance of extinction. But if N is large enough and the transmission probability is significantly above $1/N$, the persistence of the disease goes on for a long time, exponential in N.)

4.9   Percolation applications

Ground water percolation: Consider a sandy soil, so that sometimes the space between grains is large enough for water to flow through and sometimes it is not. You can imagine that the bigger spaces are nodes with the (potential) passageways being the thinner spaces between them. See [55], where water flow through cracked rocks is considered. In neither case is there a regular array, but one can still idealize the problem in terms of abstract percolation.

Galaxies: Star formation in one region of a galaxy is often the result of a supernova pressure wave from a large star that was born in a previous star formation episode a few million years earlier. Thus "star formation" can be thought of as a persistent epidemic in a galaxy. For details see Schulman and Seiden [204].

Neurology: Excitation of one neuron can set off excitation in others. Here things are seriously more complicated because there are also inhibitory signals. Another important factor is that there is a refractory period: after excitation the neuron will be quiescent for a while. This is also present in the galaxy model, and in the language of percolitis can be incorporated by assuming partial immunity for a period of some days. The phenomenon to be explained is certain oscillations in the brain, which (sometimes) turn out to be a consequence of the interplay of percolation and refractoriness. For details see Traub and Miles [224]. It's amusing though that the time scales are so different (millions of years, days, milliseconds), despite the similarity of mechanisms.

Another application of neurology—surprising to me—is in thought experiments associated with a "balanced" cortex. This is the work of Shew and Plenz [210]. They imagined 100 coupled neurons and concluded that criticality, and a balanced cortex, occurs when the probability of excitation (translate: transmission) is 1/100, that is, $1/N$ with N the number of neurons. For more on their work, see Sec. 10.7.

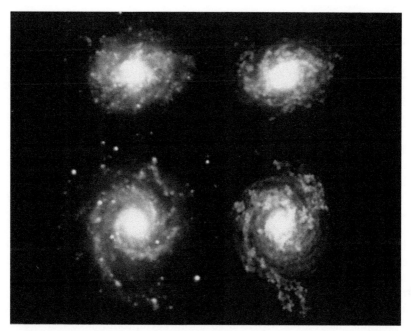

Fig. 4.11 On the left are actual galaxy images. On the right are the simulation results. These are based on random processes, so there is no reason to have exact correspondence. It's the qualitative similarity that's of interest. Taken from [204], which was a collaboration with Phil Seiden.

4.10 True epidemiology

In Footnote 5 we discussed true epidemiology, and with the coronavirus (COVID-19) rampant (by the time this is published hopefully that will be a memory) it is worth discussing what it would take to make our percolitis model relevant to actual diseases. The objective is to reduce deaths and disability, not necessarily the incidence of the disease. This is an important distinction. At least two epidemiologists disagree (as of 2020—they may have changed their minds) with many of the objectives of those who are trying to reduce the incidence of the disease (Knut Wittkowski and John Ioannidis). Both point out that, although this version of the virus may have a higher fatality rate, had they been treated as variants of the flu they would have been background noise.[31] Of particular interest is the concept of "herd immunity," which is a situation where enough people have had the disease that it's no longer likely to be transmitted. A figure of 80% was mentioned as signifying herd immunity. That seems to me to be low and the "herd" is likely to have local variation (think about recent experience with measles). The "herd" in this case would be school children, who are (usually) asymptomatic but most carry immunity once infected.[32] Ioannidis's survey

---

[31]Yes, "noise" represents many deaths. But epidemiologists are, by necessity, a hard-hearted lot, like traffic engineers (see the end of Sec. 9.8.1).

[32]The question of immunity to COVID-19 is a difficult one. What seems to be the case is that most people develop at least temporary immunity—not everyone, but most—and how long it lasts is

of a county in California has been disputed, and he seems to be in a minority (to say nothing of Wittkowski). The truth? Maybe by the time this book is published it will be known, but, right now, no one can say. (I'm not sure how they would account for initial experience in Italy and New York City, but I'll leave it to them to explain.)

What would it take to make percolitis more realistic? There are two features that must be taken into account. The first is immunity. As indicated, presently there rage disputes over this issue (for the coronavirus), but let's suppose that there is immunity and it lasts long enough for the disease to disappear. But the parts that make percolitis difficult are the latency period and asymptomatic carriers. There is a period, for COVID-19 this is typically 5 days, during which people who will ultimately come down with the disease can transmit it but are themselves asymptomatic. From the perspective of a potential victim this means that everyone (except those who have already had the disease) is a potential carrier. From the perspective of a percolitis modeler, you must keep track of individuals. This changes to a problem in which individuals carry a label. In addition there is the matter of local communities, but that is a weakness of percolitis in general. Nevertheless, if one is willing to keep track of individuals the percolitis model is adaptable.

It turns out that although individuals matter, if there is overall immunity, only the total number at each stage of the disease needs to be counted. I've written a simple program that allows a person to be an asymptomatic carrier but with total immunity. Not surprisingly the higher the transmission probability, the more people get infected. However, it is unrealistic to imagine that one can confine the population to relatively small N. People travel. The version of the virus that is plaguing us these days started in China (and lots of travelers from there brought it to the west coast of the United States). However, it also seems that the version that hit New York (where there were many deaths) apparently went through Europe first. In any case, here's my (MATLAB™) adaptation of percolitis:

```
function out=percolitis_enhanced(in)
% Percolitis with enhancements: latency and persistence.
%      Immunity is permanent.
if ~exist('in'),in=struct;end     % If program is run without
                                  % alternative data, defaults are used.
                                  % But to change from defaults a
                                  % "structure," "in," must be created.
[N,in]=setdefault(in,'N',1e3);    % Number of people
[T,in]=setdefault(in,'T',2e3);    % Total time of observation
[tP,in]=setdefault(in,'tP',5);    % Time people remain asymptomatic
[tS,in]=setdefault(in,'tS',6);    % Time people can have symptoms.
                                  % Follows tP.
[p,in]=setdefault(in,'p',100/N);  % Transmission probability.
                                  % Could have different values
                                  % before and after symptoms appear
```

unknown. Nevertheless, there does seem to be enough so that, if the infection occurrence is sufficiently widespread, herd immunity, however temporary, would emerge.

```
                           % (but are = in present program).
[seed,in]=setdefault(in,'seed',1204); % Random seed
[show,in]=setdefault(in,'show',1);     % Whether or not to show figures.
[sickfrac,in]=setdefault(in,'sickfrac',.01); % Fraction initially sick
out.input=in;
rng(seed)                              % "rng" is a matlab function fixing
                                       % the random number generator.
s0=ceil(sickfrac*N);                   % "ceil" rounds up (if necessary)
                                       % to give an integer.
s=zeros(1,T+1); s(1)=s0;
status=zeros(1,N); status(1:s0)=1; infected=s0;
for t=1:T,
   firstzero=find(status==0,1);   % The "1" at the end means find the 1st
   if numel(firstzero)==0, firstzero=N+1; end
   status(1:(firstzero-1))=1+status(1:(firstzero-1));
   carriers=find(status(1:firstzero-1)<tP+tS);
                                  % (have already added 1)
   ncarriers=length(carriers);
   infect=(rand(1,ncarriers)<p); % (only step using randomness)
   ninfect=length(find(infect>0));
   status(firstzero:min(N,firstzero+ninfect-1))=1;
   newlyinfected=length(find(status==1));
   infected=infected+newlyinfected;
   s(t+1)=length(find(status>0 & status<(tP+tS)));
end
if show==1,
   figure(100),plot(1:(T+1),s),axset(1,1,2,T+1,3,0)
   range=(tP+tS+1):(T+1);
   figure(110),plot(range,s(range)),axset(1,1,2,T+1,3,0)
end
out.sickness=s;
out.infected=infected;


%-----------------------------------------
function [a,in]=setdefault(in,a_string,a_default);
% "in" is a structure. a_string is a string variable.
% The variable to be set is "a", appearing as a_string.
% That is, a_string='a'
if isfield(in,a_string),
   eval(['a=in.',a_string,';']);
else
   a=a_default;
   if nargout==2,
      eval(['in.',a_string,'=a;'])
```

```
      end
end
```

Explanation: The variable `status` keeps track of who does not and who does have the disease and how long they've had it. Those who do *not* are the ones who have zeros, starting with `firstzero`, and includes others at higher index values. For those who have the disease, `status` keeps track of how long ago they got it, and ceases to consider them either sick or potential transmitters after `tS+tP` time steps. `status` is (automatically) a variable with declining values as a function of index. s keeps track of the total sick at any given time. `setdefault` is a program that I use again and again. It gives standard values to variables. To change them (to values other than the default, e.g., $N = 500$) use the command line `in.N=500; out=percolitis_enhanced(in);` (in the "command" window). Possibly it will be necessary to precede this with `clear in;`, in case the variable (or "structure" in MATLAB™ terminology) `in` has already been defined.

It is also possible to run percolitis_enhanced.m for several p values and several seeds. There follows a program (unimaginatively called run_percolitis_enhanced.m) to do this. Again, I use MATLAB™:

```
function out=run_percolitis_enhanced(in)
if ~exist('in'),in=struct;end
% parameters for the runs:
[N,in]=setdefault(in,'N',500);            % # of people
[T,in]=setdefault(in,'T',300);            % Time of observation
[tP,in]=setdefault(in,'tP',5);            % Time asymptomatic
[tS,in]=setdefault(in,'tS',7);            % Time sick
[ps,in]=setdefault(in,'ps',[10:5:70]/N);% Probabilities of transmission
                                          % Testing several values
                                          % (the p's are called ps)
[seed,in]=setdefault(in,'seed',222);      % Random number generator
[Nrepeat,in]=setdefault(in,'Nrepeat',16);% # times each value is checked
% Giving the input structure (inp) for running various p's a different
% name, to distinguish it from the original input structure (in)
inp.N=N;inp.T=T;inp.tP=tP; inp.tS=tS;
inp.show=0;                               % input to percolitis_enhanced is inp.
rng(seed)  % fixes random number generator to make results reproducible.
np=length(ps);
infect=zeros(np,2);
for kp=1:np,
    inp.p=ps(kp); % The p value for each of the Nrepeat runs.
    infected=zeros(1,Nrepeat);
    for k=1:Nrepeat,
        inp.seed=seed+10*k;
                % A different seed for the random generator for each run.
        out=percolitis_enhanced(inp);
```

```
      infected(k)=out.infected;
   end
   infect(kp,1)=mean(infected);
   infect(kp,2)=std(infected);
end


% Following info is from
% https://www.cdc.gov/nchs/nvss/vsrr/covid19/index.htm
% Accessed on 20200423 (Date: yyyymmdd)
% Purpose: to establish the number of people who died.
bracket=[1,4,14,24,34,44,54,64,74,84,120];
                  % Never used. Upper value of each age range.
profile=[3848208,15962067,41075169, 42970800,45697774,...
    41277888,41631699,42272636,30492316,15394374,6544503];
                  % dots are continuation of a line
deaths=[0,2,1,21,183,462,1257,2993,5093,6429,6917];
prob_death=sum((profile/sum(profile)).*(deaths./profile));
death=infect(:,1)'.*prob_death;
out.dead=death;
NYC_deaths=8072;
NYC_population=8.4e6;
NYC_dead=(NYC_deaths/NYC_population)*infect(:,1)';
out.NYC_dead=NYC_dead;
```

For large p (more than $20/N$ with $N = 500$), *everyone* is infected; the overall death rate for New York City (statistics as of April 23, 2020) is about 1 in 1000—that's the fraction who die (although all are infected). This does seem to be worse than ordinary infection and "social distancing" over a long period of time can be difficult. In any case, all I'm offering is a program with obvious deficiencies.

5

Ferromagnetism

Applications discussed in this chapter: Ferromagnetism, Collective behavior, Fluctuations: $N^{1/2}$ and $N^{1/4}$, Curie–Weiss theory

---

Ferromagnetism is a classical topic in statistical physics and Onsager's solution[1] of the two-dimensional Ising model (to be defined) led the way to the modern understanding of phase transitions. My reason for including magnetism here is that it will give us the opportunity to see *non-square-root-N* fluctuations. The fluctuations at the critical point turn out to be much larger than elsewhere, and this is ascribed to significant cooperation among the degrees of freedom. Does this happen in other contexts? Absolutely. Large fluctuations worry everyone from swimmers to insurance companies.

One reason why it will be easy to demonstrate the large fluctuations here is that, rather than study a stochastic process, we will use a *partition function*. Along the way we'll define a stochastic process, but the actual calculation will involve the partition function.

First, a description of the physical system: Electrons carry an intrinsic magnetic moment. A single such magnet is like a compass needle when placed in an external field. However, a pair of electrons inside atoms may behave differently from a pair of compass needles. Macroscopic magnets will have their lowest energy when they are pointed opposite to one another, but for the electrons, say in iron or nickel, the opposite is the case: they tend to align. This is because of "exchange forces" related to the fact that electrons have anti-symmetric wave functions.[2] The details need not

---

[1] Lars Onsager [171], Norwegian, 1903–1976, won the Nobel Prize. Taught at Johns Hopkins where he was fired for being unable to teach, at Brown, where the same thing happened. Went to Yale, where they (apparently) didn't care. But they did care that he should have a PhD, so he sent a thesis back to Trondheim (where he'd studied). Getting the degree took a while since no one could understand the thesis. (Perhaps I exaggerate—but I was told this by a Norwegian scientist in Trondheim!) I can personally testify to Onsager's deficiency as a speaker, having listened to him for an hour's worth of lecture.

[2] The Pauli principle is the requirement that electrons (and other particles with half-integer spin) must not be in the same state. It is supported by relativistic field theoretic proofs. For Fe, Ni and a number of other elements, it is important that the spatial states of a pair of electrons be different, which means that the spin states are the same—and parallel. In other words, there's a bigger loss of energy by having the spatial states identical than by having the spins (hence the magnetic moment) point in the same direction.

When Things Grow Many. Lawrence S. Schulman, Oxford University Press.
© Lawrence S. Schulman (2022). DOI: 10.1093/oso/9780198861881.003.0005

concern us here (and are subtle). Suffice to say that one model of such a system is the Ising model, in which the spins can only take two values—plus or minus one. There is the further assumption that the magnetization of each spin is the spin itself. Any units that *should* be there are absorbed into the coupling constants (J snd h), which have units of energy. The total energy of an array of spins is given by

$$E = -J \sum_{\langle ij \rangle} \sigma_i \sigma_j - h \sum_i \sigma_i . \tag{5.1}$$

Here σ_i ($= \pm 1$) is the spin of particle i, which is assumed to be arranged on some sort of lattice. The first sum is over "$\langle ij \rangle$," meaning pairs ij, such that the spins interact with one another, for example they might be nearest neighbors on a two-dimensional lattice. J is a coupling constant, which when positive can yield ferromagnetism (spins tending to point in the same direction, as in iron). The second sum is simply the effect of an external field, h. This is a classical model that captures some, but certainly not all, features of actual magnets.

Let the overall configuration (on the entire lattice) be designated $\tilde{\sigma}$; thus $\tilde{\sigma} = (\sigma_1, \ldots, \sigma_N)$ gives the value of each spin,[3] so that one can write $E = E(\tilde{\sigma})$. At this point various kinds of equilibrium can be contemplated. There's the *microcanonical* ensemble in which each state—each configuration, $\tilde{\sigma}$—has equal probability. Then there is the *canonical* ensemble, in which the system is in contact with a reservoir of temperature T; it can exchange energy with this reservoir, but nothing else. A basic postulate[4] of statistical mechanics assigns probability to each state under these circumstances. This distribution, known as the Boltzmann distribution, is

$$\Pr(\tilde{\sigma}) = \frac{1}{Z} \exp(-E(\tilde{\sigma})/k_B T) , \tag{5.2}$$

with Z a normalization constant, so that $\sum_{\tilde{\sigma}} \exp(-E(\tilde{\sigma})/k_B T) = Z$. (The Z stands for *Zustandssumme* in Boltzmann's lingua franca. See below for more information on Z.)

Remark: Mean field theory. As a warmup (and for later use), let's apply the Boltzmann distribution to a single spin. In that case $E = -Bm_0\sigma$ with $\sigma = \pm 1$, where B is an external field and m_0 is the magnetization of a single spin. (So the h of Eq. (5.1) is Bm_0.) In that case the probability that spin is up relative to the field is $\Pr(\mathsf{up}) = \frac{1}{Z}e^{-\beta E_{\mathrm{up}}} = \frac{1}{Z}e^{\beta Bm_0}$. Similarly $\Pr(\mathsf{down}) = \frac{1}{Z}e^{-\beta Bm_0}$. (As usual, $\beta = 1/k_B T$.) The partition function, Z, is the sum of these, so that

$$\langle m \rangle = m_0 \frac{\exp(\beta Bm_0) - \exp(-\beta Bm_0)}{\exp(\beta Bm_0) + \exp(-\beta Bm_0)} = m_0 \tanh(\beta Bm_0) . \tag{5.3}$$

---

[3]Note that, although the spins are labeled 1 to N, they need not be on a straight line. Thus on a square lattice σ_i and σ_{i+M} might be nearest neighbors, where M is the length of a side of the lattice.

[4]The original assumption—that all states are equally likely in the microcanonical ensemble—and the Boltzmann distribution for the canonical ensemble "derived" from it (see Appendix B), are basic postulates of statistical mechanics, and, like many postulates, they are fairy tales. They work; they give accurate predictions, but are they completely true? Maybe not.

Now do mean field theory for many spins, for the energy function, Eq. (5.1). The magnetization of a single spin (w.l.o.g.[5]) is taken to be $m_0 = 1$ and the given spin (whose magnetization we wish to determine) is taken to be #i. Then we replace the "interacting" spins (not #i) by $\mu = \langle \sigma \rangle$. μ will then be determined by self-consistency. The energy, Eq. (5.1), is thus replaced by

$$E = -zJ \sum_i \sigma_i \mu - h \sum_i \sigma_i, \qquad (5.4)$$

with z the number of neighbors of the given spin. (The number z is lower case. It is not the partition function and is not complex—but it is the usual notation.) The energy of a single spin in an effective field is therefore

$$E_1 = -\sigma(zJ\mu + h), \qquad (5.5)$$

for an effective field of $zJ\mu + h$ and with $\mu \equiv \langle \sigma \rangle$. (The i has been dropped, since at this stage all spins are equivalent.) This gives the self-consistent equation for the mean field spin value as

$$\langle \sigma \rangle \equiv \mu = \tanh(\beta(zJ\mu + h)). \qquad (5.6)$$

This material is also given in Sec. 8.1.1 with slight differences in notation. See Sec. 5.2 for some consequences of Eq. (5.6).

Now for a dynamical calculation. The way things stand we've defined an energy and a distribution function, but no dynamics. How can these spins go from one configuration to another?

It turns out that an algorithm originally used to optimize numerical integration [151] provides a non-unique answer. This algorithm is a stochastic process that finds the most significant contributions to the integral, and is a plausible set of transitions from one configuration to another. Of course one should ask what "plausible" means. They are transitions which could occur if the system moved irreversibly to its most likely set of configurations, namely the Boltzmann distribution. I'll be precise.

For a given lattice (and N) the space of configurations consists of all possible values of $\tilde{\sigma}$ (so X, on which the transitions are defined, is of cardinality 2^N). For a stochastic process I need to define a transition probability from each $\tilde{\sigma}_0$ to each possible target $\tilde{\sigma}_1$. In other words, I need a matrix

$$R(\tilde{\sigma}_1, \tilde{\sigma}_0) = \Pr(\tilde{\sigma}_1 \leftarrow \tilde{\sigma}_0). \qquad (5.7)$$

Note that $R_{\tilde{\sigma}_1 \tilde{\sigma}_0}$ represents a transition from right to left, from $\tilde{\sigma}_0$ to $\tilde{\sigma}_1$. (Caution: Some authors write the transition matrix, (what I call) R, the other way.) The matrix R will provide the dynamics; it will tell you with what likelihood $\tilde{\sigma}_0$ transitions to $\tilde{\sigma}_1$. To be sure, it's a *stochastic* dynamics, with a level of randomness. As such you can

---

[5]w.l.o.g. = Without Loss Of Generality. This is a common abbreviation.

talk about the probability that a system has reached some configuration. Let $p(\tilde{\sigma}, t)$ be the probability that at time-t the system's configuration is $\tilde{\sigma}$. Then the matrix R gives the distribution at time-$(t+1)$ as $p(\tilde{\sigma}, t+1) = \sum_{\tilde{\sigma}'} R(\tilde{\sigma}, \tilde{\sigma}') p(\tilde{\sigma}', t)$.

There is considerable choice in giving values to R. Here is one way to define $R(\tilde{\sigma}_1, \tilde{\sigma}_0)$:

- Pick a spin, say k, taking some value (plus or minus 1) in $\tilde{\sigma}_0$.
- Let $\tilde{\sigma}_1$ be the configuration in which σ_k has a value opposite to what it has in $\tilde{\sigma}_0$; all other spins are the same as in $\tilde{\sigma}_0$.
- Evaluate both $E(\tilde{\sigma}_0)$ and $E(\tilde{\sigma}_1)$ and calculate $\Delta E \equiv E(\tilde{\sigma}_1) - E(\tilde{\sigma}_0)$.
- If $\Delta E < 0$ *always* perform the transition, that is, $R(\tilde{\sigma}_1, \tilde{\sigma}_0) = 1$.
- If $\Delta E \geq 0$ *sometimes* perform the transition. This is the key possibility. Sometimes you do the "wrong" thing. You do this randomly, as I now describe.
- There's a parameter T, the temperature. Define an inverse temperature $\beta \equiv 1/k_B T$;[6] the only purpose in defining beta is to make the writing simpler. Now evaluate the quantity $\exp(-\beta \Delta E)$ (which is ≤ 1, since the case $\Delta E < 0$ is already accounted for). This number is the transition probability; that is, $R(\tilde{\sigma}_1, \tilde{\sigma}_0) = \exp(-\beta \Delta E)$ and $R(\tilde{\sigma}_0, \tilde{\sigma}_0)$ is 1 minus this quantity.
- All other matrix elements of R are zero.
- Definition: A matrix R_{mn} is *stochastic* if for each (m, n) $R_{mn} \geq 0$ and $\sum_m R_{mn} = 1$. (For some basic facts on stochastic matrices, see Appendix D.5.)
- The matrix (as defined so far) may not be stochastic because column sums may add to more than 1. Take the largest column sum and divide all columns by it. Then, for those columns that now add to less than 1, increase R's diagonal to make all column sums 1.

In MATLAB™ the commands would be

```
M=max(sum(R));
R=(1/M)*R;
R=R+eye(size(R))-diag(sum(R));
```

eye is the identity matrix, ones on the diagonal, zeros elsewhere. U=diag(u) is the matrix with $U_{ii} = u_i$, $U_{ij} = 0$, $i, j = 1, \ldots, N$, $j \neq i$ and R is 2^N by 2^N.

Exercise 14 Show that with these rules the Boltzmann distribution (Eq. (5.2) or Eq. (5.9)) is recovered. (This is done below, ignoring the last step above. For this proof, include the last step.)

---

[6] k_B is the Boltzmann constant. It is about 1.38×10^{-23} joules per kelvin. Its use is purely historical— it would make more sense to define temperature in energy units. But in a world that cannot agree on Fahrenheit vs. Celsius (a.k.a. centigrade) vs. kelvin, who could expect reasonable choices. For this algorithm one usually just defines β, and doesn't worry about T or k_B. Moreover, even when T makes an appearance (as in the derivation of Eq. (5.24)) one often takes units such that k_B is one.

Exercise 15 Propose other stochastic matrix choices for $R(\tilde{\sigma}_1, \tilde{\sigma}_0)$. Allow multiple spin flips as well. In all cases you need to satisfy what I prove below for the matrix just defined: it should give the Boltzmann distribution as its stationary state.

For the transition matrix I've defined there's a special probability distribution, call it $p_0(\tilde{\sigma})$, the stationary state. It has the property

$$p_0 = R p_0 . \tag{5.8}$$

Now comes the punch line: I claim (as anticipated by Eq. (5.2)) that the stationary state can be written

$$p_0(\tilde{\sigma}) = \frac{1}{Z} \exp\left(-\beta E(\tilde{\sigma})\right) , \text{ with } Z = \sum_{\tilde{\sigma}'} \exp\left(-\beta E(\tilde{\sigma}')\right), \text{ a normalization factor.} \tag{5.9}$$

That is, it's the Boltzmann distribution. As such it is fundamental to thermal physics. To prove that this is the stationary state and is invariant under the action of R consider[7] two probabilities, $p(\tilde{\sigma}_1)$ and $p(\tilde{\sigma}_0)$, whose states differ by the value of a single spin, say spin-k. For definiteness let $\tilde{\sigma}_0$ have that spin be +1 (and for $\tilde{\sigma}_1$ −1). Also for definiteness, let $E_+ > E_-$ and rename $p_+ \equiv p(\tilde{\sigma}_0)$ and $p_- \equiv p(\tilde{\sigma}_1)$. Then, with prime indicating "after" and no prime "before," we have, with obvious notation,

$$\begin{aligned} p'_+ &= R_{++}p_+ + R_{+-}p_- \\ p'_- &= R_{-+}p_+ + R_{--}p_- . \end{aligned} \tag{5.10}$$

By the rules just given, since $E_+ > E_-$, $R_{++} = 0$ and $R_{+-} = \exp\left(-\beta\left(E_+ - E_-\right)\right)$. Similarly $R_{-+} = 1$ and $R_{--} = 1 - \exp\left(-\beta\left(E_+ - E_-\right)\right)$. Therefore the new value of the ratio p_+/p_- is

$$\left(\frac{p'_+}{p'_-}\right)_{\text{after}} = \frac{\exp\left(-\beta\left(E_+ - E_-\right)\right)}{1 + \left(\frac{p_+}{p_-}\right)_{\text{before}} - \exp\left(-\beta\left(E_+ - E_-\right)\right)} . \tag{5.11}$$

If $\left(\frac{p_+}{p_-}\right)_{\text{before}} = \exp\left(-\beta\left(E_+ - E_-\right)\right)$ then the ratio is unchanged, showing that p_0 is indeed an eigenstate of eigenvalue 1.

Next we turn to the innocent-looking "normalization factor," which by virtue of its actual importance carries its own name, the *partition function*:

$$Z \equiv \sum_{\tilde{\sigma}} \exp\left(-\frac{E(\tilde{\sigma})}{k_B T}\right) . \tag{5.12}$$

It will turn out that if you can calculate Z you pretty much know everything about the system.[8]

---

[7] Here I follow the appendix in [135].
[8] Z is related to the free energy, F, by $Z = \exp\left(-\beta F\right)$.

But I must add a cautionary note. The dynamics I've defined is probabilistic. You don't definitely go from one configuration to another, as you would if you had Hamiltonian dynamics (classical or quantum). Perhaps you've coarse grained, or perhaps you are willing to settle for less. Stochastic dynamics is not in general reversible, as most forms of Hamiltonian dynamics are. And unless all eigenvalues are of norm 1, you've already put in an arrow of time: for irreducible dynamics there is a single eigenvalue 1; and in the absence of eigenvalues of norm 1, all transients will disappear.

5.1 Curie–Weiss ferromagnets

To allow an exact solution we go over to a slightly different model,[9] characterized by the energy

$$E = -\frac{J}{N} \sum_{\langle ij \rangle} \sigma_i \sigma_j - h \sum_j \sigma_j , \qquad (5.13)$$

where $\langle ij \rangle$ now means sum over *all* pairs[10] and J is now divided by N (so that energy is extensive[11]). What characterizes this energy function is that we ignore the geometry from the beginning and assume that each atom interacts equally with every other atom. This is like percolitis in which universal mutual daily contact was assumed. And, like percolitis, mean field theory becomes exact.

As above, the system configuration is described by an N-tuple, $\tilde{\sigma} \equiv (\sigma_1, \ldots, \sigma_N)$. Following my allegations of its importance, we calculate the partition function

$$Z(T, h) \equiv \sum_{\tilde{\sigma}} \exp(-\beta E(\tilde{\sigma})) , \qquad (5.14)$$

where the dependence of Z on J is suppressed and $\beta = 1/k_B T$. There are 2^N contributions to this sum.

It turns out that there's a wonderful simplification in this model. Let $M \equiv \sum_i \sigma_i$, the total magnetization. Then obviously the second term in E (of Eq. (5.13)) depends only on M, but the first term also has that property. This is because

$$M^2 = \sum_i \sigma_i \sum_j \sigma_j = \sum_k \sigma_k^2 + \sum_{\substack{i \neq j}} \sigma_i \sigma_j = N + 2 \sum_{i<j} \sigma_i \sigma_j . \qquad (5.15)$$

It follows that

$$E = -\frac{J}{2N} M^2 - hM + \text{const} \longrightarrow -\frac{J}{2N} M^2 - hM , \qquad (5.16)$$

---

[9] When you're studying magnetism for its own sake you usually start with a nearest-neighbor model and talk your way into a *mean field theory*, which gives the same results as the Curie–Weiss model.

[10] "Pairs" means that each pair, i–j, is counted only once and that there is no diagonal. Sometimes this is written $i < j$ under the summation symbol.

[11] An *extensive* variable scales with system size. An example is energy: if you double the volume (and keep other parameters the same), you double the energy. Other examples of extensive quantities are volume, number of particles, and entropy. By contrast, *intensive* variables do not change. Examples are pressure, temperature, and chemical potential.

where the arrow indicates that we drop the constant. Adding a constant to the energy changes nothing physical: when you calculate probabilities from $\exp(-\beta E)/Z$ the constant cancels.

In evaluating Z we had many, many (2^N) contributions, but now we have a variable, M, that takes only $N+1$ values, from $-N$ to $+N$ (in steps of 2). This means that typical values of M are associated with many configurations, that is, many sequences of the form $(\pm 1, \pm 1, \dots)$.

There next follow a number of algebraic steps whose verification I leave as an exercise. Define A to be the number of spins "up," that is, the number of i's for which $\sigma_i = +1$. Define the per-particle magnetization to be $\mu \equiv M/N$ and the fraction pointing up to be $\alpha \equiv A/N$. Then it follows that $\mu = 2\alpha - 1$ and $\alpha = (1 + \mu)/2$.

Exercise 16   Do the algebra.

Next the density of states: How many configurations are associated with a given A? It's the number of ways of picking out A spins to be up, from among the entire collection of N of them. That is, it is the combinatorial coefficient $\binom{N}{A} = \frac{N!}{A!(N-A)!}$. The partition function is therefore

$$Z = \sum_{A=0}^{N} \binom{N}{A} \exp\left(N\frac{\frac{J}{2}\mu^2 + h\mu}{k_B T}\right), \tag{5.17}$$

where I sum over *all* the indicated integer values of A. We use Stirling's approximation, $n! \sim \sqrt{2\pi n}\left(\frac{n}{e}\right)^n$, at a slightly lower precision (dropping the $\sqrt{2\pi n}$) to obtain (again omitting details) for the combinatorial coefficient

$$\binom{N}{A} \sim \exp[N\phi(\mu)]$$

$$\text{with} \quad \phi(\mu) = -\frac{1-\mu}{2}\log\left(\frac{1-\mu}{2}\right) - \frac{1+\mu}{2}\log\left(\frac{1+\mu}{2}\right). \tag{5.18}$$

For convenience in working with the energy, let $\epsilon(\mu) \equiv -(J/2)\mu^2 - h\mu$. Writing, as usual, $\beta \equiv 1/k_B T$, and converting the sum to an integral, the partition function becomes[12,13]

$$Z = \sum_A \exp\{N[\phi(\mu) - \beta\epsilon(\mu)]\} \longrightarrow \frac{N}{2}\int_{-1}^{+1} d\mu \exp\{N[\phi(\mu) - \beta\epsilon(\mu)]\}. \tag{5.19}$$

The mischief in this integral (and what makes it easy to get physical answers) is the N in the exponent. As usual it means that whichever peak (as a function of μ)

---

[12]The half in $N/2$ comes because changes in "M" are in units of 2.

[13]I said earlier that this model has an "exact" solution. Going from a sum to an integral is obviously not exact, but its deviation is a small issue; it is not a qualitative change in the results, as, say, mean field theory is for the nearest neighbor Ising model.

occurs in the difference of the functions ϕ and $\beta\epsilon$, it will be enormously magnified, so that we only need to look at that peak (although we'll look at the spread around the peak for finer questions).

We therefore seek the maximum of the integrand.[14] The argument of the exponent in Eq. (5.19) is the difference of $\phi(\mu)$ and $\beta\epsilon(\mu)$, so that to find the maximum we need to calculate the derivative of each function with respect to μ. These derivatives are

$$\frac{d\phi}{d\mu} = \frac{1}{2}\log\frac{1-\mu}{1+\mu} \quad \text{and} \quad \frac{d(\beta\epsilon)}{d\mu} = -\beta\,(J\mu + h)\,. \tag{5.20}$$

The equilibrium value of μ—where the integrand is maximal—is gotten by setting these expressions equal. With a little algebra this implies

$$\mu = \tanh\beta(J\mu + h). \tag{5.21}$$

This is a remarkably simple result, and coincides with the usual mean field theory of ferromagnetism.[15] There are also roots of Eq. (5.21) with an imaginary part to μ, but they are not physical.

5.2 Magnetization

For the sake of completeness I'll talk briefly about just what kind of phase transition occurs here. Our real focus is on fluctuations, but we might as well cover a bit of physics.

Like the percolitis equation for s_0, Eq. (5.21) represents a relation that μ must satisfy[16] if there is to be a stationary point of the function $\phi(\mu) - \beta\epsilon(\mu)$. To find a solution of Eq. (5.21) use the same method as for percolitis, except that now there are two free variables, T (or β) and h. (J scales with the other variables and I henceforth set it equal to 1. Ditto for k_B, although sometimes it will be written to allow comparisons.) We'll study[17] the case $h = 0$ and plot the left- and right-hand sides of Eq. (5.21) as a function of μ. See Fig. 5.1.

By drawing such figures we reach the following conclusion. For $T > 1$ (remember $J = 1$), the only intersection is at $\mu = 0$; in other words, the system is not magnetized

[14]Again there is an approximation to this "exact" answer. This time it's the neglect of the integral over a finite range. And once again it changes nothing. For example, the integral of Eq. (5.19) only runs from -1 to $+1$ and not from 1 to ∞ nor from $-\infty$ to -1. Including the integral over a range not present in the original integral obviously introduces an error. However, the error is bounded by $\exp(-N\text{const})\times$ something small relative to $\phi - \beta\epsilon$ at its maximum (because the maximum as a function of μ occurs in $[-1, 1]$). As such it decreases exponentially with N and is negligible. Then there are the approximations due to looking only at the first few moments of the integral. What's important is that these corrections go to zero like $\exp(-N \cdot \text{positive const})$ as $N \to \infty$.

[15]See also Sec. 8.1.1. This is the standard mean field theory result, given above (z is absorbed into J), and for example is found in Ch. 15 of the 7th edition of Kittel's book [126]. What corresponds to Eq. (5.21), with zero external field, is Kittel's Eq. (8) or (9). (You may have a different edition of the book—there are eight of them.)

[16]Implicit in the definition of μ is the requirement that it be real. There are many complex roots to the transcendental equation for μ, but they are not relevant. (An example: for $T = 1.5$ there is $\mu \approx 0 + i \cdot 1.4511$.)

[17]Concerning $h \neq 0$, see the remarks at the end of this subsection.

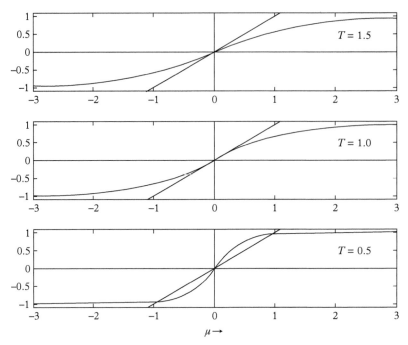

Fig. 5.1 Finding μ. For each value of T (1.5, 1 and 0.5) two curves are drawn. One is the same for all 3: μ as a function of μ, in other words, a straight line. The second curve is $\tanh(\mu/T)$. Intersections of the curves are solutions to Eq. (5.21), with $h = 0$, i.e., $\mu = \tanh\mu/T$ (with $k_B = 1$).

(also recall that $h = 0$). For $T < 1$ there are three intersections. To see their meaning we recall the integrand in Eq. (5.19) and for convenience define $f(\mu) \equiv \phi(\mu) - \beta\epsilon(\mu)$. This is the function that has an extremum, and which will ultimately be promoted to an *extreme* extremum by the N multiplying it. But to know if this is physically relevant we must also know that it's a maximum. For that we'll need the second derivative. From Eq. (5.20) it follows that

$$\frac{d}{d\mu}[\phi(\mu) - \beta\epsilon(\mu)] = \frac{df}{d\mu} = \frac{1}{2}\log\frac{1-\mu}{1+\mu} + \beta(\mu + h) , \tag{5.22}$$

and thus

$$\frac{d^2 f}{d\mu^2} = \beta - \frac{1}{1-\mu^2}. \tag{5.23}$$

Continuing the assumption that $h = 0$, for $T > T_c = 1$, β is less than 1, while $\mu = 0$, so that f'' is negative and indeed there is a maximum. For $T < T_c = 1$ there are three values of μ (satisfying Eq. (5.21)), and we examine each. The middle one on the plot, $\mu = 0$, gives a value of f'' that is $\beta - 1$. Now however $\beta > 1$, so this is no longer a maximum but a minimum, and therefore this state is physically excluded. For the other two roots it is again not trivial to show that f'' is negative, but indeed

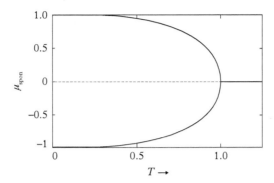

Fig. 5.2 Value of the spontaneous magnetization as a function of temperature (solid lines). Like the solution of the percolitis equation, these values have been attained by iterating $\mu' = \tanh \beta\mu$, *except that now there is no a priori reason for this limit to provide a solution, since there is no dynamical context.* Nevertheless, the technique works, and, as for percolitis, the number of iterations necessary to get the process to converge in the neighborhood of the critical point is large.

it is.[18] Thus the two non-zero μ values are the physical ones, and represent the fact that the system becomes magnetized despite there being no external field present. This corresponds to the victory of energy over entropy—the non-zero μ states have lower energy and the temperature is low enough that energy considerations win the day. Note that there are *two* states and the system doesn't get to be in both of them at once. This turns out to be like the rubber band game—the choice of *which* state depends on a number of factors, some of them random, but there *must* be a choice.[19] The critical point is $T_c = 1$, for which there is only one solution, but it is actually a multiple root. Here we anticipate interesting behavior and our hopes will be realized, especially when we get to the subject of fluctuations. The way to see analytically that indeed $T_c = 1$ is by comparing the slopes of the lines μ and $\tanh \beta\mu$, which match when $\beta = 1$ (or $T \equiv 1/\beta = 1$, recalling that $k_B = 1$).

The results of the curve plotting are summarized in Fig. 5.2. As you might have guessed already, the order parameter in this case is the magnetization and indeed it attains non-zero values for $T < 1$.

Recall that when I displayed the Guggenheim plot (Fig. 1.1) I asserted that the order parameter grew as (approximately) the cube root of the difference in the temperature from its critical value. (The order parameter there was the difference in

---

[18] As for percolitis (see Footnote 11, Chapter 4), it is easy to show positivity near the critical point and more difficult further away, because of the transcendental nature of the condition. Near $T = 1$, $h = 0$ (the critical point), we show below that for $T = 1-\epsilon$, $\mu = (\sqrt{3\epsilon})$. Therefore $f'' = 1/\beta - 1/(1-\mu^2) = 1/(1-\epsilon) - 1/(1-3\epsilon) \approx (1+\epsilon) - (1+3\epsilon) = -2\epsilon < 0$, so that, as claimed, f has a maximum. Away from the critical point it is easiest to invert the transcendental equation and take f'' as a function of the solved-for variable, μ. Thus from Eq. (5.21) one has $\beta = (\text{arctanh}\,\mu)/\mu$ and $f'' = (\text{arctanh}\,\mu)/\mu - 1/(1-\mu^2)$. This is in fact negative for all $|\mu| < 1$, except at zero, where it approaches zero.

[19] In a way, this is analogous to the fact that black holes are black (and white holes white). As for the present case, although the equations are symmetric, a choice must be made. For a similar result, involving a choice in analytic continuation for holes (black and white), see [61].

density between the liquid and gas phase.) What is the corresponding growth for our mean field ferromagnet? For this we need to solve Eq. (5.21) with $h = 0$ (and $J = 1$). The equation determining μ is $\mu = \tanh(\mu/(T_c - \epsilon))$, where $T_c = 1$ and ϵ (> 0) is the deviation (below) from criticality. Use the expansion of the hyperbolic tangent: $\tanh x = x - x^3/3 + 2x^5/15 + \dots$. Near the critical point μ is small, so we drop terms higher than cubic. In the remaining equation we want the dependence of the *non-zero* value of μ on $\epsilon \equiv T_c - T$, so we can divide by μ. This yields $1 = 1/(1 - \epsilon) - \mu^2/3(1 - \epsilon)^2$ and, to lowest order in ϵ, the spontaneous magnetization is

$$\mu = \pm\sqrt{3\epsilon}, \text{ which is to say that } |\mu| \sim (T_c - T)^{1/2}. \qquad (5.24)$$

So the mean field Curie–Weiss model has an order parameter that goes to zero as the 1/2 power of the temperature away from criticality, which is different from the (approximately) 1/3 found for the liquid–gas phase transition. Mean field theory is not the whole story.

Remark: Percolitis comparison: Recall that for percolitis we found that the order parameter is *linear* in the deviation of the control parameter from its critical value. This is the message of Eq. (4.7), where s_0—the order parameter—grows linearly as r—the control parameter—goes away from $r_c = 1$. This shows that although many behaviors (exponents, in this case) are similar to one another, some are truly different. Those in Guggenheim's plot are all the same and are said to fall into the same *universality class*; in other words, the way they approach the critical point is, in a sense, universal. Clearly percolitis and mean field ferromagnetism are in different universality classes.

Exercise 17   Use Eq. (5.24) to show that, for T below (but near) T_c, $d^2 f/d\mu^2 < 0$ for non-zero μ_{spon}. Hint: For the solution, see Footnote 18.

    There's another kind of phase diagram that brings out the features of magnetism. One can study magnetization as a function of external field (h). Again one uses graphical methods to understand qualitative behavior. One can also get some understanding of the phenomenon of *hysteresis*. However, this topic is a bit of a digression from our main goal and will not be pursued here.

Exercise 18   Calculate $d\mu/dh$ for h near 0 and $T = 1$. This is the magnetic susceptibility and is analogous to the specific heat. Determine its behavior as $h \to 0$.

5.3   Fluctuations greater than $\sqrt{N}$

How large are the fluctuations of the magnetization, μ? Let's make this easy on ourselves by looking only at $h = 0$ and $T \geq T_c$ (or $\beta \leq \beta_c$), where the equilibrium state is $\mu = 0$. That way the fluctuations, $\Delta\mu \equiv \sqrt{\langle(\mu - \langle\mu\rangle)^2\rangle}$, are simply $\sqrt{\langle\mu^2\rangle}$.

As usual the answer is found using the partition function. From Eq. (5.19)

$$Z = \frac{N}{2} \int_{-1}^{+1} d\mu \, \exp\left[N f(\mu)\right], \quad \text{with } f(\mu) = \phi(\mu) - \beta\epsilon(\mu). \tag{5.25}$$

We already know two derivatives of f, but now we must go further. To establish our in-principle results there is no need to get involved in a lot of algebra, and I'll approximate f by the first terms in its power series around the special value of μ satisfying Eq. (5.21), which is zero. Recall that

$$\frac{d^2 f}{d\mu^2} = \beta - \frac{1}{1 - \mu^2} \tag{5.26}$$

and that, at zero μ, $df/d\mu = 0$. The next two derivatives of f are

$$\left.\begin{aligned} \frac{d^3 f}{d\mu^3} &= -\frac{2\mu}{(1 - \mu^2)^2}, \\ \frac{d^4 f}{d\mu^4} &= -\frac{2}{(1 - \mu^2)^2} + \mu \times \text{stuff}. \end{aligned}\right\} \tag{5.27}$$

Since we are only interested in the values at $\mu = 0$, we have $f''(0) = \beta - 1$, $f^{(3)}(0) = 0$, and $f^{(4)}(0) = -2$. The partition function is thus

$$Z = e^{N f(0)} \frac{N}{2} \int_{-1}^{+1} d\mu \, e^{N[(\beta-1)\mu^2/2 - 2\mu^4/4!]}, \quad (\beta \le 1). \tag{5.28}$$

Recall how the partition function gives the answer. The desired expectation value is

$$\langle \mu^2 \rangle = \sum_\mu \mu^2 \Pr(\mu) = \frac{1}{Z} e^{N f(0)} \frac{N}{2} \int_{-1}^{+1} d\mu \, \mu^2 \exp\left[N\left(\frac{1}{2}(\beta - 1)\mu^2 - \frac{1}{12}\mu^4\right)\right], \tag{5.29}$$

since the partition function is after all made up of Boltzmann factors. There are two cases, $\beta < 1$ and $\beta = 1$. I will examine them separately.

$\beta < 1$. **Away from the critical point.** In this case the fourth-power term in the integral can be dropped. The integral can now be done explicitly, but this is not necessary. Define a new variable $x = \mu\sqrt{N(1 - \beta)}$. The new limits of integration are $\pm\sqrt{N(1 - \beta)}$, which we set to $\pm\infty$. Then

$$\langle \mu^2 \rangle = \frac{1}{Z} e^{N f(0)} \frac{N}{2} \int_{-\infty}^{\infty} \frac{dx}{\sqrt{N(1 - \beta)}} \frac{x^2}{N(1 - \beta)} \exp\left(-x^2/2\right). \tag{5.30}$$

The $1/Z$ makes a lot of things cancel: first all factors to the left of the integral, and then the $1/\sqrt{N(1 - \beta)}$ due to the change of variable. We get

$$\langle \mu^2 \rangle = \frac{\int_{-\infty}^{\infty} dx \, \frac{x^2}{N(1-\beta)} \exp\left(-x^2/2\right)}{\int_{-\infty}^{\infty} dx \, \exp\left(-x^2/2\right)} = \frac{1}{N(1 - \beta)} \cdot \text{constant of order unity}. \tag{5.31}$$

The "constant of order unity" arises from the ratio of the scale-free integrals, which, generally speaking, is made of 2's and π's and such like. In this particular case it actually happens to be 1, but that's not important for our general conclusions.

This is our usual result. Fluctuations are of order $1/\sqrt{N}$.[20] Note though that as $\beta \to 1$ the fluctuations will grow, apparently becoming infinite. We'll see how that gets taken care of.

$\beta = 1$. At the critical point.   Now the quadratic term in the integrand drops out and we have

$$\langle \mu^2 \rangle = \frac{1}{Z} e^{Nf(0)} \frac{N}{2} \int_{-1}^{+1} d\mu\, \mu^2 \exp\left(-N\mu^4/12\right). \tag{5.32}$$

Again we change variables, this time letting $x = \mu N^{1/4}$. Going through the same steps that led to Eq. (5.31), we have

$$\langle \mu^2 \rangle = \frac{\int_{-\infty}^{\infty} dx\, \frac{x^2}{\sqrt{N}} \exp\left(-x^4/12\right)}{\int_{-\infty}^{\infty} dx\, \exp\left(-x^4/12\right)} = \frac{1}{\sqrt{N}} \cdot \text{constant of order unity}. \tag{5.33}$$

Now fluctuations ($\sqrt{\langle \mu^2 \rangle}$) are on the order of the *fourth* root of unity.[21] *Finally we have gone beyond the famous square-root-of-N.* At criticality a system enjoys large collective and coherent motions. This can affect experiments and is also related to other phenomena, such as universality and the appearance of fractals.

Exercise 19   I'll call this an exercise, but actually I don't know the answer, so maybe I should call it a project. Here's the question: We've found that per-particle fluctuations go like $N^{-1/4}$. The reason is that correlations among the underlying variables have become significant. You don't really have N independent participants, because the dynamics has them doing things in a correlated way. How many independent motions are there? I would guess there are $\sqrt{N}$; I say that because if this number of degrees of freedom acted independently, they would give rise to per-particle fluctuations of order $1/\sqrt{\sqrt{N}}$. So the question is, what are these $\sqrt{N}$ degrees of freedom? Or maybe show something a bit easier, that—effectively—there *are* that many independent degrees of freedom, even if you can't explicitly identify them.

Other critical exponents can also be obtained, for example let $\beta = 1$ but vary h. (If you plan to check this, don't forget that $\langle \mu \rangle = (3h)^{1/3}$ for $\beta = 1$ (which is easy to show).) What's not possible is anything to do with space, since (as in percolitis) there's no space involved.

---

[20]This is the per-particle fluctuation. The macroscopic magnetization M will be N times as large.
[21]Now the "constant of order unity" is 6, again irrelevant.

6

Maximum entropy methods

Applications discussed in this chapter: Information theory, Maximum entropy, Neural networks, Supreme Court voting, Flocking

This is a way of selecting a probability distribution based on available information—and nothing more! In other words, you are given some data and want the "most random" probability distribution that could have given rise to those data. "Most random" is interpreted as using the least "information." The method goes by the name "maximum entropy" [116; 117], since maximum entropy is identified with minimum information. To proceed though I'll have to say exactly what I mean by information [209].[1]

6.1 Information

Suppose a ball is equally likely to be in one of n boxes, but it is not known which. Call the *missing* information I, or, since it should depend on n, $I(n)$. Clearly, the more boxes there are the less you know of the ball's whereabouts, so the function $I(n)$ increases with increasing n. If $n = 1$, no information is missing, so $I(1) = 0$.

Suppose the n boxes are in a row, and there are m such rows. Then there are mn boxes. If there is a single ball in just one of the boxes, the information missing is $I(mn)$, since there are mn boxes. On the other hand, the location of the ball could be given by specifying the row and column, so

$$I(mn) = I(m) + I(n). \qquad (6.1)$$

We'll now look for a function, I, for which Eq. (6.1) holds for all real numbers (not just integers, m and n). This is more stringent than what I have justified so far and one can give arguments to support this generalization [123], but we'll simply assume that Eq. (6.1) can be generalized to

$$I(xy) = I(x) + I(y), \qquad (6.2)$$

[1]Claude E. Shannon, author of [209], 1916–2001, American, information theory, cryptanalysis, a descendant of an early settler of New Jersey.

When Things Grow Many. Lawrence S. Schulman, Oxford University Press.
© Lawrence S. Schulman (2022). DOI: 10.1093/oso/9780198861881.003.0006

for x, y strictly positive real numbers. Moreover, we'll also assume that the function I is both continuous and differentiable. Again, these assumptions can be somewhat loosened at the expense of a lot more verbiage.[2]

Now take the derivative with respect to y and set $y = 1$. This yields

$$xI'(x) = I'(1) \equiv k,\tag{6.3}$$

where the prime indicates a derivative and I have defined $k \equiv I'(1)$. This leads to $dI/dx = k/x$, which integrates to

$$I(x) = \int_1^x \frac{k}{u}\, du + I(1) = k\left[\log x\right]\Big|_1^x = k\log x.\tag{6.4}$$

In a physics context, one usually takes $k = k_B$, the Boltzmann constant, while in information theory $k = 1/\log 2$, so that if you have two boxes the missing information would be 1, 1 "bit."

Now suppose there are a lot of "boxes" but they don't all have the same probability of having a ball in them. Label the boxes $\alpha = 1, \ldots, N$, and let the probability that the ball is in box α be p_α. Of course $\sum_\alpha p_\alpha = 1$.

To reduce this problem to one we've already solved, imagine many copies of the system, say M of them, such that every $p_\alpha M$ is close to an integer. Thus we set[3] $n_\alpha \equiv M p_\alpha$ with the n_α's integers.

Call n_1 of the M boxes "container #1," n_2 of the boxes "container #2" and so on. Now take M balls and distribute them among the boxes. In this distribution you put n_1 into container #1, n_2 into container #2 and so on. Thus if you grab some particular ball its probability of being in container #k is p_k, as stipulated.

Next we need a measure of ignorance. Imagine that the balls have labels, 1, 2, ..., M. You will have $M!$ ways of putting them in the M boxes. (M choices for the first, $M - 1$ for the second, etc.) However, all you know is which container the ball is in, and not if it got there before another ball in the same container. Therefore, the $M!$ over-counts by $n_1!$, which is the number of ways container #1 could have been filled. This is true for all containers. Therefore the original $M!$ needs to be divided by each $n_k!$ for $k = 1, \ldots, N$. Thus the multiplicity of distinct ways to fill the M boxes is

$$\mathcal{M} = \frac{M!}{n_{\alpha_1}!\, n_{\alpha_2}! \ldots n_{\alpha_N}!},\tag{6.5}$$

and this is the measure of your ignorance. (Note the different M's: the calligraphic M ($\mathcal{M}$) is the multiplicity of distinct ways of filling the boxes, while the plain (italic)

---

[2]Eq. (6.2) is related to the Cauchy (functional) equation, $f(u + v) = f(u) + f(v)$, under the transformation $u = \log(x)$ and $v = \log(y)$. The solution we've found for Eq. (6.2) is equivalent to having linear solutions to the Cauchy equation. However, there exist other solutions *if* you accept the axiom of choice and allow non-Lebesgue measurable functions. See the Wikipedia article at `https://en.wikipedia.org/wiki/Cauchy%27s_functional_equation` (accessed July 2020) for "Cauchy's functional equation."

[3]At this point, you might want to show that, given some error, taking M large enough can affect the ultimate answer (Eq. (6.8)) by less than that error. For example, rational approximations to the $\{p_\alpha\}$ and the set of α's being finite (i.e., $N < \infty$) will make our argument exact, so that it would be easy to make δ-ε-type arguments for the approximations and the result, Eq. (6.8). Similarly (and almost in the same way), you can allow $N \to \infty$.

M (M) is the number of copies of the original system.) What you don't know—even when you've specified the numbers $(n_{\alpha_1}, n_{\alpha_2}, \dots)$—is which among these $\mathcal{M}$ arrangements actually occurred. It follows that your missing information is $k \log \mathcal{M}$. But the important quantity is not this; rather it is the missing information *per copy of the original system*. The notation and definition are:

$$I\{p_\alpha\} \equiv \lim_{M \to \infty} k\frac{1}{M} \log \mathcal{M}. \tag{6.6}$$

We evaluate the quantities in Eq. (6.6) for finite M, and then let $M \to \infty$. To do this, make use of Stirling's approximation ($\log(n!) \approx n(\log n - 1)$ is sufficiently accurate) and recall that $n_\alpha = Mp_\alpha$:

$$\frac{M}{k} I\{p_\alpha\} = \log \mathcal{M} = M \log M - M - \sum_\alpha Mp_\alpha \left[\log(Mp_\alpha) - 1\right]$$

$$= M \left\{ \log M - 1 - \sum_\alpha p_\alpha \left[\log M + \log p_\alpha - 1\right]\right\}$$

$$= M \left\{ \log M - 1 - \sum_\alpha p_\alpha (\log M - 1) - \sum_\alpha p_\alpha \log p_\alpha \right\}. \tag{6.7}$$

Using the fact that $\sum_\alpha p_\alpha = 1$, we obtain

$$\boxed{I\{p_\alpha\} = -k \sum_\alpha p_\alpha \log p_\alpha}. \tag{6.8}$$

Eq. (6.8) is of fundamental importance in information theory and in understanding the statistical properties of matter. It is the entropy per particle.

Exercise 20 Suppose that a binary signal—a "0" or a "1"—has probability p of being transmitted correctly and probability q of having an error $(q + p = 1)$.

What is the missing information for each transmitted signal?

Now use redundancy to enhance reliability: For each bit (0 or 1) in your bit stream you send three signals. At the receiving end they have the same reliability as in the one-signal case. But, you now take a vote. If two or three are ones, you take the original signal to have been 1; otherwise, you take it to have been 0.

What is the probability that your answer is correct?

Show that for $1/2 < p < 1$ there is improved reliability.

What is the missing information? For small q, what is the leading term in q? in q^2? If the leading term in q^2 is 0, find the next term.

What level of redundancy (i.e., how many bits for each initial bit), along with a voting rule, is necessary to ensure one part in 8 million accuracy (so a megabyte file is likely to be copied correctly)?

For numerical answers assume $q = \exp(-6) \approx 1/400$.

6.2 Maximum entropy

In using "maximum entropy estimates" one makes what might be called philosophical assumptions. The method, or philosophy, is this: You know what you know. For all the rest you assume a probability distribution that maximizes the entropy. There are two reasons to like this method: it sounds right and it can't be wrong. What I mean by the latter is that, if its predictions are wrong, you've discovered a new feature of the system you're studying, some new fact that if you're clever you'll phrase as a constraint to get a probability distribution that takes into account everything you *now* know. This method was first, to my knowledge, advocated by Jaynes [116; 117].

Let's look at an example. Suppose there's an isolated "ideal gas," meaning you can assume that inter-particle interactions are negligible. Suppose you know the expected energy, E, of each particle and the mass, m. First define your variables. Position? Doesn't matter, we'll assume the whole thing is contained in some large volume, V, and anywhere within V is equally likely. The only other variable in Newtonian dynamics is velocity, so what we're looking for is $\Pr(v)$, the probability that the particle has velocity v. But we'll also assume isotropy, so the probability only depends on v (the magnitude of the vector, v). Making use of the statistical independence of each particle (of the *ideal* gas) we have only two constraints:

$$E = \sum_v \frac{1}{2} m v^2 \Pr(v),\qquad(6.9)$$

$$1 = \sum_v \Pr(v).\qquad(6.10)$$

The second condition is part of the definition of probability, but we list it separately since we plan to find the "most random" distribution by varying the probabilities and this condition needs to be separately enforced.

Now comes the main step. We earlier defined information and, following Eq. (6.8), identified it as (the negative of) entropy. With this perspective, "most random" means highest entropy. Thus we wish to maximize $S \equiv -k_B \sum_i \Pr(v) \log \Pr(v)$, subject to Eqs. (6.9) and (6.10).

To deal with the constraints, we use the method of Lagrange multipliers (see Appendix G). Instead of S, we find stationary points of

$$W(\Pr(v)) \equiv S - \lambda_0 \left(\sum_v \Pr(v) - 1 \right) - \lambda \left(\sum_v \frac{1}{2} m v^2 \Pr(v) - E \right),\qquad(6.11)$$

where (besides $\Pr(v)$) λ and λ_0 are considered variables, to be fixed by the variational principle. Let me change notation slightly to emphasize that $\Pr(v)$ is just another function. We'll call it $f(v)$ ($\equiv \Pr(v)$). Next we want to take the functional derivative. To do this we consider $f(v)$ a separate variable for each v. The derivative is thus

$$\frac{\partial W(f(v))}{\partial f(v)} = \frac{\partial}{\partial f(v)} \left[-k_B \sum_i f(v) \log f(v) - \lambda_0 \left(\sum_v f(v) - 1 \right) - \lambda \left(\sum_v \frac{1}{2} m v^2 f(v) - E \right) \right]$$

$$= -k_B \log f(v) - k_B - \lambda_0 - \lambda \frac{1}{2} m v^2.\qquad(6.12)$$

Setting the left-hand side to zero obviously implies that

$$f(v) = \exp\left(-1 - \frac{\lambda_0}{k_B}\right)\exp\left(-\frac{\lambda}{k_B}\frac{1}{2}mv^2\right). \tag{6.13}$$

Using this form of f we can now adjust the parameters λ_0 and λ to satisfy Eqs. (6.9) and (6.10). For these values, maximizing S is the same as maximizing W. To write the result in a familiar form we set $\lambda = 1/T$ and $\exp(1 + \lambda_0/k_B) = Z(T)$, since λ_0 is implicitly a function of λ and λ is implicitly a function of T. Z now clearly acts as a normalization and the energy, E, determines the value of (what we now recognize to be) the temperature T by Eq. (6.9) using

$$\Pr(v) = \frac{1}{Z(T)}e^{-mv^2/2k_B T}, \tag{6.14}$$

which is the Maxwell–Boltzmann distribution.[4]

Exercise 21 There's an additional constraint on the probabilities, namely that they all be non-negative. Why didn't we have to enforce it? (And it would be difficult to enforce, being non-holonomic.)

Exercise 22 Check that $\frac{1}{2}m\langle v^2\rangle = \frac{3}{2}k_B T$. Remember that v is three-dimensional and use the appropriate integration weight.

That's the derivation, but let's review what went into it. We did no dynamics, no little balls bouncing around, no larger volume of which V was a small part. We only said that there's *some* probability function, gave it one observational constraint (expected energy), and found the distribution that used the least information consistent with that observation. That's the method of maximum entropy. It has been found to be useful in many applications— physical, biological, image processing, and others.

Later we will go into detail for the (many) specific applications of these methods, but, just to set the stage on the kind of problems met, we briefly mention one additional example. There is considerable interest in the connectivity of neural networks. The work by Schneidman et al. [191] examining the connectivity of salamander retinas uses the techniques we have just described. What they can measure is both the average current through a neuron and the correlations; in other words, which neurons tend to fire together. For tractability these things are usually quantized: either there's a spike (a neuron fires) or not, which we'll characterize as +1 if it fires and −1 if it doesn't. (No mention is made in that paper of inhibitory neurons.) Similarly the time interval for response is quantized; for the study in [191] the time interval was taken to be 20 ms.

---

[4]In three dimensions $Z(T) = (2\pi k_B T/m)^{3/2}$.

Using the notation $\sigma_i = \pm 1$, the information available was thus $\langle \sigma_i \rangle$ and $\langle \sigma_i \sigma_j \rangle$. The maximum entropy function given this kind of information (using Lagrange multipliers, which have been renamed Z, h_i and J_{ij}) is

$$\Pr(\sigma_1, \sigma_2, \ldots, \sigma_N) = \frac{1}{Z} \exp\left(\sum_i h_i \sigma_i + \frac{1}{2} \sum_{i \neq j} J_{ij} \sigma_i \sigma_j \right). \tag{6.15}$$

So we're in the land of Ising models—making physicists (or maybe mathematicians) happy. But it's not quite the same. Now the coupling constant is not a single number ("J," as we had in Eq. (5.1)), but is a symmetric matrix whose values are to be fixed by experiment.[5] In fact Eq. (6.15) is close to a related well-known model: the spin glass. In that system the individual coupling constants, $\{J_{ij}\}$, are themselves random variables, sometimes taken to be ± 1, sometimes continuum-valued. This model was originally proposed because its long relaxation time resembled the behavior of glass. Hidden in the folds of its configuration space are many long lived metastable states in which the system can be caught. The study of this system has led to many developments (e.g., simulated annealing), although its resemblance to glass has been challenged. It has also served as a model for memory storage [7, 8].

Where our system differs from a spin glass is that in those studies $\{J_{ij}\}$ is given, whereas here our objective is to find out what the $\{J_{ij}\}$ are, that is, the connectivity of the neurons. (The same is true for h and h_i.)

Remark: Another application of the maximum entropy method treated in this book is the work of John Harte in ecology. See Sec. 10.6.

6.2.1 Mathematical details*

Next I address the mathematical question of how to maximize entropy while maintaining constraints; in other words, how do you find the probabilities that accomplish this? Once one has probabilities it's a matter of curve fitting to determine the coefficients in Eq. (6.15).

I'll follow a paper by Darroch and Ratcliffe[6] [47] that is cited in [191]. It is called the "log-linear" method, for reasons to be explained below. However, unless you're interested in the methods themselves you can skip these details and use packaged programs—which exist—for doing this. I will also simplify [47] to focus on the problem at hand. In an example in Sec. 6.3, we do a maximum entropy problem with a relatively small number of constraints and don't use the method described here. The method that *is* used relies (as just anticipated) on packaged "solution" programs in MATLAB™. Nevertheless, it's a good idea to see how this can work, in principle.

---

[5]Others have looked at non-symmetric matrices. For more literature see [3; 188; 192].

[6]See [47]. This is an early method, presumably used in [191], but there exist other techniques that have been developed since this work.

You have a collection of possible states. Call the space of these states X;[7] in the neural network case [191] the space is the set of N-tuples, strings of length N, consisting of plus and minus ones, so the cardinality is 2^N. You are looking for a probability distribution on X, some function $p(x)$, for $x \in X$. You have a guess as to what the probability distribution might be—call it $\pi(x)$. It doesn't need to be a very good guess. We do require $\pi(x) > 0 \ \forall x$ [8] as well as $\sum_x \pi(x) = 1$.[9]

Just as for our probability distribution above you have a number of constraints on the probabilities—you know d of their expectation values. That is, you have quantities B_s, with coefficients b_{sx}, $s = 1, \dots, d$, such that you require (of $p(x)$)

$$\langle B_s \rangle = \sum_x p(x) b_{sx} = \beta_s \,, \tag{6.16}$$

for given numbers, β_s. For convenience we define $B_0(x) \equiv 1$ and the requirement $\sum_x p(x) = 1$ is included as $\langle B_0 \rangle = \sum_x p(x) = \beta_0 = 1$ (so that one should take $b_{0x} = 1$). Thus Eq. (6.16) is assumed to hold for $s = 0, 1, \dots, d$. What is shown in [47] is that you can find numbers $(\mu_0, \mu_1, \dots, \mu_d)$, such that

$$p(x) = \pi(x) \prod_{s=0}^{d} \mu_s^{b_{sx}} \,, \tag{6.17}$$

and for which—connecting this to what we have been doing—the resultant $\{p(x)\}$ *maximizes* $S = -\sum_x p(x) \log (p(x))$. And the method is constructive—it actually *shows how to find these numbers.*

Note that one can define a vector

$$w(x) \equiv \log \frac{p(x)}{\pi(x)} = \sum_s b_{sx} \log \mu_s \,, \quad \text{or} \quad W = \sum_s B_s \log \mu_s \,. \tag{6.18}$$

This equation explains the term "log-linear": the *logarithm* of $p(x)/\pi(x)$ is a *linear* function on X, a sum of the vectors B_s, $s = 0, 1, \dots, d$ (or b_{sx}).

Now consider the relative entropy (a.k.a. the Kullback–Leibler divergence)

$$S(p|\pi) \equiv \sum_x p(x) \log \left(\frac{p(x)}{\pi(x)} \right) . \tag{6.19}$$

This is known to be non-negative and can only be zero if $\pi = p$ [44].

---

[7]I've changed notation from that of [47]. Here is the translation:

- $X \leftrightarrow I$, $x \leftrightarrow i$.

- $S \leftrightarrow H$, entropy.

- $\beta_s \leftrightarrow k_s$, $\gamma_s \leftrightarrow h_s$.

- New quantities are also defined. B_s and A_s are vectors with components b_{sx} and a_{sx}. For B, $0 \leq s \leq d$, and for A, $1 \leq s \leq c$ ($\equiv d + 1$).

[8]$\forall$ means "for all."

[9]In [47] the more general condition $\sum_x \pi(x) \leq 1$ is allowed.

Suppose $q(x)$ is another probability distribution satisfying the constraints[10] in Eq. (6.16). Then

$$
\begin{aligned}
S(p|\pi) &= \sum_x p(x) \sum_s b_{sx} \log \mu_s = \sum_s \log \mu_s \sum_x p(x) b_{sx} \\
&= \sum_s \log \mu_s \sum_x q(x) b_{sx} = \sum_x q(x) \log \left(\frac{p(x)}{\pi(x)} \right) \\
&= \sum_x q(x) \left[\log \left(\frac{q(x)}{\pi(x)} \right) - \log \left(\frac{q(x)}{p(x)} \right) \right] = S(q|\pi) - S(q|p).
\end{aligned}
\tag{6.20}
$$

This says that $S(q|\pi) - S(p|\pi) = S(q|p)$. Since $S(q|p) \geq 0$ this shows that p minimizes the relative entropy (with respect to π), and that any q that also minimizes the relative entropy must in fact be p (since $S(q|p) = 0$ only if $p = q$).

This then is what we were looking for. From the definition of relative entropy (and of entropy) $S(p) = S(\pi) - S(p|\pi)$. For any fixed π, $S(\pi)$ is just a number, so minimizing $S(p|\pi)$ is the same as maximizing $S(p)$.

6.2.2   Constructing the probabilities*

Next we turn to the construction providing the desired probability distribution. It's a bit of a mess, since the way it's done in [47] is via a complicated change of variables. Define two new $c \equiv (d + 1)$-dimensional vectors, u_s and t_s. They are chosen to define new variables, A_s and γ_s:

$$
A_s = t_s(u_s + B_s) \quad \text{and} \quad \gamma_s = t_s(u_s + \beta_s).
\tag{6.21}
$$

You want to pick $\{t_s\}$ and $\{u_s\}$ so that the new quantities are positive and sum (for $s \leq d$) to less than 1. For example you could make the following choices:

$$
u_s = 1 + |\min(\beta_s, b_{sx})| \quad \text{and} \quad t_s = \frac{1}{(N+1)} \frac{1}{(1 + |\max(\beta_s, b_{sx})|)},
\tag{6.22}
$$

with N the cardinality of X. You then define

$$
A_c \equiv B_0 - \sum_{s=1}^d A_s \quad \text{and} \quad \gamma_c \equiv 1 - \sum_{s=1}^d \gamma_s.
\tag{6.23}
$$

With these definitions, the constraint equations and the form of p become

$$
\langle A_s \rangle = \gamma_s,
\tag{6.24}
$$

$$
p(x) = \pi(x) \prod_{r=1}^c \lambda_r^{A_r(x)},
\tag{6.25}
$$

with $\lambda_s = \nu \mu_s^{1/t_s}$, $s = 1, \ldots, d$, $\lambda_c = \nu$ and $\nu = \mu_0 \prod_1^d \mu_s^{-u_s}$.

---

[10]We are assuming the existence of *some* probability distribution satisfying the given constraints. Moreover, if some $p(x)$ is zero, let $0 \log 0 = 0$, so this can be ignored.

Given the new notation, the iterative procedure for calculating the probabilities is the following: Let $p^{(0)} = \pi$. Let $p^{(n+1)}(x) = p^{(n)}(x) \prod_{r=1}^{c} \left(\frac{\gamma_r}{\gamma_r^{(n)}} \right)^{A_r(x)}$, where $\gamma_r^{(n)} = \langle A_r \rangle_{p^{(n)}}$, the last expectation evaluated with respect to $p^{(n)}$ (as indicated).

If you would like to see a proof that the iteration converges to the desired result, see [47]. As I mentioned there are probably computational packages to do this, using [47] or other methods.

6.2.3 Using the probabilities

Once you have the probabilities, you can fit them to the form Eq. (6.15) to deduce the coupling coefficients, which here have the significance of connectivity.

You can read the results of these calculations in [191]. These authors worked with 40 cells of a salamander's retina. Their first question was: Do you need more than second-order correlations to fit the data? Their conclusion was that you do not. This was considered important since higher-order correlations would tremendously expand the size of the space they needed to consider. However, they also emphasized the need for second-order information. In particular, the firing of 10 neurons at once, using only firing rates (lowest-order information), was extremely unlikely, whereas in fact it was seen. Yet another result was the behavior of the coefficients, J_{ij}, as the network size increased. You'll recall that in our "Curie–Weiss" model we scaled the attraction by the system size (the $1/N$ in Eq. (5.13), as opposed to its absence in Eq. (5.1)). So you might expect that for large collections of neurons the J_{ij} values, the "coupling," would become smaller—but according to [191] it does not.

6.3 Using maximum entropy to study Supreme Court voting

This subsection follows a particular article [136], which, I have to confess, I found to be an amusing exercise, although to legal scholars the subject is deadly (occasionally) serious. You could phrase the problem as: Given the voting records of the Supreme Court of the United States, how predictable is the next decision? Or, more mundanely, what is the probability distribution for the voting of the justices?

The first thing to notice is that a lot of decisions are unanimous. (According to [136], these and other data are taken from Spaeth et al. [217]). This implies correlations among the justices (surprise, surprise). If justices voted with no correlations, the probability of a unanimous decision would be $2 \times 2^{-9} \approx 0.004$, 0.4%, in contrast to the observed (about) 30%.[11] So you need correlations.[12] First-order correlations, that is, the expected vote of any particular justice, are useless in this case, since a justice must vote "yes" or "no" but the question is often so convoluted that "yes" or "no" may be liberal, conservative, or undefinable.

The formal (mathematical) step is to assign each justice ("i") a voting record, $\sigma_i(t)$, with t labeling the decisions (so t is integer valued) and σ being plus or minus

---

[11] This estimate holds if all h's are zero. With small values (as explained earlier), it's almost true. In any case, it's nowhere near the actual values.

[12] This is reminiscent of [191].

Court	Percentage unanimous		Court	Percentage unanimous
1946–1949	34%		1972–1975	33%
1949–1953	28%		1975–1981	32%
1955–1956	42%		1981–1986	38%
1957–1958	30%		1986–1987	30%
1958–1962	36%		1988–1990	38%
1962–1965	40%		1990–1991	40%
1965–1967	35%		1991–1993	42%
1967–1969	33%		1993–1994	42%
1970–1971	37%		1994–2005	43%

Table 6.1 Votes of the Supreme Court, broken into "natural courts," meaning that there is no change of personnel during that period. The second column is the percentage of the votes that was unanimous, usually exceeding 30%.

one, depending on whether the vote is positive or negative. As remarked, due to the nature of the questions posed, "yes" or "no" means little, and we expect $\langle \sigma_i \rangle$ to be about zero. So our study will involve at least 2-point correlations, $\langle \sigma_i(t)\sigma_j(t) \rangle$, the relation between the votes of justice i and those of justice j. It will turn out that (1-and) 2-point correlations almost tell the whole story, and that's what's about to be demonstrated.

The problem can now be defined as follows. We seek $\Pr(\sigma_1, \sigma_2, \ldots, \sigma_9)$, the probability distribution for votes by all the justices on any particular issue. We impose the following constraints: $\langle \sigma_i \rangle = \langle \sigma_i \rangle_{\text{actual}} \equiv b_i$, $\langle \sigma_i\sigma_j \rangle = \langle \sigma_i\sigma_j \rangle_{\text{actual}} \equiv C_{ij}$, where $i, j = 1, \ldots, 9$. Note that each pair of indices is counted only once and there is no diagonal, so the C's constitute $9 \times 8/2 = 36$ constraints. And of course the probabilities must sum to one. The question behind this is: Is this enough? That is, given only second-order correlations, can you accurately forecast the vote? If you took all *correlations* into account you'd surely recover the precise voting record, but we'd like an easier problem, truncating at 2. How good will the prediction be?

The mathematical problem is to find stationary points of

$$W = S(P) - \lambda_0 \left(\sum P(\{\sigma_i\}) - 1 \right) - \sum_i \lambda_i \left(\langle \sigma_i \rangle_P - b_i \right) - \sum_{i<j} \lambda_{ij} \left(\langle \sigma_i\sigma_j \rangle_P - C_{ij} \right), \quad (6.26)$$

with the subscript "P" meaning that the expectation is evaluated with the probability distribution P, and $S(P)$ is the entropy associated with the distribution P (the usual "$-\sum p \log p$"). P is thus a function of $\{\sigma_j\}$, that is, P is supposed to give the likelihood of each decision by each justice. (The $\{b_i\}$ and $\{C_{ij}\}$ are fixed by the data and are the quantities to be matched.) Finding the variation of S with constraints is most easily accomplished through the method of Lagrange multipliers (as described in Sec. 6.2 and Appendix G). You enforce the constraints through selection of the λ's. W is supposed to be an extremum with respect to variation of all variables, the P's as well

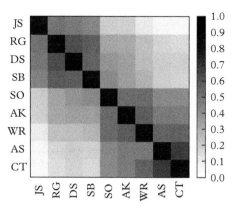

Fig. 6.1 Correlations among the justices in voting during the period of the Rehnquist court (1994–2005). The letters are initials (JS, John Paul Stevens; RG, Ruth Bader Ginsberg; DS, David Souter; SB, Stephen Breyer; SO, Sandra Day O'Connor; AK, Anthony Kennedy; WR, William Rehnquist; AS, Antonin Scalia; CT, Clarence Thomas). The scale is at the right, with darker colors indicating greater correlation. Note that all correlations are positive. Source: [136].

as the λ's, but the derivatives with respect to the λ's simply enforce the constraints. The derivative with respect to P yields

$$\delta W = \delta P \left\{ \log P - 1 - \lambda_0 - \sum_i \lambda_i \sigma_i - \sum_{i<j} \lambda_{ij} \sigma_i \sigma_j \right\} . \tag{6.27}$$

Since δW is to be zero and δP is arbitrary, it's clear that the solution is of the form $P \sim \exp(-\sum_i \lambda_i \sigma_i - \sum_{ij} \lambda_{ij} \sigma_i \sigma_j)$. However, in order to bring our notation in line with that of [136] we will write this as

$$P(\{\sigma_i\}) = \tfrac{1}{Z} \exp(-E\{\sigma_i\}) ,$$
$$E(\{\sigma_i\}) = -\sum_i h_i \sigma_i - \tfrac{1}{2} \sum_{ij} J_{ij} \sigma_i \sigma_j , \tag{6.28}$$
$$Z = \sum_{\{\sigma_i\}} \exp(-E\{\sigma_i\}) .$$

This is only a change in notation, from λ's to Z, h_i and J_{ij}. (A similar transformation was carried out in the paper by Schneidman et al. [191]. See Sec. 6.) The convention of [136] is that the double sum over i and j above is over both of them running from 1 to 9, but with only 36 independent J's, since $J_{ij} = J_{ji}$ and $J_{ii} = 0$. There are thus $36 + 9 + 1 = 46$ constants to obtain from a corresponding number of equations.

Finding the actual numbers for the h's and J's can be done with standard programs. In [136] the authors suggest the `fsolve` routines of MATLAB™.[13] As anticipated the h's are close to zero, so most of the information resides in the J's. The data apply to the Rehnquist court, an extremely prolific one with 895 decisions and which functioned from 1994 to 2005. The raw correlation data are shown in Fig. 6.1.

---

[13]`fsolve` is a MATLAB™ program that takes a function of several variables, sets it to zero, and finds the parameter values of the function that makes that happen.

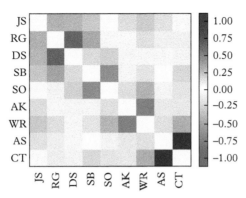

Fig. 6.2 Values of the matrix J of Eq. (6.27). Thus if J_{ij} is close to 1, the "energy" E is lowered and it is more likely that both justices vote the same way. A blue square has the opposite implication. None of the matrix elements approaches the level -1. (If you are viewing this in monochrome, darker colors generally indicate more correlation, since there is only one anti-correlation in the -0.25 to -0.50 range, WR and JS.) Source: [136].

As you can see Thomas and Scalia often vote together, but the surprise is in the analysis. Despite the fact that all correlations are positive, it turns out that a positive vote by some justices is likely to coincide with a *negative* vote by certain others. This can be seen in Fig. 6.2. The two deepest blues $(J \sim -1/2)$ are the pairs WR–JS and SB–CT (note that the matrix is inherently symmetric, so there are four squares with this color). We'll talk politics in a little while, but certainly the color scheme in Fig. 6.2 suggests that there are major differences of opinion. To see this first check Fig. 6.1. The justices are in a political order: liberal to conservative, JS being the most liberal (according to [217]), Thomas the most conservative. This accounts for the high level of agreement among the justices near the upper left and lower right of Figs. 6.1 and 6.2. It is also evident in the color scheme of Fig. 6.2, with lots of red and orange along the diagonal, and blue along the edges.

Note, by the way, that the probability function (Eq. (6.28)) is similar in fact and in principle to that of Eq. (6.15). In particular it looks like the Ising model for a spin glass, but, as for the other example, it's the J's that you seek, rather than the other way round.

Now we can take up our original question: How often is it important that three justices be taken into account? There are order-3 correlations and so far we have neglected them. Does this matter? Various tests can be applied to the probabilities of Eq. (6.28) and we'll use a rather simple criterion. How often are there zero, one and so on dissents? The probability of a single dissent is $P(1,1,1,1,1,1,1,1,-1) + P(1,1,1,1,1,1,1,-1,1) + \ldots + P(-1,-1,-1,-1,-1,-1,-1,-1,1) + \ldots$, which is directly calculable from Eq. (6.28). Our "P" uses (first-and) second-order correlations (call it $P^{(2)}$); one can also define $P^{(1)}$, the probabilities obtained using only first-order correlations as constraints. And now the predictions of various splits can be calculated using the two derived probabilities and comparing them with the actual results. This is illustrated in Fig. 6.3. As is

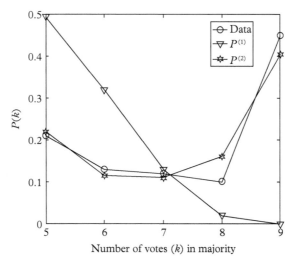

Fig. 6.3 Probability of a k-vote majority. $k = 9$ represents a unanimous decision and majorities can form with as few as five. The actual votes are shown with circles. In the figure decisions have not been classified as liberal or conservative and the mean value of each $\langle \sigma \rangle$ is nearly zero. In both the values indicated by $P^{(1)}$ (using a triangle symbol) use only first-order correlations and the values designated $P^{(2)}$ (star symbol) are those shown in Eq. (6.28), involving second-order correlations. Adapted from: [136].

evident the predictions using second-order correlations are excellent and are reliably wrong only for the one-dissent case.

Now let's talk politics. By labeling each vote as liberal or conservative and calling (say) the liberal vote $\sigma = -1$ and the conservative vote $+1$, one loses the plus/minus symmetry of the previous version of the (same) data. Because judges tend to be consistent in their liberal/conservative perspective, the values of the h's will be considerably larger. The h's act like fields, so that the liberal justices will feel one direction of the field (tending to make them vote -1) and the conservative justices will feel the opposite field. With this change the (new) function $P^{(1)}$ does a much better job of prediction. This is not evident in the record of unanimous votes, but is particularly significant when there's a near tie.

What I like about this model (and which is remarked upon in [136]) is that the predilections of the justices emerge without an a priori classification. You don't need to know that JS is liberal and CT conservative; the data speak for themselves. Of course our intuitive (or maybe educated, but not in the maximum entropy sense) classification is based on the same data. All that's happened is that names are given to the two extremes.

7
Power laws

Applications discussed in this chapter: Linguistics (Zipf's law), Sociology, Mathematics (urns), Computer science (graphs), Economics, Geology, Biology (species, genera)

A power law is a function,

$$f(x) = ax^{-b}, \tag{7.1}$$

relating x and f, with two parameters, a and b.[1] If you take the logarithm of each side you get $\log f(x) = \log a - b \log x$, so doing a log-log plot would give a straight line of slope $-b$ and intercept $\log a$. We'll see that a great deal of data fit this pattern, in economics, in linguistics, in physics, in sociology, and on and on—even in music. We'll also see that with finite data (limiting the variable x, and with error bars) you can't be *sure* you have a power law, since other relations can look like a power law over quite a few decades of data.

Here are a few examples of data sets that are usually considered power laws—some of them will be discussed in greater detail in this and other sections.

- Zipf's law in linguistics. In some large book, plot the number of times a word appears (the dependent variable f above) versus its rank (e.g., if the most used word is "the" its rank would be 1). The power ($-b$ in Eq. (7.1)) is usually near -1.
- Zipf's law for almost everything. As well as many others, G. K. Zipf, a Harvard professor, looked at lots of data sets, like cites and their populations, the number of people watching a TV channel, and more. Not only does Zipf have successors, but he also had predecessors (notably Auerbach [13] and Estoup [56]). Of possible interest to readers of the present material, citations of scientific papers also follow a power law (cf. [184]).
- $1/f$ noise. Suppose you measure the voltage through a resistor, paying special attention to the fluctuations. Then you take the spectrum of those fluctuations, that is, you measure the strength of each $f$requency component in the autocorrelation

---

[1]There are also cutoffs, that is, limits to the range of x. If f is to be normalizable there is at least one, possibly two (depending on b), cutoffs.

When Things Grow Many. Lawrence S. Schulman, Oxford University Press.
© Lawrence S. Schulman (2022). DOI: 10.1093/oso/9780198861881.003.0007

function (that's the f in $1/f$).[2] When you do this, you find that the intensity drops almost like $1/f^\alpha$ with α close to 1. Music (as opposed to speech) also has [140, 229, 230] this pattern.[3]

- Magnitude of earthquakes. You plot magnitude (on the Richter scale this is already logarithmic) against frequency of occurrence.
- Family names in the United States. Again, you plot the number of people having a given family name versus the rank. But here there's a catch. For other cultures this may not be the case. In Korea there seems to be an exponential distribution.
- Features of proteins and genes. As mentioned in the introduction I found that in the Tangled Nature model the likelihood of rare genes does not decrease exponentially but rather like a power law. This is also reflected in Nature. There are several papers and books [77, 99, 128, 129, 183, 208, 225] that study both genomes and protein features. They find that rare structures obey a power law.

This is a small sample of distributions that *seem* to follow a power law.

To explain the italicized "seem" in the last sentence, consider what's known as the log-normal distribution. That distribution is

$$p(x) = \frac{1}{x\sigma\sqrt{2\pi}} \exp\left(-\frac{(\log x - \log x_0)^2}{2\sigma^2}\right) \tag{7.2}$$

for constants x_0 and σ. As a numerical experiment I took the power law

$$f = 30 \exp\left(-x^{1.1}\right), \qquad \text{for } x = 1, 2, \ldots, 1000,$$

and optimized a multiple of the log-normal distribution to fit the function f. It turns out that there's no point in showing the figures: for a factor of about 2000 in the arguments of the functions and three orders of magnitude in the arguments, you can't tell the two logarithmic curves apart. This is true for powers starting at around 0.75 and as far as 2 (I haven't checked beyond that). Obviously, the two functions (power law and log-normal) are not the same and differ both for powers beyond those I've indicated and for exponents outside the close-to-1 range. But the data that you're trying to fit often are only valid for a limited range. Moreover, they are typically noisy, especially for small values of the function f. The conclusion: for values of exponents close to 1, log-normal is a stiff competitor and the less data you have the less confidence there can be in using a power law.

---

[2]To be precise, let the signal, the fluctuating voltage, be $V(t)$. You first find its autocorrelation function, $R(\tau) = E\left((V(t+\tau))(V(t))\right)$. (This assumes V is stationary and of mean zero.) Then you Fourier transform this function and find the intensity as a function of frequency. ("E" is expectation, $E(x) = \int dx\, xp(x)$. See Appendix A, on notation.)

[3]It's probable that music does follow a power law. I'm somewhat convinced by Levitin et al. [140] and by Voss and Clarke [229, 230] as well. My doubts were aroused by the following: Some years ago I sat next to John Clarke (one of the discoverers of this feature) at lunch and reported that my students had Fourier analyzed various pieces of music and had not found his effect. (Voss and Clarke [229, 230] had used analog methods, whereas my students used Fourier transforms by computer of the actual signal.) Their project was to see if the call of the humpback whale also had a $1/f$ pattern, but first I wanted them to check their methods on ordinary music. Clarke did not respond.

7.1   Power laws are scale free

Power laws have an additional attraction. They are scale free. This means that they look the same at all magnifications. This ties into the theme of fractals (see Appendix C). In equations, "looking the same at all sizes" translates into a property of the probability distribution $p(x)$, namely

$$p(\gamma x) = h(\gamma)p(x),\tag{7.3}$$

with γ some constant and h a function, at this point seemingly arbitrary. In words, if you look at the the function $p(x)$, not over some values of x but over a multiple (γ) of those values, the function will be the same, possibly multiplied by a function dependent on γ. Obviously if $\gamma = 1$ there cannot be any change, so we know that $h(1) = 1$. If everything is assumed differentiable then by taking the derivative with respect to γ and setting $\gamma = 1$ we have $xp'(x) = h'(1)p(x)$. This immediately yields $p(x) = ax^{-b}$ with $b = -h'(1)$. (The constant a is fixed by normalization; see below.) This also fixes the function h to be $h(\gamma) = \gamma^{-b}$. (The x dependence, as well as a, cancels.) So all scale-free functions are power laws (at least if they're differentiable).

The property of being scale free can be contrasted with distributions *not* having this property, that is, they *do* have particular dimension(s), *scale(s)*. Suppose you have N nuclei at time zero. Then at later time, t (>0), the average number that survive is $N\exp(-t/\tau)$ for some constant τ (ignoring very small and very large times[4]). This constant τ could be called the lifetime of the nucleus and is the characteristic time scale for the decay.[5] Another distribution with a scale is the normal distribution; for example to a good approximation the probability that an American male has height between h and $h + dh$ is given by $p(h)dh$ with $p(h) = (\sigma\sqrt{2\pi})^{-1}\exp\left(-(h - h_0)^2/2\sigma^2\right)$.[6] The average of this distribution is h_0 and the spread is σ, both of which have dimensions of length and which establish the scale of the distribution. Also note that the log-normal distribution, which might not be distinguished from a power law for some data sets, is also excluded.

Although a power law has no scale, it does have a range of validity. The function $p(x) = ax^{-b}$ is a probability density, so it must satisfy $\int p(x)dx = 1$; but if x ranges from $-\infty$ to $+\infty$ this would be impossible. So if $b > 0$, there must be a cutoff at some minimum value, that is, $p(x) = ax^{-b}$ can only be valid for $x > x_{min}$ (x_{min} is

---

[4]For very short times you get the quantum Zeno effect, a slow down and possible speed up of decay, and for long times there is a power law. Both are difficult to observe experimentally. I associate the name L. A. Khalfin, a Russian, with discovery of both long and short time effects, although the Zeno (paradox) allusion came later. Khalfin told me that V. A. Fock, his advisor, at first did not believe him, but eventually came around. Landau never came around.

[5]For nuclei one often gives instead the "half-life," the time at which half the nuclei (of some particular sort) have decayed. Thus $\exp\left(-t_{1/2}/\tau\right) = 1/2$ with $t_{1/2} = \tau\log 2$ the half-life.

[6]Of course this is only approximate, but it's a very good approximation. In particular $h < 0$ is meaningless although $p(h)$ is not zero, just very small for negative h. If the mean is about 180 cm and the standard deviation is about 20 cm, then the probability at zero (or even at small positive values) is essentially zero (the integral from 40 cm to $-\infty$ is 1/[more than the human population of the planet], by several orders of magnitude). Achondroplasia, gigantism and similar conditions are not part of this distribution.

that minimum).[7] At the other end, $x = \infty$, if $b > 1$ there is generally no mathematical reason for a cutoff, although in fitting data things usually get noisy for large "x." The normalization condition, $\int_{x_{min}}^{\infty} p(x) = 1$, implies that $a = (b-1)x_{min}^{b-1}$. Clearly this can only make sense if $b > 1$, even though on occasion the data may call for smaller values of b. In the latter case, there is generally an implicit or explicit assumption that there is also a maximum x value.

For the moment then, let's assume we have a power law. The question is, why?

As you can imagine, this is a broad question, because it not only applies to linguistics but also to the various other distributions that may (or may not) exhibit power law distributions. Zipf himself [244] had some fuzzy ideas about this distribution minimizing speakers' and listeners' efforts (he brings Freud into the story). Calling them fuzzy is already a judgement on my part: some people (e.g., [16, 181]) take his ideas seriously, while others find them to be mostly nonsense [127, 220]. Instead, I will discuss a number of possible mechanisms, more from the mathematical point of view than the psychological. For details see Newman's excellent[8] article [163].

It turns out that lots of situations can lead either to power laws or to distributions that resemble them for an extended range of powers (what I've been calling "b").

7.2 Diffusion

Suppose a particle is confined to one dimension and starts at a point $x_0 > 0$. Thereafter it diffuses. This means it undergoes a random walk and will only cover a distance proportional to the square root of the time it's diffusing.[9] We add a complication: There's an absorber at $x = 0$. We now ask how much is left (in $0 < x < \infty$) after some time t? This could be your personal fortune at a casino. It's just a matter of time until you go broke (unless you leave when you're ahead—but the odds are against that in "well-run" casinos). *If you are not interested in the details of the derivation, skip to Eq. (7.12).* At time-0 the particle is at x_0, which is to say its density function is a (Dirac)-delta function

$$\rho(x,0) = \delta(x - x_0),\tag{7.4}$$

where ρ is the density. This means that $\rho(x,t)\,dx$ is the probability that the particle will be found between x and $x + dx$. It is known (e.g., [226]) that ρ satisfies the *diffusion equation*

$$\frac{\partial \rho}{\partial t} = D\frac{\partial^2 \rho}{\partial x^2},\tag{7.5}$$

where D is a constant measuring the propensity to diffuse. The solution to Eq. (7.5) can be given in terms of a propagator, K, namely

---

[7]Correspondingly if $b < 0$ one requires that there be some maximum x value, and if $b = 0$ there need to be two limits. Usually, but not always, it's more convenient to define the distribution with a negative b.

[8]... but watch out for typos and clerical errors.

[9]The contrasting situation is *ballistic* motion, in which a particle moves without interference in a wave-like fashion. For this motion distance is proportional to time—the proportionality constant being velocity.

$$\rho(x, t+t_0) = \int dy \, K(x, t; y) \, \rho(y, t_0) \,. \tag{7.6}$$

If there were no absorber and the particle could travel freely on the entire line, the propagator would be $G(x, t; y) \equiv \frac{1}{\sqrt{4\pi Dt}} \exp\left(-\frac{(x-y)^2}{4Dt}\right).$[10] However, there *is* a wall and one can construct the propagator using the method of images

$$K(x, t; y) = G(x, t; y) - G(x, t; -y) \,. \tag{7.7}$$

That this is the solution can be seen as follows. Both $G(x, t; y)$ and $G(x, t; -y)$ satisfy the differential equation, Eq. (7.5), for $x \geq 0$. Moreover, the particular combination in Eq. (7.7) vanishes at $x = 0$. And finally, $K(x, t; y)$ has a delta-function singularity at $y = 0$ and $t = 0$. That's all it takes for the proof.[11]

The power law shows up when you ask what is the probability that the diffuser has *not* been absorbed at time-t. This is equal to the total probability in the region of positive x, namely

$$P(t) = \int_0^\infty dx \int_0^\infty dy \, K(x, t; y)\rho(y, 0) \qquad \|\text{Recall } \rho(y, 0) = \delta(y - x_0) \text{ for the next equality}$$

$$= \frac{1}{\sqrt{4\pi Dt}} \int_0^\infty dx \left[\exp\left(-\frac{(x - x_0)^2}{4Dt}\right) - \exp\left(-\frac{(x + x_0)^2}{4Dt}\right) \right] \,. \tag{7.8}$$

We are interested in this quantity for long times. "Long" must be defined in terms of only fixed, dimensional quantities, since x takes both large and small values. The relevant quantity is

$$\lambda \equiv \frac{x_0^2}{4Dt} \to 0 \,, \tag{7.9}$$

which is dimensionally correct since D has the dimension distance squared over time (making λ dimensionless). This naturally leads to a change of (integration) variable to $u \equiv x/x_0$, so that (doing a bit of algebra) we get

$$P(t) = \sqrt{\frac{\lambda}{\pi}} \int_0^\infty du \left(e^{-\lambda(u-1)^2} - e^{-\lambda(u+1)^2} \right) \qquad \text{with} \quad t \sim \frac{1}{\lambda} \quad (\text{hence } \lambda \to 0) \,. \tag{7.10}$$

We do asymptotics directly on this expression. The integrand can be rewritten as

$$\exp(-\lambda) \exp(-\lambda u^2) 2 \sinh(2u\lambda) \,. \tag{7.11}$$

The first term, $\exp(-\lambda)$, disappears as $\lambda \to 0$. The gaussian kills the integrand as $u \to \infty$ while the sinh has the opposite behavior, so there is a single peak along the way. To find

---

[10]My own background is more in the realm of quantum physics, where you have the one-dimensional Schrödinger equation for a free particle, $i\hbar \partial \psi / \partial t = -(\hbar^2/2m)\partial^2 \psi / \partial x^2$, and the propagator is (e.g., [194], Eq. 6.5) $\sqrt{m/2\pi i\hbar t} \exp\left((im/2\hbar t)(x - y)^2\right)$. To get the propagator for the diffusion equation (Eq. (7.5)) make the following transformation (I'll call the new time variable τ, but in the text (e.g., Eq. (7.5)) it's called t): $\tau = it$ and $D = \hbar/2m$. When there's an absorbing wall, say at zero, one uses the method of images (e.g., [194], Eq. 6.43) and the propagator is that given in the main text.

[11]For the Schrödinger equation there's no absorption and one often takes this (Dirichlet) boundary condition to be the same as having an infinite potential for $x \leq 0$.

that, we differentiate and set the derivative to zero: for $\phi(u) \equiv \exp(-\lambda u^2)\sinh(2u\lambda)$, $\phi'(u) = 0$ implies $u \sinh(2u\lambda) = \cosh(2u\lambda)$, whose solution (using $\tanh w \sim w$ for w near zero) is $u = 1/\sqrt{2\lambda}$.[12] Not going beyond lowest order asymptotics and replacing u by $1/\sqrt{2\lambda}$ in Eq. (7.11), we get

$$P(t) \sim \sqrt{\frac{\lambda}{\pi}} e^{-1/2} 2 \sinh \frac{2\lambda}{\sqrt{2\lambda}} = \lambda \sqrt{\frac{4}{e\pi}} = \text{const} \cdot \frac{1}{t}. \qquad (7.12)$$

To summarize, the probability that a diffusing particle starting somewhere to the right of the absorber has not hit the absorber by time-t drops off like one over the time. This was found by knowing the probability density at each point as a function of time and integrating over the positive real line. The power in this power law is 1, a value that is often found in practice. The expected value of the time for absorption would be $\int dt\, t \Pr(\text{the particle is absorbed at time-}t) = \int dt\, t(-1)\frac{dP}{dt} \sim \int dt\, t(1/t^2) \sim \log t_{\max} = \infty$; this expectation does not exist.

Returning to the casino example, the slow dropoff ($1/t$) of the expected time to extinction is part of the reason casinos make money: you don't go broke all that fast—and maybe you even enjoy yourself. But before you run off to the casino, bear in mind that the above example is missing an important feature of casinos: they have "drift." For the idealized diffusion equation, Eq. (7.5), it is equally likely for a step to go further from 0 or closer to zero. In casinos the house always[13] has the odds; it's always more likely that you'll lose money than that you'd gain.

7.3 Preferential attachment (the rich get richer)

The term "preferential attachment" is a bit dry, and calling the versions of the power law yielding processes after the inventors, Yule [239], Simon [212], and possibly others, doesn't convey information. Perhaps it should be called "the rich get richer" process.

Here is one variant of the process: You have N urns, with k_ℓ balls in urn #ℓ (ℓ running from 1 to N). You now randomly pick a ball—not an urn—and create another ball in the same urn. In this way urns with lots of balls get even more balls faster than urns with fewer balls. So the rich get richer.

The first model of this sort was a little different. Yule [239] was interested in the distribution of species: How many species are in a given genus? You start with some distribution of genera, each containing a number of species, and you add species to a genus in proportion to the number of species already present. So far it's the same as the urn model. But Yule added a wrinkle: at every M steps you would start a new genus, creating a single species for that genus. You let this process go on for a long

---

[12]Note that $u \to \infty$, but all we required in our approximation of the hyperbolic tangent is that $u\lambda$ be small. For that quantity, $u\lambda \sim (1/\sqrt{2\lambda})\lambda \sim \sqrt{\lambda} \to 0$.

[13]My statement is *almost* true. With card counting—keeping track of what cards have already appeared—the game of "21" can favor the player. This effect is weakened if many decks of cards are used in the play. Casinos, however, do not rely on this alone: they keep an eye on the players (since players generally use some device to help count) and bar from playing those deemed to be counting cards. For an exhaustive discussion, including aspects of the history (which existed before I lost \$10 in Reno), see https://en.wikipedia.org/wiki/Card_counting (accessed November 2020).

time and ask for the distribution function p_k, the probability that a given species is in a genus with k species. As stated this process ignores species extinction, but appears to give good results nevertheless.

So far I've talked about biology (species, genera) and urns, but there are many versions of the same process: computer scientists like to talk about networks and graphs, and students of scientific literature note that citations follow this pattern, with the data favoring a power law distribution, presumably the result of a preferential attachment process. Oh yes, and money. The Pareto distribution (about which more, later) of wealth and income shows tremendous disparities in the amount of money different individuals have or acquire.

Deriving the actual power in the power law can be a bit messy, so I'll only derive one version, motivate the general notion and present the results.

The Albert and Barabási [5] algorithm is a way of modeling preferential attachment. It does so in the language of networks, which I'll follow here. A network is a collection of nodes. Some of these nodes are connected to others, so that each node has a *degree*, the number of nodes it connects to. Its connections can be directed or undirected, depending on what the node is supposed to be modeling (i.e., for nodes A and B, you might want either A→B or A↔B). We will talk about undirected connections, to keep things simpler (so we're only looking at A↔B). At each time step a new node is added to the network. Each new node is, upon its creation, connected to an existing node. The connection rule depends both on the existing connectivity and on a probability p. With probability p the node is randomly connected with equal likelihood to *any* node, irrespective of that node's connectivity. So with this probability the rich do not get richer. But, with probability $(1-p)$ they do: specifically, with probability $(1-p)$, attachment to a given node is proportional to that node's degree, to the number of other nodes already attached.

We'll work out the distribution in the case $p=0$. This is the extreme case of the rich get richer.

Let $D(n, t)$ be the degree of node n at time-t. According to the rules specified

$$\langle D(n, t+1) \rangle = D(n, t) + \frac{D(n, t)}{\sum_m D(m, t)} = D(n, t) \left(1 + \frac{1}{2t + \#\text{original nodes}} \right). \quad (7.13)$$

I'll explain this equation, term by term. The angular brackets on the left represent an expectation with respect to the random choices made above. To the right of the (first) equal sign is $D(n, t)$, which is the number of nodes already present, and to its right the probability of adding a node; the latter is the heart of the model. That probability is proportional to the relative number of nodes already connected. Finally (in the second equality) the denominator in that expression is the total number of connections altogether, namely the original nodes plus two connections for each time step. (The "two" is because the connections, often called "edges," go both ways.) So far, everything is exact. Next we massage this a bit, taking into account the fact that we are only interested in asymptotics, large times, when there are already many, many nodes and most are quite old. This means that we can dispense with the number of original nodes since it only adds a fixed constant to the total degree.

Moreover, since we are only interested in long time asymptotics, we can drop the brackets (on $D(n, t+1)$) and turn Eq. (7.13) into a differential equation. For sufficiently large times $D(n, t+1) - D(n, t)$ is approximated by $\frac{d}{dt} D(n, t)$, so that

$$\frac{d}{dt} D(n, t) = \frac{D(n, t)}{2t}. \tag{7.14}$$

Note that after dividing by $D(n, t)$ there is no n dependence on the right-hand side (which is $1/2t$). The value of n will only depend on the integral limits, specifically on t_i (to be defined below). Eq. (7.14), when divided by $D(n, t)$, can be immediately integrated to yield

$$\log D(n, t) - \log D(n, t_i) = \frac{1}{2} \left(\log t - \log t_i \right) = \log D(n, t), \tag{7.15}$$

where t_i is the *initial* time (the time at which node n is created) and $\log D(n, t_i) = 0$, by definition. It follows that

$$D(n, t) = \sqrt{\frac{t}{t_i}}. \tag{7.16}$$

Our interest is in asymptotics of the degree distribution, so we ask for the probability that the degree exceeds some (large) number, k. This probability is

$$\Pr\left(D(n, t) > k \right) = \Pr\left(\sqrt{\frac{t}{t_i}} > k \right) = \Pr\left(\frac{t_i}{t} < \frac{1}{k^2} \right) = \frac{\text{const}}{k^2}, \tag{7.17}$$

by the uniformity of the t_i distribution within $[0, t]$.

One last step remains. The probability calculated was for D exceeding k. To know the probability of getting to k exactly we must take the negative of the derivative of the foregoing expression, so that finally

$$\Pr\left(D(n, t) = k \right) \propto -\frac{d}{dk} \frac{\text{const}}{k^2} = \frac{\text{const}}{k^3} \tag{7.18}$$

(with a different positive constant).

The number of highly connected nodes follows a power law, dropping off like the cube of that number. If p were non-zero the dropoff would be like $\frac{3-p}{1-p}$ and for directed graphs this would be replaced by $\frac{2-p}{1-p}$.

> **Exercise 23** For the directed network with $p = 0$, show that the dropoff indeed has power two.

For the Yule process the result is the following: the asymptotic probability that there are n species in a genus goes like $p_n \sim \text{const}/n^\alpha$ with $\alpha = 2 + \frac{1}{m}$, where m is the interval between the creation of new genera. The method of proving this involves stochastic differential equations and is given in Newman's paper [162], but is a bit involved for a derivation to be presented here. (And you thought the proof of Eq. (7.18) was messy.)

For other forms of the attachment process slightly different exponents are given, but generally speaking they are more than 2, although functions of these quantities (e.g., the integral) can be close to 1.

7.4   Exponential functions of exponential distributions

Consider a variable that is exponentially distributed: $p(u) = \alpha e^{\beta u}$, with u the cutoff at some value (so β can be positive or negative). Lots of things have this distribution.

Here's an example of the exponential distribution: Suppose there is a certain rate, r, for something to happen, maybe for customers to come to a store; then (assuming independence) it's known that the probability that n will be in the store simultaneously is $\Pr(n) = (r^n/n!)e^{-r}$; that's the Poisson distribution. But what is the probability that there is a wait of length t between customers? That can be written as the probability (density) that no one came up to time-t, at which point someone did come in. That would be $\Pr(\text{wait between customers} = t) = \lim_{\Delta t \to 0}(1 - r\,\Delta t)^{t/\Delta t}r$ since for the first $(t/\Delta t)$ time steps no one came, which has probability $1 - r\,\Delta t$ for each Δt, and probability $r\Delta t$ that someone (finally?) entered between t and $t + \Delta t$ (and we divide by Δt since this is a probability *density*). Letting $t/\Delta t = N$ we need to evaluate $\Pr(\text{wait time}) = \lim_{N \to \infty}(1 - rt/N)^N r = re^{-rt}$.[14]

Now imagine x to be a variable that depends exponentially on u, so that $x = \gamma e^{\delta u}$. It will have some probability distribution and the two must be related by

$$|p_x(x)dx| = |p_u(u)du|, \tag{7.19}$$

so that

$$p_x(x) = p_u(u)\left|\frac{du}{dx}\right| = \alpha e^{\beta u}\frac{1}{|\delta|x} = \frac{\alpha}{|\delta|}\frac{1}{x}\left(\frac{x}{\gamma}\right)^{\frac{\beta}{\delta}} = \mu\frac{1}{x^\sigma}, \tag{7.20}$$

with $\sigma = 1 - \frac{\beta}{\delta}$ (and μ also fixed).

To continue the store example, suppose you are the storekeeper and your level of boredom increases exponentially with the wait time (so "x" is a measure of your boredom). Now "β" $= -r$ so your boredom will follow a power law with exponent greater than one. Still, it is quite likely (because of the slow dropoff of a power law) that, at some point of the day, you will be sufficiently bored to sneak a peek at the book you've been keeping under the counter.[15]

A more serious example of such a distribution is a randomly killed population [185]. Suppose the number alive after a time interval t is $N = N_0 e^{rt}$, where r is the logarithm of the reproductive rate. Unfortunately these creatures are substantially wiped out from time to time, having to start over again with N_0 individuals. Let

---

[14]A similar identity for the exponential was proved in Footnote 3, Chapter 4, except that now the limit $N \to \infty$ truly occurs.

[15]The variable "u" in this case would be t so that r has the unit of inverse time. In this case δ has the same units, and all is consistent. What units are used to measure boredom? Who knows? But whatever it is, γ takes care of that. (Cf. the definition of x.)

the time for this reduction be exponentially distributed, $p(t) = p_0 e^{-Kt}$. Then the population distribution will be a power law $\Pr(N) \sim N^{-1-K/r}$ (correspondence: $\beta \to -K$ and $\delta \to r$).

7.5 Superposition of exponentials

Another source of power laws arises from a superposition of exponentials. Thus

$$t^{-1-\alpha} = \frac{1}{\Gamma(\alpha+1)} \int_0^\infty dx \exp(-xt)x^\alpha, \qquad (7.21)$$

as is clear from the change of variables $u = xt$. In Sec. 11.1 I give an example of an apparent power law arising from a superposition of exponentials. Since the example is from physics it has been put elsewhere, but it does serve to illustrate this phenomenon. Similarly in Sec. 10.7. Those are *fakes*, but they're close. Eq. (7.21) is a continuous integral which is a bona fide power law. For more on the gamma function, $\Gamma(z)$, see Appendix D.

7.6 Critical phenomena and self-organized criticality (SOC)

For percolitis the approximate iteration, $s' = 1 - \exp(-rs)$ (Eq. (4.4)), leads to a time of extinction—*at the critical point* $(r = 1)$—of the disease that drops off like a power law (Sec. 4.4). The power is one. The "1" is particular to percolitis, but the occurrence of a power law is general. Moreover, unlike percolitis there are often many quantities having such behavior, for example correlation lengths and specific heats. Thus for the critical point in a liquid–vapor (i.e., fluid) transition you get the phenomenon of *critical opalescence*. Away from the critical point the correlation length is small. Say you're above the transition temperature and are talking about an Ising model. Then if a particular spin is up, it's likely that its close neighbors are also up. But once you go a few lattice spaces away it's as likely to be up as down. Not so as you approach criticality. This *correlation* grows, so that more and more distant spins are also up. In a fluid, near criticality correlation lengths grow tremendously and, instead of being at most a few molecular distances (perhaps less than a nanometer) in length, they grow and grow, reaching at some point hundreds of nanometers. But that's the wavelength of light! When light hits an object whose size is comparable to its wavelength there is enhanced scattering—hence critical opalescence, in which the substance strongly scatters light. In mathematical terms, you go from an exponential to a power law.

Could this be the source of the many observed power laws? Most likely, for some, yes, for some, no. The problem is that critical points are not generic: they require particular values of the control parameters temperature, pressure, transmission probability and so on. For percolitis, although the probability of transmission (p) varies from 0 to 1, it must be close to 0 (1/population) to achieve criticality. For a liquid–gas transition, say the critical point of water, temperature and pressure could have a variety of values, but criticality *only* occurs when the temperature is 647 kelvin and the pressure is 218.3 atmospheres (approximately).

Self-organized criticality—SOC—is the idea that a system will respond to external forces in such a way that it approaches a critical state—*on its own*. An incomplete version of this was seen in the galaxy model, in which star formation depended on a percolation-like phenomenon, but if there was too much star formation it inhibited further activity. Basically this was negative feedback, something also seen in neuronal models (there is a refractory period after a neuron fires). The system settled down close to the critical value, leading to relatively long spiral arms for the galaxies. But this was not the original metaphor for SOC; and in the original metaphor you got 100% criticality, not *almost* criticality. Other examples are discussed in Secs. 4.6 and 4.7 in connection with a percolitis or chemical model. The case, in chemical terms, of 3S→2S+I, where S and I are particular chemicals, turned out to be SOC, including the effect of fluctuations. This is an example of mean field theory giving rise to SOC.

The original metaphor is the sand pile model, proposed by Bak, Tang and Wiesenfeld [19]. Imagine a flat surface covered with sand. You have a further supply of sand and it comes from a small region above your surface, landing on a similarly small area of the sand below. For a while the sand piles up, forming a conical heap, with the cone eventually reaching a critical angle, known as the angle of repose. (I've seen this phenomenon in piles of slate. It can occur for a variety of granular materials.) At some point the pile becomes unstable and an avalanche ensues. This avalanche is scale free, that is, its possible size follows a power law. What's happened is that you started with arbitrary—non-critical—initial conditions and drove the system until it reached criticality, at which point characteristic features of criticality emerged, in particular a power law.

That's the metaphor, but there two problems: sand does not behave this way and this is not the model actually studied in the original paper. Neither problem is a game-changer since the model is a metaphor; presumably there are real systems that behave this way, even if sand does not.

There is a second paper [18] describing the "sand pile" model by the same authors; it is less telegraphic than the first, hence more comprehensible. Start with the one-dimensional case. On one side of the one-dimensional model the integer height is maximal (the value of which is also given) and on the other zero. Whenever there is a height jump of more than some (selected) critical value the sand falls to the lower region. This does *not* produce anything interesting since in short order the system

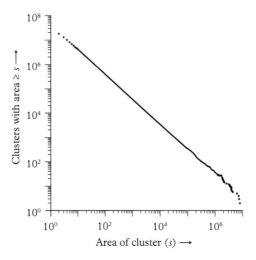

Fig. 7.1 Sizes of clusters for forest fire percolation. From [163.] Source: *Physical Review.*

reaches a unique minimally stable state and any perturbation leads back to the same state. But in two or more dimensions (for the sand configuration) you do see criticality. A perturbation has a wide variety of possible consequences, in fact a power law over many decades. What's happening is that, in contrast to the one-dimensional case, the perturbation spreads and ultimately affects more distant sites that do not simply go towards a unique stationary (but unstable) state.

There's a picture of a forest fire that also gives rise to SOC. This is described in [50] and simplified in [163]. Whether it will save any houses in California or Australia (where there are real fires burning) is unlikely but it's a percolation model of sorts. You have a two-dimensional square lattice, representing a land area in which there might be forest. In this land area an occupied square means that there's a tree that could be burnt. At first you begin with all squares unoccupied, but then with probability p you randomly pick sites to occupy. So far it's a standard percolation problem, eventually leading to a globally occupied cluster. (A *cluster* is a collection of occupied squares, each one of which has at least one occupied nearest neighbor. *Nearest neighbor* means one of the four squares touching the square in question.) But now you introduce lightning. This can strike anywhere, at any time—with some probability, call it p'. If it strikes an empty square (no tree) that's the end of it. If it hits an occupied square then that tree burns as well as the entire cluster that it belongs to. As the number of unburnt trees increases a spanning cluster eventually forms, but just as surely gets hit by lightning. This lowers the number of unburnt trees and increases the empty sites. Once again the empty sites fill and again a spanning cluster is eradicated. The point is that under this process the spanning clusters are themselves quite thin. This means that it doesn't take much to produce a spanning cluster, nor to destroy it. The result of this activity is a percolation system that hovers around criticality. It will therefore satisfy power laws (e.g., how many in the spanning cluster) and have fractal structure (of the spanning cluster). Apparently this model cannot be solved exactly in more than one dimension, but there is a nice power law. In Fig. 7.1 I show the results, which have a slope [85] (i.e., power) of about 1.19.

Besides the examples given, SOC has spread to many areas. In geology there is the Gutenberg–Richter [94] relation giving the number of earthquakes of various magnitudes during a given time period. And—guess what—it's a power law. This is an area studied by Bak and Tang [17] and in which the model of [19] was relevant. (Nevertheless, as remarked in [235], a similar model, not for sand but for blocks and springs sliding about, had been developed much earlier by Burridge and Knopoff [35].)

There is much to recommend this idea. It does seem to be one of many that can give rise to power laws, but there seems to me to be more involved, for example spatial fractality. In any case we'll encounter many systems that have been analyzed and explained in this way—social (e.g., [222]), economic, and many others.

A word of caution: If the data show a power law, even over three or four decades, and even if the slope is close to 1, you should not assume that SOC is at work. I have tried Ising and percolation models that give power laws which are not at criticality. To claim SOC one should have some argument suggesting criticality and good straight lines over sevaral decades (in the log-log plot) before jumping to conclusions.

8

Universality, renormalization and critical phenomena

Applications discussed in this chapter: Ising model (of ferromagnetism), Phase transitions, Universality, Renormalization

---

The behavior of systems near a critical point is *universal*, by which I mean that interaction details don't matter and the critical exponents depend only on gross features of the transition. The example I've given until now is the density change in a number of gases/liquids, as depicted in the Guggenheim plot in Fig. 1.1. Similarly we've seen that the growth of the order parameter as we move away from the critical temperature in both ferromagnets and in the van der Waals gas is the same, $\sqrt{T_{\text{critical}} - T}$. But wait: in the Guggenheim plot the order parameter grew (approximately) like the cube root of the difference of the temperature from its critical value, while for mean field theories it grew like the square root. On top of which in percolitis, another mean field theory, the order parameter was linear in the control parameter, that is, $s \sim r - 1$ (using mean field percolitis notation).

So "universal" may not be the right word, but it's what's used. It turns out that you need to add qualifiers. A phase transition is universal within its "universality class," of which there are several. Which universality class a given transition belongs to depends on the dimension of the space in which the transition occurs, the number of components of the order parameter, the range of the coupling force and other properties. However, it is remarkable that with all those qualifiers there are still many apparently disparate substances that fall into the same universality class. For them the short-range features of the interactions don't matter and the critical exponents are the same. Apropos critical exponents, these are the powers with which certain properties behave as the critical point is approached. I've already given examples of contrasting powers of the order parameter as the control parameter approaches the critical value. But there are many such critical exponents, applied to specific heat,

When Things Grow Many. Lawrence S. Schulman, Oxford University Press.
© Lawrence S. Schulman (2022). DOI: 10.1093/oso/9780198861881.003.0008

magnetization, and so on, properties associated with the critical point, which often go to zero or infinity.

One of the most important exponents pertains to the correlation length. This length approaches infinity as the relevant parameter goes to its critical value, but the power (of the control parameter's deviation from criticality) with which it does so can vary from one universality class to another, but within any class it is fixed.

The model we'll study has advantages and disadvantages. The advantage is that there's a bona fide real space renormalization; the disadvantage is that the phase transition is a bit artificial.

8.1   The nearest neighbor one-dimensional Ising model

The system is the one-dimensional Ising model with nearest neighbor interactions. The Hamiltonian is

$$\mathcal{H} = -J \sum_{k=1}^{N} \sigma_k \sigma_{k+1} - h \sum_{k=1}^{N} \sigma_k , \tag{8.1}$$

with J a positive coupling constant and h an external field. Note that, unlike the Curie–Weiss model, we now have only nearest neighbor interactions. We take $N = 2^m$ with m a large positive integer and periodic boundary conditions (#N is the same as #0). For reference purposes (and for general education) we solve it three ways—one incorrect, the second for the record and the third using the renormalization group—to show what is behind the Guggenheim plot and criticality in general.

8.1.1   Mean field treatment

The incorrect "solution" uses mean field theory and is what was done above—see Eq. (5.21) and [126]. The difficult part of Eq. (8.1) is the product, $\sigma_k \sigma_{k+1}$. The trick is to say that spin-k feels an *average* or *mean* field due to all the others, so we replace spin-$(k+1)$ and spin-$(k-1)$ by that average, namely $\mu \equiv \langle \sigma_j \rangle$, which does not depend on j: it's the same for all spins. Then you can use a one-particle Hamiltonian and look at only $\mathcal{H}_k = -\sigma_k(2J\mu + h)$, which is a single spin in a field $(2J\mu + h)$ (other spins, $\sigma_{k\pm1}$, appear two times in the sum, hence the "2"). This has a simple solution since it involves only non-interacting spins. We can focus on a single spin, since the calculation is the same for all of them (and allows us to drop the subscript k). First evaluate the partition function for a non-interacting spin in a field $h + 2J\mu$:

$$Z = \sum_{\sigma = \pm 1} \exp(\beta\sigma(2J\mu + h)) = 2\cosh(\beta(2J\mu + h)) . \tag{8.2}$$

(As usual, $\beta = 1/k_B T$.) Self-consistency then demands that the expectation of σ (a function of magnetization) is given by

$$\mu \equiv \langle \sigma \rangle = (+1) \Pr(\sigma = +1) + (-1) \Pr(\sigma = -1)$$

$$= \frac{1}{Z} \left[(+1) e^{\beta(+1)(2J\mu+h)} + (-1) e^{\beta(-1)(2J\mu+h)} \right]$$

$$= \frac{e^{\beta(2J\mu+h)} - e^{-\beta(2J\mu+h)}}{e^{\beta(2J\mu+h)} + e^{-\beta(2J\mu+h)}} = \tanh \beta(2J\mu + h). \qquad (8.3)$$

This is exactly the same equation as Eq. (5.21) (with "z," the number of neighbors, equal to 2). These equations predict a critical point at $\beta = 1/2J$ (or $T = 2J/k_B > 0$), $h = 0$; in the context of the Curie–Weiss model, that's what happens.[1] Here, for the nearest neighbor Ising model in one dimension, it's a lie[2]—as we'll see.

8.1.2 Transfer matrix treatment*

Using the transfer matrix (to be defined in a moment) we obtain an exact treatment of this model. This is often taught as a warmup for the corresponding (transfer matrix) treatment of the two-dimensional model, but since we won't be covering that problem our reason for presenting this is to be able to compare our results with the renormalization treatment, and to see the falsity of the mean field theory.

First some abbreviations. Let $\mathcal{J} \equiv \beta J$ and $H \equiv \beta h$. So $\beta \mathcal{H} = -\mathcal{J} \sum_k \sigma_k \sigma_{k+1} - H \sum_k \sigma_k$. (Note the distinction between $\mathcal{H}$, the Hamiltonian, and H, proportional to the external field. Also, calligraphic $\mathcal{J}$ is not the same as italic, capital J.) Define the matrix

$$L \equiv \begin{pmatrix} e^{\mathcal{J}+H} & e^{-\mathcal{J}} \\ e^{-\mathcal{J}} & e^{\mathcal{J}-H} \end{pmatrix}, \qquad (8.4)$$

which can also be written $L(\sigma, \sigma') = \exp \left(\mathcal{J} \sigma \sigma' + \frac{H}{2}(\sigma + \sigma') \right)$, with the labeling scheme $\left(\begin{smallmatrix} ++ & +- \\ -+ & -- \end{smallmatrix} \right)$. L is the transfer matrix. Using L and the definition of the matrix product we have

$$Z = \sum_{\substack{\text{all } k \\ \{\sigma_k = \pm 1\}}} \exp \left(\mathcal{J} \sum \sigma_k \sigma_{k+1} + H \sum \sigma_k \right) = \sum_{\substack{\text{all } k \\ \{\sigma_k = \pm 1\}}} L(\sigma_1, \sigma_2) L(\sigma_2, \sigma_3) \dots L(\sigma_N, \sigma_1)$$

$$= \operatorname{Tr} L^N = \lambda_+^N + \lambda_-^N \qquad (8.5)$$

---

[1] ... aside from a multiplicative change in the definition of J.

[2] Even in two and three dimensions it's still a lie. It's true that the Ising model has a critical point at a finite temperature in those cases (as predicted by mean field theory), but the behavior of the order parameter is wrong. The order parameter is the magnetization and is usually written $\mu \sim (1 - T/T_c)^\beta$ with β the critical exponent. (The notation is lousy, with β also meaning $1/k_B T$—but there are only so many letters in the Greek and Roman alphabets. Besides, I didn't invent it.) Mean field theory predicts $\beta = 1/2$ for all dimensions, whereas for two dimensions $\beta = 1/8$, for three it's approximately $0.326\,419(\pm 3)$. By the way, the three-dimensional value of β is also believed to apply to the physical liquid–gas transition. Note that, although it is close to $1/3$, it is significantly different, implying that the $1/3$ that one would read off the graph in Fig. 1.1 is not quite correct (and that there isn't a simple argument giving $\beta = 1/3$).

(spin-0 is the same as spin-N because of the periodic boundary conditions). In the last equality we have made use of $\text{Tr}\,A^n = \sum a_k^n$, where $\{a_k\}$ are the eigenvalues of A. Thus $\lambda_\pm$ are the eigenvalues of L and are given by (left as an exercise, based on Eq. (8.4))

$$\lambda_\pm = e^{\mathcal{J}}\left[\cosh H \pm \sqrt{\sinh^2 H + e^{-4\mathcal{J}}}\right]. \tag{8.6}$$

Instead of evaluating $\langle\sigma\rangle$, I'll cut to the chase and use the relation $Z = \exp(-\beta F)$ with F the usual free energy. This function contains the significant information about the system, and in particular the expectation of the spin value can be obtained from it, as we do below. First we look at the per-spin free energy, which is

$$f^{(N)} \equiv \frac{F^{(N)}}{N} = (-k_BT)\frac{1}{N}\log Z = (-k_BT)\left[\log\lambda_+ + \frac{1}{N}\log\left(1 + \left(\frac{\lambda_-}{\lambda_+}\right)^N\right)\right], \tag{8.7}$$

where the superscript N indicates the number of spins involved and our interest is in the limit $N \to \infty$.

Exercise 24 Show that $\text{Tr}\,A^n = \sum a_k^n$, where $\{a_k\}$ are the eigenvalues of A.

Exercise 25 Find the eigenvalues of L (given in Eq. (8.6)).

Now comes the punch line. For any non-zero $\mathcal{J}$, even if $H=0$, the ratio λ_-/λ_+ is bounded away from zero—in fact (for $H=0$), it's $\tanh\mathcal{J}$. (Note that as $N\to\infty$ we also have that $\frac{\lambda_-}{\lambda_+}$ does not approach zero.) Therefore the $N\to\infty$ limit of its Nth power is zero. It follows that

$$f \equiv \lim_{N\to\infty} f^{(N)} = (-k_BT)\log\lambda_+ = \frac{-1}{\beta}\log\left[\left(e^{\mathcal{J}}\right)\left(\cosh H + \sqrt{\sinh^2 H + e^{-4\mathcal{J}}}\right)\right]$$
$$= -J - \frac{1}{\beta}\log\left(\cosh\beta h + \sqrt{\sinh^2\beta h + e^{-4\beta J}}\right). \tag{8.8}$$

Using f we can also calculate the expected value of the spin, which is[3]

$$\mu \equiv \langle\sigma\rangle = -\frac{\partial f}{\partial h} = \frac{\sinh\beta h + \frac{\sinh\beta h\cosh\beta h}{\sqrt{\sinh^2\beta h + e^{-4\beta J}}}}{\cosh\beta h + \sqrt{\sinh^2\beta h + e^{-4\beta J}}}. \tag{8.9}$$

For any finite β this is a smooth function of h and vanishes for $h=0$: there is *no* spontaneous magnetization and no phase transition.

---

[3]The basic equation is $F/N = (-1/N\beta)\log Z$, with $Z = \sum\exp(\beta J\sum\sigma_i\sigma_j + \beta h\sum\sigma_j)$ and F the free energy. Thus $\mu \equiv \langle\sigma\rangle = (1/N)(1/Z)\sum\sigma_j\exp(\beta J\sum\sigma_i\sigma_j + \beta h\sum\sigma_j)$. It follows that $\mu = (1/NZ)(1/\beta)\partial\exp(\beta J\sum\sigma_i\sigma_j+\beta h\sum\sigma_j)/\partial h$. Going back to $f\equiv F/N$, this implies the relation between f and μ in Eq. (8.9).

That's the bad news.

The good news is that you can consider this a zero-temperature transition. As $T \to 0$ the value of $\exp(-2\beta J)$ becomes smaller and smaller, so if you'd take a small positive value of h and a small negative value, the difference in μ would be nearly 2.[4] Only when h went below the exponentially small $e^{-2\beta J}$ would things smooth out.

This makes sense from the standpoint of correlation length. The formalism developed here does allow a direct calculation (see below), but it's clear intuitively that as $T \to 0$ the cost of a change in spin between two spins next to each other becomes high $(= 2J/k_B T)$ and it is increasingly unlikely that such a break occurs. In other words, we expect long stretches of all pluses, followed by a long stretch of minuses and so on. The length of such a uniform stretch is (essentially) the correlation length, which one can immediately see goes to infinity as $T \to 0$. We'll see precisely how that happens in Secs. 8.1.3 and 8.1.4.

8.1.3 Spatial correlations*

If you know that spin #0 is "up," what do you know about spin #k, where k is large, but still small compared with N?[5] Such a property is measured by correlations. Thus[6]

$$C(k) \equiv \langle \sigma_j \sigma_{j+k} \rangle, \tag{8.10}$$

where the brackets indicate averaging over the Boltzmann distribution (as we will shortly do explicitly). Periodic boundary conditions guarantee that C does not depend on j, but even with other boundary conditions, if N is large enough and $j+k$ and j are not too close to the ends, the value of j becomes irrelevant. If the function $C(k)$ drops off like $\exp(-k/\xi)$ for some positive number ξ, we say that ξ is the correlation length. If you had a bona fide phase transition, for example in the two-dimensional Ising model at low temperature, then this quantity actually goes to a finite constant. This is because if (say) the spontaneous magnetization is up (and $T < T_c$ for there to be any spontaneous magnetization) then there is some positive probability that σ_0 is up. In this case the correlation function subtracts this effect.

Let us see what happens in one dimension.

---

[4] As $T \to 0$, $\beta \to \infty$, so $\exp(-2\beta J) \to 0$. Moreover, in the same limit, $\sinh \beta h \to \pm\exp(\beta|h|)/2$ and $\cosh \beta h \to \exp(\beta|h|)/2$. If $\exp(-2\beta J) \ll |\exp(\beta h)|$ then the right-hand side of Eq. (8.8) is
$$\frac{\pm\exp(\beta|h|) + \frac{\pm(\exp(\beta|h|)/2)(\exp(\beta|h|)/2)}{\sqrt{\exp(2\beta|h|)/4}}}{\exp(\beta|h|)/2 + \sqrt{\exp(2\beta|h|)/4}}. \text{ This in turn equals } \text{sgn}(h) \text{ (for } \beta \to \infty).$$

[5] Spin "#0" in the one-dimensional case coincides with spin #N, but the numbering is irrelevant. I could as easily talk about spin #5 and #$(k+5)$. Similarly for the two-dimensional case some mapping is assumed between pairs of numbers (which could be coordinates on a lattice) and the integers.

[6] For any temperature above zero, $\langle \sigma \rangle = 0$. When this quantity is not zero one defines the correlation as $C(k) = \langle \sigma_j \sigma_{j+k} \rangle - \langle \sigma_j \rangle^2$.

We need to evaluate

$$C(k) \equiv \langle \sigma_j \sigma_{j+k} \rangle = \frac{1}{Z} \sum_{\sigma_1 = \pm 1, \ldots, \sigma_N = \pm 1} \sigma_j \sigma_{j+k} \exp \left(\mathcal{J} \sum_{k=0}^{N-1} \sigma_k \sigma_{k+1} + H \sum_{k=0}^{N-1} \sigma_k \right), \quad (8.11)$$

where I have used the fact that the terms excluding the factor $\sigma_j \sigma_{j+k}$ are the Boltz-mann probabilities. Now we make use of the transfer matrix, L, to write this as

$$C(k) = \frac{1}{Z} \sum_{\{\sigma\}} L(\sigma_1, \sigma_2) \ldots L(\sigma_{j-1}, \sigma_j) \, \sigma_j \, L(\sigma_j, \sigma_{j+1}) \ldots$$

$$\times \ldots L(\sigma_{j+k-1}, \sigma_{j+k}) \, \sigma_{j+k} \, L(\sigma_{j+k}, \sigma_{j+k+1}) \ldots L(\sigma_{N-1}, \sigma_0). \quad (8.12)$$

In matrix form this is

$$C(k) = \frac{1}{Z} \operatorname{Tr} \left(L^j \tilde{\sigma} L^k \tilde{\sigma} L^{N-j-k} \right), \quad (8.13)$$

where $\tilde{\sigma}$ is itself a 2-by-2 matrix that does the job of giving a plus 1 in the sum when needed and a minus 1 when that's needed. Let's examine that claim. The matrix $\tilde{\sigma}$ is $\tilde{\sigma}(\sigma, \sigma') = (\sigma + \sigma')/2$ $(\sigma, \sigma' = \pm 1)$, or in matrix form

$$\tilde{\sigma} = \begin{pmatrix} 1 & 0 \\ 0 & -1 \end{pmatrix} \quad (8.14)$$

(which incidentally is the Pauli σ_z). In Eq. (8.12) there's a sum over σ_k and σ_{k+j}. When you take $\sigma_k = 1$ you get the upper term on the diagonal of $\tilde{\sigma}$, +1, and when it's -1, $\tilde{\sigma}$ also does the job. In Eq. (8.13) there are now two extra matrix multiplications; that's the entire change.

Examining Eq. (8.13) shows that there's an L^j on the left (inside the trace opera-tion) and there's an L^{-j} on the right. But, for a trace, $\operatorname{Tr}(AB) = \operatorname{Tr}(BA)$ (just write out the indices for A and B). So the two terms involving these plus and minus jth powers of L can be dropped.

Let's make things a bit easier on ourselves with two additional simplifications. First, there's translational invariance: there's nothing special about $j = 0$ or $j = 17$ or whatever, so we simply take it to be 0. Second (and here we lose some information) let's only look at $h = 0$, which is the same as $H = 0$. We now have

$$C(k) = \operatorname{Tr} \left(\tilde{\sigma} L^k \tilde{\sigma} L^{N-k} \right) / \operatorname{Tr} L^N, \quad (8.15)$$

with

$$L = \begin{pmatrix} e^{\mathcal{J}} & e^{-\mathcal{J}} \\ e^{-\mathcal{J}} & e^{\mathcal{J}} \end{pmatrix}. \quad (8.16)$$

At this point I don't know how to avoid more matrix technique. But the same thing, only more complicated, also comes up in quantum mechanics, so here goes.

The matrix L has a *spectral decomposition*[7] (or spectral expansion). This is true for far more than 2-by-2 matrices, so let me state things more generally (but I'll use the notation L to avoid confusion).[8] Let the normalized eigenvectors of L be ϕ_α, so that

$$L\phi_\alpha = \lambda_\alpha \phi_\alpha \qquad \text{and} \qquad \phi_\alpha^\dagger \phi_\alpha = 1, \tag{8.17}$$

Where α is the eigenvalue and eigenvector label and the dagger is the adjoint. So if you write ϕ as $|\phi\rangle$, $\phi^\dagger$ would be $\langle\phi|$. Let's suppose there is no degeneracy (all λ's are distinct); then it is true that[9]

$$L = \sum_\alpha \lambda_\alpha \phi_\alpha \phi_\alpha^\dagger. \tag{8.18}$$

In interpreting this equation, each ϕ_α should be considered a (2-row, 1-column) column vector, and the adjoints, $\phi_\alpha^\dagger$, row vectors (1-row, 2-columns). It is clear that L^n is given by Eq. (8.18) with only the replacement of λ by λ^n. (If it's not clear, note that $P \equiv \phi\phi^\dagger$ for ϕ a normalized eigenvector is a *projection*, which among other things means that $P^2 = P$.)

We rewrite Eq. (8.15) as

$$C(k) = \text{Tr}\left(\tilde{\sigma}\left[\lambda_1^k \phi_1 \phi_1^\dagger + \lambda_2^k \phi_2 \phi_2^\dagger\right]\tilde{\sigma}\left[\lambda_1^{N-k}\phi_1\phi_1^\dagger + \lambda_2^{N-k}\phi_2\phi_2^\dagger\right]\right)/\text{Tr}\,L^N. \tag{8.19}$$

Enough abstraction: time to get to work and evaluate the ϕ's. But that turns out to be easy, because, for $H = 0$, L is simple. The answers are

$$\lambda_1 = 2\cosh\mathcal{J}, \qquad \phi_1 = \frac{1}{\sqrt{2}}\begin{pmatrix}1\\1\end{pmatrix}, \tag{8.20}$$

$$\lambda_2 = 2\sinh\mathcal{J}, \qquad \phi_2 = \frac{1}{\sqrt{2}}\begin{pmatrix}1\\-1\end{pmatrix}. \tag{8.21}$$

---

[7]For a matrix to have a spectral decomposition (or expansion) it needs to be diagonalizable. In quantum mechanics you may already have seen this. If you have a Hamiltonian H with eigenvalues E_n and eigenvectors u_n or $|n\rangle$, then $H = \sum_n E_n u_n u_n^\dagger = \sum E_n |n\rangle\langle n|$. This is a spectral decomposition. The set of eigenvalues is the *spectrum* and the (projection) operators $|n\rangle\langle n|$ allow the decomposition. It can become more complicated when there is degeneracy (equal eigenvalues). Hermitian operators encountered in quantum mechanics are always diagonalizable. When the matrices are infinite dimensional you get involved in questions of self-adjointness. But even for finite dimensional matrices one can encounter matrices requiring a Jordan form (e.g., the raising operator $\begin{pmatrix}0&1\\0&0\end{pmatrix}$) that are not diagonalizable. However, those issues don't arise in the present application.

[8]The statements made about the spectral decomposition refer to Hermitian matrices. For, say, stochastic matrices (cf. Appendix H) you may get different left and right eigenvectors, although Eq. (8.18) still holds. However, in some cases you may need a Jordan form.

[9]Degeneracy doesn't complicate things seriously; however, in this case the two eigenvalues are distinct so we don't get into that.

Exercise 26   For these *normalized* eigenvectors, check that $\langle n|n'\rangle = \delta_{nn'} = \text{Kronecker}$ delta $= 1$ when $n = n'$ and 0 otherwise $(n, n' \in \{1, 2\}$ in this case).

Exercise 27   Check the spectral expansion. See that you recover L, as in Eq. (8.18).

Now there's a lot of arithmetic to do. There are four matrix products in Eq. (8.19) and a lot of λ's that go with each. Let's do them systematically. It's useful to indicate all matrices:

$$P_1 \equiv \phi_1 \phi_1^\dagger = \frac{1}{2}\begin{pmatrix} 1 & 1 \\ 1 & 1 \end{pmatrix}, \tag{8.22}$$

$$P_2 \equiv \phi_2 \phi_2^\dagger = \frac{1}{2}\begin{pmatrix} 1 & -1 \\ -1 & 1 \end{pmatrix}, \tag{8.23}$$

$$\tilde{\sigma} = \begin{pmatrix} 1 & 0 \\ 0 & -1 \end{pmatrix}. \tag{8.24}$$

What's nice is that you encounter another peculiarity of matrices. If you look at $\tilde{\sigma}P_1\tilde{\sigma}P_1$ it's obviously a square of $\tilde{\sigma}P_1$. But when you evaluate $\tilde{\sigma}P_1$ and square it, you get zero.[10] Ditto for P_2. So only two terms need be kept. But actually it's even better, since they're equal. So you get

$$\tilde{\sigma}P_1\tilde{\sigma}P_2 = \tilde{\sigma}P_2\tilde{\sigma}P_1 = \frac{1}{2}\begin{pmatrix} 1 & -1 \\ -1 & 1 \end{pmatrix}, \tag{8.25}$$

which is the same as P_2. Putting it all together

$$C(k) = \text{Tr}\, P_2 \left(\lambda_1^k \lambda_2^{N-k} + \lambda_1^{N-k}\lambda_2^k\right)/\lambda_1^N, \tag{8.26}$$

where I've used our previous result for Z. The trace of P_2 (which is the only matrix in the product in Eq. (8.26)) is 1; the term with λ_2^{N-k} is overwhelmed by the denominator and all that survives is

$$C(k) = \left(\frac{\lambda_2}{\lambda_1}\right)^k. \tag{8.27}$$

This gives the correlation length as $\xi = -1/\log(\tanh \mathcal{J})$. This is never infinite, but as temperature goes to zero it grows. In particular, as $\mathcal{J} \to \infty$, $\tanh \mathcal{J} = \frac{1-e^{-2\mathcal{J}}}{1+e^{-2\mathcal{J}}} \approx 1 - 2e^{-2\mathcal{J}}$ so that $\xi \approx \frac{1}{2}e^{2\mathcal{J}} = \exp(2J/k_BT)/2$, a rather rapid growth.

---

[10]The peculiarity is that you can square something that is not zero and get zero. This is not true for ordinary numbers.

8.1.4 Renormalization group treatment

Recall the following definitions from Sec. 8.1.2: $\mathcal{J} \equiv \beta J$ and $\mathcal{H} \equiv \beta h$. For this subsection we take $H = 0$, so that we want to evaluate

$$Z = \sum_{\{\sigma_k = \pm 1\}} \exp\left(\mathcal{J} \sum \sigma_k \sigma_{k+1}\right). \tag{8.28}$$

The basic idea of this method is that, since N is a power of 2, if we were to sum over half the spins (say the even-numbered spins) we would have a problem with only $N/2$ spins. Sounds good, but you want more. You want the new $N/2$-spin problem to look very much like the old one, meaning that it should still have the form of Eq. (8.28), possibly with a different $\mathcal{J}$. In fact this can be done, which makes the one-dimensional Ising model a poster child for the (real space[11]) renormalization group. The method is often called "decimation," because of the explicit process of successively getting rid of half the spins.

Let's see how to eliminate σ_2, and ultimately all the other even-numbered spins. The only place σ_2 appears is

$$\sum_{\sigma_2 = \pm 1} \exp(\mathcal{J}(\sigma_1 \sigma_2 + \sigma_2 \sigma_3)). \tag{8.29}$$

The sum is easy, but we want to write it as $f(\mathcal{J}) \exp(\mathcal{J}' \sigma_1 \sigma_3)$ for some choice[12] of a function f and a number $\mathcal{J}'$. Can this be done? What makes this work is that the σ's can only take two values, ± 1. The requirement is

$$\sum_{\sigma_2 = \pm} e^{(\mathcal{J}(\sigma_1 \sigma_2 + \sigma_2 \sigma_3))} = 2\cosh(\mathcal{J}(\sigma_1 + \sigma_3))$$

$$\overset{?}{=} f(\mathcal{J}) \exp(\mathcal{J}' \sigma_1 \sigma_3) = \begin{cases} f(\mathcal{J}) \exp(\mathcal{J}'), & \text{if } \sigma_1 = \sigma_3 = \pm 1, \\ f(\mathcal{J}) \exp(-\mathcal{J}'), & \text{if } \sigma_1 = -\sigma_3 = \pm 1. \end{cases}$$

$$\tag{8.30}$$

Remarkably, equality can be achieved. For $\sigma_1 = -\sigma_3 = \pm 1$, the argument of the hyperbolic cosine on the left-hand side is obviously zero, so the lower equality in Eq. (8.30) can be satisfied. But this is also true for the case where the σ's are equal, whether they're both +1 or both −1. This is because the hyperbolic cosine is an even function and takes the same value whether $\sigma_1 + \sigma_3$ is plus 2 or minus 2. So, indeed, when you

---

[11]"Real space" means that we are summing over the spins in the usual coordinate space. Actually this is often a frustrating exercise and more reliable results are obtained by Fourier transforming and integrating there—you start by eliminating the shortest distance scales and work your way to longer ones, just as we are doing here, but in a conjugate space.

[12]This "f" is not that same as that which appeared in Sec. 8.1.2.

sum over σ_2 you get something that looks the same as you had before (for σ_1 and σ_3), but with a change of coupling constant $(\mathcal{J}')$ and a factor (f).

It follows, by examining the two cases, that

$$\left.\begin{array}{c} \exp(2\mathcal{J}) + \exp(-2\mathcal{J}) = f(\mathcal{J})\exp(\mathcal{J}'), \\ 2 = f(\mathcal{J})\exp(-\mathcal{J}'). \end{array}\right\} \tag{8.31}$$

Multiplying and dividing these two equations implies

$$f(\mathcal{J}) = 2\sqrt{\cosh(2\mathcal{J})}, \tag{8.32}$$

$$\mathcal{J}' = \frac{1}{2}\log(\cosh(2\mathcal{J})). \tag{8.33}$$

Now imagine that you've done this for every even-numbered spin. The result is

$$\sum_{\text{all } \sigma\text{'s}} \exp\left(\mathcal{J}\sum\sigma_k\sigma_{k+1}\right) = Z(N,\mathcal{J}) =$$

$$= \sum_{\text{odd } \sigma\text{'s}} f(\mathcal{J})^{N/2} \exp\left(\mathcal{J}'\sum\sigma_k\sigma_{k+2}\right) = f(\mathcal{J})^{N/2}Z(N/2,\mathcal{J}'). \tag{8.34}$$

In words: the partition function for N spins with coupling $\mathcal{J}$ is the same as the partition function for $N/2$ spins with coupling $\mathcal{J}'$, multiplied by the function $f(\mathcal{J})^{N/2}$. Now we make use of additional knowledge. The free energy is extensive, that is, if you double the system size you double F. This means that $\log Z$ is proportional to the number of spins, in other words there exists a function ζ, *independent* of N, such that $\log Z(N,\mathcal{J}) = N\zeta(\mathcal{J})$. Using Eq. (8.34), this implies

$$\zeta(\mathcal{J}) = \frac{1}{N}\log Z(N,\mathcal{J}) = \frac{1}{N}\log\left(f(\mathcal{J})^{N/2}Z(N/2,\mathcal{J}')\right) = \frac{1}{2}\log f(\mathcal{J}) + \frac{1}{2}\zeta(\mathcal{J}'), \tag{8.35}$$

or, to turn things around,

$$\zeta(\mathcal{J}') = 2\zeta(\mathcal{J}) - \log\left(2\sqrt{\cosh(2\mathcal{J})}\right). \tag{8.36}$$

Note from Eq. (8.33) that not only is $\mathcal{J}' < \mathcal{J}$, but also with successive iterations the effective coupling constant goes to zero; in other words, calling this quantity $\mathcal{J}_n$ (after the nth iteration), $\mathcal{J}_n \to 0$.[13]

Keep going. Suppose you've cut N by enough powers of 2 that you've arrived at a single spin, that is, your effective "N" is 1. If the original number of spins was large enough $\mathcal{J}$ has gone to zero, and $\zeta(0) = \log(e^0 + e^0) = \log 2$.

---

[13] To see that $K' < K$, exponentiate Eq. (8.33) to get $\cosh 2K = \exp(2K')$ or $\exp(2K)+\exp(-2K) = 2\exp(2K')$. To check the further claim—that successive values of K go to zero (rather than merely decreasing)—take the derivative $dK'/dK = \tanh K$. No matter how large K is, this number is always less than 1, so that after n iterations K_n is less than $\tanh^n K \to 0$. In fact this is an underestimate since the tanh factor also shrinks as K is reduced.

This steady reduction of $\mathcal{J}$ can be looked on as a "flow." $\mathcal{J}$ is the coupling constant over the temperature and as you progress to larger and larger distance scales it flows. The flow is toward smaller and smaller values of $\mathcal{J}$, which can also be looked upon as a steady increase in temperature. In other words, as you go to bigger scales you move further away from the critical point, which in this case is $T = 0$, zero temperature.

The behavior of the correlation length is also clear. Suppose that at some $\mathcal{J}$ the length is ξ. (It's not necessary to define it precisely, but, just for the record, it's defined by $\langle \sigma_0 \sigma_n \rangle \sim \exp(-n/\xi)$.) If you drop half the spins the correlation length is cut in half. So

$$\xi(\mathcal{J}) = 2\xi(\mathcal{J}').$$
(8.37)

Thus as $\mathcal{J} \to 0$, $T \to \infty$, the correlation length goes to zero. Conversely, as one goes in the opposite direction, $\mathcal{J}$ grows, increasing by a factor 2 on each iteration. Thus as $\mathcal{J} \to \infty$ or $T \to 0$ the correlation length grows larger and larger, tending itself to infinity at the critical point.

Exercise 28 We earlier found that the correlation length is given by $\xi = -1/\log(\tanh \mathcal{J})$. Now we find that with $\mathcal{J}'$, a particular function of $\mathcal{J}$ (given by Eq. (8.33)), $\xi' = \frac{1}{2}\xi$. Are these consistent?

Recovery of the free energy.   The transfer matrix method immediately gave values for the free energy. Supposedly we have now solved the same problem using the renormalization group. How does one recover the free energy?

Suppose you've let $\mathcal{J}$ "flow" till it's quite small, say $\mathcal{J} = 0.01$. Then you have two free spins (supposing that initially you had 2^{m+1} spins and have taken m steps). The small value of $\mathcal{J}$ means that they are essentially free, so $Z \approx 2^2$ and $\zeta(0.01) \approx \log 2$. Now "unflow," go back to the larger $\mathcal{J}$ using Eq. (8.33) in the form $\mathcal{J} = (1/2) \cosh^{-1}\left(e^{2\mathcal{J}'}\right)$. It is possible to do this arithmetic and find $\mathcal{J} \approx 0.100\,334$. The value of ζ at this $\mathcal{J}$ value can be obtained from Eq. (8.36) and is $\zeta(\mathcal{J}) \approx 0.698\,147$. This process can be continued and the results compared with what you would get from (say) the transfer matrix. From [148] we take a table of the results of continuing this process, Table 8.1. As you can see, this process gives excellent results.

Indifference to details.   The issue of "indifference," in some sense, is the main point of the renormalization group, the fact that, as distance scales grow, many systems fall into a single universality class: certain details don't count. But showing this on the

$\mathcal{J}$	$\zeta(\mathcal{J})$ RG	$\zeta(\mathcal{J})$ exact
0.01	log 2	0.693 197
0.100 334	0.698 147	0.698 172
0.327 447	0.745 814	0.745 827
0.636 247	0.883 204	0.883 210
0.972 710	1.106 299	1.106 302
1.316 710	1.386 078	1.386 080
1.662 637	1.697 968	1.697 968
2.009 049	2.026 876	2.026 877
2.355 582	2.364 536	2.364 537
2.702 146	2.706 633	2.706 634

Table 8.1 $\zeta(\mathcal{J}) \equiv Z(N, \mathcal{J})/N$ for various $\mathcal{J}$ flowing from the near-zero value of 0.01. The exact value of ζ is given in the right-most column, while the value given by the renormalization group (RG) is in the middle column.

one-dimensional Ising model can be difficult, perhaps because the decimation trick was so easy.

To show this "main point" for at least one kind of "detail," consider the following Hamiltonian:

$$\beta \cdot \text{Hamiltonian} = -\mathcal{J}_a \sum_j \sigma_{2j}\sigma_{2j+1} - \mathcal{J}_b \sum_j \sigma_{2j+1}\sigma_{2j+2}. \tag{8.38}$$

There are two different bond strengths, with $\mathcal{J}_a \neq \mathcal{J}_b$; bonds from even numbers to odd numbers to their right have strength $\mathcal{J}_a$, while those to their left have strength $\mathcal{J}_b$. (I put lower numbered sites on the left, higher numbers on the right, except that #N is the same as #0.) Periodicity and the number of spins are the same as before. This Hamiltonian differs in its detailed behavior from that considered earlier. Does it matter?

To find out we attempt the first step in the decimation procedure, summing over (say) σ_2. Focus only on that sum:

$$\sum_{\sigma_2 = \pm 1} \exp\left(\mathcal{J}_a\sigma_2\sigma_3 + \mathcal{J}_b\sigma_2\sigma_1\right) = \exp\left(\mathcal{J}_a\sigma_3 + \mathcal{J}_b\sigma_1\right) + \exp\left(-\mathcal{J}_a\sigma_3 - \mathcal{J}_b\sigma_1\right). \tag{8.39}$$

We try to follow our previous steps in the decimation. We (attempt) to set

$$2\cosh(\mathcal{J}_b\sigma_1 + \mathcal{J}_a\sigma_3) = f\exp(\mathcal{J}'\sigma_1\sigma_3) = \begin{cases} f\exp(\mathcal{J}'), & \text{if } \sigma_1 = \sigma_3 = \pm 1, \\ f\exp(-\mathcal{J}'), & \text{if } \sigma_1 = -\sigma_3 = \pm 1. \end{cases} \tag{8.40}$$

where f can now depend on both coupling strengths, $f = f(\mathcal{J}_a, \mathcal{J}_b)$. There are again two equations to satisfy, but now there are three variables. Is this a problem? It turns out that it is not. The equations are

$$2\cosh(\mathcal{J}_a + \mathcal{J}_b) = fe^{\mathcal{J}'},\tag{8.41}$$

$$2\cosh(\mathcal{J}_a - \mathcal{J}_b) = fe^{-\mathcal{J}'}.\tag{8.42}$$

(Note that cosh is an even function, so, for example, $\cosh(\mathcal{J}_a - \mathcal{J}_b) = \cosh(\mathcal{J}_b - \mathcal{J}_a)$.) Solving these is done exactly as before, the only differences are that $\mathcal{J}_a + \mathcal{J}_b$ has replaced $2\mathcal{J}$ and $\mathcal{J}_a - \mathcal{J}_b$ has replaced 0. That is, we again divide and multiply (plus taking a log):

$$\mathcal{J}' = \frac{1}{2}\log\left(\frac{\cosh(\mathcal{J}_a + \mathcal{J}_b)}{\cosh(\mathcal{J}_a - \mathcal{J}_b)}\right),\tag{8.43}$$

$$f(\mathcal{J}_a, \mathcal{J}_b) = 2\sqrt{\cosh(\mathcal{J}_a + \mathcal{J}_b)\cosh(\mathcal{J}_a - \mathcal{J}_b)}.\tag{8.44}$$

It now follows that on the *next* step there is only one coupling constant, "$\mathcal{J}$" (equal to the $\mathcal{J}'$ of Eq. (8.43)).

The point, an example of what I referred to earlier as the "main point," is that, having done a different decimation once, the problem has now been reduced to the previous one. From this stage on, there is only one "$\mathcal{J}$" and we can proceed exactly as before. The details of the interaction do not matter as far as critical exponents are concerned, although the actual critical values will be a function of both $\mathcal{J}$'s.

Exercise 29 Can you find a different example of "indifference to details?" I've given something simple in which taking one decimation goes from $\beta \cdot \text{Hamiltonian} = -\mathcal{J}_a \sum_j \sigma_{2j}\sigma_{2j+1} - \mathcal{J}_b \sum_j \sigma_{2j+1}\sigma_{2j+2}$ to $-\mathcal{J}' \sum_j \sigma_j \sigma_{j+1}$. Perhaps if the interaction involved a next nearest neighbor or a distribution of coupling constants (with a mean and standard deviation) the scheme would work.

Remark: The renormalization group has become a staple of the working statistical mechanician as well as other physicists, from those who worry about condensed matter to those who specialize in particle physics. In fact the word "renormalization" was first (to my knowledge) used in taming the infinities in QED, quantum electrodynamics. The method is often (affectionately?) referred to as the "RG," a recognition of its common use.

Nevertheless, and despite the Nobel Prizes[14] on the subject, I don't like it.[15] It represents a truth, but, perhaps more than other techniques in physics, it works when it works—but you don't know ahead of time when it's dependable.[16] There are few rigorous calculations using this technique.[17] The truth that it represents is the ability to ignore details, and I've shown this in a simple example. But I have to confess that I tried to develop richer examples, say with next nearest neighbor interactions (for the one-dimensional Ising model), but could not find anything simple. (See Exercise #29.) Still I believe interaction details are unimportant near criticality—it's just that I don't know how to reach that conclusion using the decimation transformation, which is simple and exact.

The two-dimensional Ising model can also be studied using real space renormalization and there's a lovely article by Maris and Kadanoff [148] that does this in both one and two dimensions. For two dimensions you can *almost* do what we did exactly in Eq. (8.30). The trouble is that additional couplings are created. You would like new variables that depend only on the four nearest neighbors, but when you try something like Eq. (8.30) you find that other couplings—not only the nearest neighbors—become involved. Kadanoff and Maris provide an ansatz to deal with this, but, as they say, it is only approximate. In particular, they get the critical exponents wrong.

It is also true that most RG work is done in Fourier space. You transform the coordinate space variables and then integrate over the short wavelength degrees of freedom. The idea is that near criticality you have many degrees of freedom cooperating with each other (cf. our discussion of fluctuations in ferromagnetism) and despair of solving any of them. So you don't solve them! You get rid of them, integrating over successively longer range degrees of freedom. You then watch the control parameters flow.

---

[14]The prize was given to Kenneth Wilson, but should have been shared by Leo Kadanoff. There was a chance to make up for the first error when Pierre-Gilles de Gennes received the award, but again the Nobel Committee erred. I should point out that I carry no special torch for Kadanoff (nor have any animus for Wilson or de Gennes), but I don't think it right that this particular committee is so powerful.

[15]This opinion puts me in a minority, but I'm good with that. I'm also in a minority with respect to views on determinism, although, there, having Einstein as company tends to assuage the pain. See also [195], preface and introduction.

[16]I exaggerate: the "replica method" is even less reliable, at least as its commonly used. This "method" takes a power of (say) the partition function (the replicas), solves it, and then lets the number of replicas go to zero, as in $\frac{1}{n}(x^n - 1) \to \log x$ as $n \to 0$. Since the partition function involves the logarithm, (some) people use it to evaluate logarithms.

[17]One of the few rigorous renormalization calculations is [162]. But this deals with a first-order transition, which in a sense is easier.

It should be mentioned that the RG method, despite the weaknesses that I have been complaining about, solved an important problem. As you approach a critical point (a second-order phase transition) degrees of freedom become coupled to one another. It is these coupled degrees of freedom that give characteristic properties—like critical exponents—to phase transitions in a particular universality class, and that care only about the overall (usually) attractive features of the transition, features that are properties of matter at large (coordinate) distance scales. And now the problem: perturbation theory fails. When you deal with a potential as a perturbation it must be cut off at some order (typically doing mean field theory on the next order). You can do first-order, second-order, even seventeenth-order perturbation theory, but you have to stop somewhere. On the other hand, critical points involve arbitrarily high orders, everything is coupled to everything (eventually). That's the problem solved by renormalization; my complaints should be taken in this light.

9
Social sciences

Applications discussed in this chapter: Economics, Finance, Crashes and bubbles, Cities, Neighborhood segregation, Elections and voters, Crowd control, Traffic

---

If you look into the history of statistical mechanics you'll mostly pick up the history of chemistry, but in fact[1] quite a bit of work was done compiling information about people, births, deaths, incomes and other information, much of it *before* Boltzmann applied these ideas to physical systems at the end of the 19th century. Of course you could argue that work on, say, the nature of heat was done before the social studies work. For example, on the website `http://history.hyperjeff.net/statmech` (accessed August 2020) there is a remark that in the 13th century Levi ben Abraham[2] connected heat and motion (and here I'd thought that Lavoisier was the one who'd rescued us from the phlogiston nonsense and that it was Count Rumford who connected heat and mechanical work (or maybe it was Joule)). The good news though is that not being a historian I don't need to decide priorities nor even worry too much about defining statistical mechanics. What I can say, as a physicist, is that it is not surprising that the methods of statistical mechanics, largely considered a branch of physics these days, includes the study of social phenomena.

One particular aspect—that of power laws—had early roots in economics, sociology and linguistics. In Chapter 7 the mathematics and a few of the applications have been

---

[1] ... as reported by Conner Herndon, in 2016 a graduate student at Georgia Tech, who spoke on this subject.

[2] 1246 to 1315, lived in France and wrote in Arabic. Studied the Bible and Talmud, but nevertheless fell afoul of certain members of the orthodox Jewish community, who excommunicated him. There's a passage in his work, *Livyat Khen*, suggesting that heat is related to motion. I contacted Prof. Haim Kreisel, who has translated some of the works of Levi ben Abraham. Prof. Kreisel told me that the writings were mainly to explain to 13th and 14th century readers the ideas of Aristotle. This in turn led me to examine Aristotle on the subject of motion, but I must confess that I could not connect his notions to modern ideas of molecular motion. All of which returns originality, for me at least, to Count Rumford and possibly Joule (nor should one forget Julius Robert von Mayer).

When Things Grow Many. Lawrence S. Schulman, Oxford University Press.
© Lawrence S. Schulman (2022). DOI: 10.1093/oso/9780198861881.003.0009

covered. Here I'll treat some of the same applications in more detail, and introduce new ones. Sometimes though the lines are blurred and topics may appear in more than one place.

9.1   Econophysics

Econophysics, which applies statistical mechanics to the study of economics, is older than the coinage of the word itself. Early work in this area was done by Pareto,[3] who noted the ubiquity of power law distributions and for whom such distributions are named. The *Pareto distribution* is the probability density function

$$\Pr(x) = \begin{cases} \frac{\alpha x_{\min}^{\alpha}}{x^{\alpha+1}}, & \text{for } x \geq x_{\min}, \\ 0, & \text{for } x < x_{\min}. \end{cases} \tag{9.1}$$

There are two parameters in this distribution, $x_{\min}$ and α. What Pareto and his successors noted was that many distributions followed this law, at least in the "tail," that is, the largest x values ("x" of Eq. (9.1)). Notable are the distributions of income and wealth. It was subsequently found that power laws appear all over the place, for example in Zipf's law, such that the word frequency (say in a long text) decreases like a power of a word's rank (relative usage). The power, α in Eq. (9.1), is often just a bit above 0. These distributions are often called "scale free" because there is no characteristic length, as there would be for (say) $\Pr(x) = \frac{1}{\xi} \exp(-x/\xi)$ (for $x > 0$, zero otherwise). (See a more precise definition in Chapter 7.) Here, ξ is the characteristic length. A second salient feature of such distributions is the absence of certain moments. Thus

$$\int_0^{\infty} \frac{x^{\gamma} dx}{(1+x)^{1+\alpha}} \tag{9.2}$$

is finite only for $\gamma < \alpha$. So if α is just above 1, there is no second moment and no central limit theorem![4]

Explaining power laws for wealth, income, words, city sizes, $1/f$-noise and much more has been the holy grail of physics and "complexity theory" for many years. See

---

[3] Vilfredo Pareto, 1848–1923, Italian (albeit born in Paris), spent most of his career in Lausanne, Switzerland.

[4] A requirement for the central limit theorem to hold for a given distribution is that there be a second moment (plus a little more) for that distribution. In this way the normal or Gaussian distribution is an attractor of distributions satisfying the central limit theorem: sums of i.i.d. distributions on the line tend to become normal. See Appendix D.3. But what if there is no second moment? Then the attractor is a Lévy distribution. The random variable X is a Lévy distribution and symmetric (with respect to the line) if the expectation of a particular function of the random variable satisfies $\langle \exp(ikX) \rangle = \exp(-\sigma^{\alpha}|k|^{\alpha})$, where $0 < \alpha < 2$ and $\sigma \geq 0$ are parameters of the distribution. (This is the Fourier transform and is also known as the "moment generating function.") If $\alpha = 2$ one has a normal distribution and $\alpha = 1$ is the Cauchy distribution. See for example [189], where both symmetric and asymmetric distributions are discussed.

Chapter 7 for a general discussion from a mathematical standpoint. Many claim to have found it, but since they are in mutual disagreement they can't *all* be right. For the holy grail at least you could exhibit a chalice; here the criterion is "intellectual satisfaction," historically not a reliable guide to scientific truth [195].

To study the wealth distribution, Ispolatov et al. [113] posit an exchange model and consider various kinds of transactions. In an exchange model one assumes that one party gives money to the other and receives goods or services in return. There is also the assumption that wealth is conserved, even when debt is involved. Of the various models considered, only one particular scheme gave the power law, and the authors modestly propose (no irony here) that this might have something to do with reality. Exchanges are of the following form. Let x be funds belonging to X and y funds belonging to Y. A pair, X and Y, is randomly selected from the population and the exchange is greedy and multiplicative. What this means is that transitions are allowed only if Y is richer than X, in which case the transaction is $(x, y) \rightarrow (x - \alpha x, y + \alpha x)$, $0 < \alpha < 1$. The simulation consists of randomly picking pairs and implementing the exchange. The result? The rich get richer and the poor get poorer, but the poor never go completely broke. Under this scheme there is no equilibrium, but the distribution has the power law seen in practice. This process is closely related to the models in Chapter 7 called "preferential attachment," a.k.a. the rich get richer, which is certainly seen in the model just described.

Remark: In [113] the authors provide an analytic equation for $N(x)$, the fraction whose wealth exceeds x and $c(x) = dN/dx$. The equation is

$$\frac{\partial c(x)}{\partial t} = -c(x) + \frac{c\left(\frac{x}{1-\alpha}\right)}{1-\alpha} N\left(\frac{x}{1-\alpha}\right) + \frac{1}{\alpha} \int_{x/(1+\alpha)}^{x} dy\, c(y) c\left(\frac{x-y}{\alpha}\right). \qquad (9.3)$$

$N(x) = -\int_{x}^{\infty} dz\, c(z)$ (as previously defined) is the population density of those whose wealth exceeds x. Eq. (9.3) is a bit of a mess, but it's nice that one can exhibit equations. (See the cited article for a derivation.) The authors show that, following $c(x)$ under this equation, the rich become richer, the poor poorer, and the process doesn't stop. What else is new. ...

Remark: It is important to distinguish *wealth* from *money*. In our system money is not conserved. Suppose I have $50,000 in my savings account. It sits there, these days earning lousy interest, but I keep it for security. The bank has lots of other accounts like mine, and let's say it has $1,000,000 in such assets. (That amount is peanuts for a bank, but I'm only giving an example.) They have to be ready for me and one or two others to suddenly decide we want our money, but they don't need to be ready for *everyone* to withdraw their money. So they need to keep (say) $100,000 around, but the other $900,000 they can use, for example, to lend as a mortgage.[5] Suppose the mortgage is for $200,000. The total amount of money has climbed to $1,200,000: my savings, everyone else's savings and John Doe's mortgage. But things don't stop there. John Doe enters into a contract with a construction firm to fix his roof and remodel his kitchen (that's why he took the mortgage) and the head of the construction firm, based on John Doe's obligation, commits herself to purchase a new car. So she signs documents requiring installment payments. That document can be sold by the car dealer, generating additional money in the economy. In the examples I've just given, debt creates new money.[6] Beyond debt, there is also the issue of money's intrinsic value. Is there such a thing? There are exchange rates, there's inflation and there are digital currencies. *The Economist* (magazine) maintains a Big Mac index, which is an exchange rate based on the cost of a Big Mac (a sandwich made by the McDonald company) in different countries. When it disagrees significantly with the official rate, it's a sign that at least one of them is unrealistic. As to inflation, what's constant about the cost of plowing the snow from your driveway? Does this mean money should have a correction factor? And what about things that become *cheaper*, like phone calls? On the other hand, what is called wealth is assumed to be constant. It may be transferred, but its total is constant.

The Ispolatov et al. [113] story leads to a more comprehensive and less tentative view put forth by Boghosian and collaborators [49, 142]. By introducing a variant of what's been described so far, plus a few parameters describing redistribution and other features, they match an economically significant curve to within 1/6 of 1% and even have a phase transition (so they come close to traditional statistical mechanics). (See also [29].)

They start from a simple model. There are N agents and everyone has some initial value of wealth. Two agents are picked randomly, say having wealth x and y. The

---

[5]The exact requirements for banks vary and 10% is not an unreasonable figure. The law does not require them to be ready for a "run on the bank" in which all parties want their money.

[6]The view expressed here is not at all idiosyncratic and can be found in classic textbooks, such as [164]. It can also be found explicitly in Wikipedia (https://en.wikipedia.org/wiki/Fractional-reserve_banking, accessed May 2020), where reserve requirements and the "creation" of money are discussed. According to Alexander Lipton and Alex Pentland (in [144]) this enlargement of the money supply can be traced to Ur (in Mesopotamia) some 5000 years ago. They also remark that, even after the loans are repaid, the interest earned by each lender remains in the system.

changes in each cancel, that is, $\Delta x = -\Delta y$, where "Δ" is the change in wealth. The rule is that $\Delta x = \eta \alpha \min(x, y)$, where α is a small constant, η is a random variable taking values ± 1 and $\Delta y = -\Delta x$. The sole asymmetry is the "min," the exchange of wealth never leaves the poorer person with zero wealth. The strange thing is that this creates an oligarch, a person (or institution) that has almost all the wealth. I've run a simple program that does this, and it's true, although it takes a *lot* of exchanges to make this happen. (In my program, involving about 150 agents and $\alpha = 0.1$, it took about 10^4 exchanges.)

This actually gives a nice relation: wealth $\approx$ const·exp($-$slope · rank) with slope ≈ 1.5, but fortunately reality is not quite as bad as that. There's another factor that must be taken into account, and this is the effect of taxation. To some extent income tax is progressive, giving the poor wealth and taking it from those who have more wealth. This is handled in a simple way, and introduces another constant, χ, such that if an agent has wealth exceeding W/N where W is total wealth, then that agent loses in proportion to $w - W/N$ (where w is that agent's wealth); the same χ governs those for whom $w - W/N$ is negative. (Actually the progressive nature of income tax is to some extent cancelled by the regressive nature of sales tax, the perks that rich people who support congresspersons get, and other factors. But the data are fit by this additional variable, so *something* is working at reducing inequality—but not by all that much. See the discussion below.)

There is a second variable, ζ in the notation of [29] (and other Boghosian references), that is related to the α given above, and that is fully defined when it comes to fitting data.

And finally they use a third variable, κ, this time accounting for *negative* wealth. Something on the order of 11% of households in the United States have less than zero, that is, money is owed to various places, credit cards, college loans, and so on, and those liabilities exceed that household's assets—meaning they have negative wealth.

When all is said and done, Boghosian and collaborators get a Fokker–Planck equation for the probability that an individual entity has a certain wealth, positive or negative. They set the time derivative to zero, a kind of adiabatic assumption. Then they establish parameter values (particular values of (χ, ζ, κ)) that fit those data. So there are a lot of assumptions that go into the model, about which I'll speak in a moment.

The first observation is that the fits are remarkable. The technique also gets the Gini coefficient right. This coefficient (named for Corrado Gini, an Italian statistician and sociologist) is a measure of inequality. You plot the cumulative share of people (from lowest to highest income) on the x-axis (the abscissa) versus the cumulative

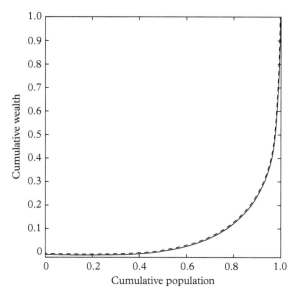

Fig. 9.1 This is a cumulative plot. On the x-axis is the total population, going from 0 to 1.0. On the y-axis is plotted cumulative wealth. Note that this goes below zero. A particular value of (χ, ζ, κ) is picked to best fit the observational curve. That curve (the solid line) is made from a United States survey of household finances [32] combined with information about the richest people [124] in the United States (who are ignored in the survey). The dashed line is the fit, for much of the range invisible. In the L^1 norm the error is about 0.1%. Based on [49], Fig. 4. Source: *SIAM Journal on Applied Mathematics.*

share of income earned (on the y-axis, or ordinate). The Gini coefficient is twice the area between the actual curve and a line from $(0,0)$ to $(1,1)$; see Fig. 9.1. This is a measure of inequality, since bending inward of the curve means that there are fewer people at the top.

Another feature is that at certain values of ζ/χ there is a phase transition, so this ratio is a kind of temperature. Specifically if this ratio is greater than one there is what they call a "partial oligarchy," meaning all wealth is concentrated in one—or a few—hands. The transition of the stationary solution of the Fokker–Planck equation exhibits this transition, although I'm not sure what this looks like in real life. *Scientific American* [29] has a graph showing that the Netherlands has no oligarchy whereas most other European countries do. I doubt if the richest person in Italy (which has $\frac{\zeta}{\chi} > 1$) is totally off the scale. The finiteness of populations makes this phase transition hard to define, even if the Fokker–Planck equation shows a discontinuity.

But I am impressed by the fit. This means that lots of features either can be ignored or cancel out. For example, people eat. So some wealth simply disappears. On the other hand, others grow food. Does that mean they create wealth? Another aspect that the quality of the fit seems to belie is unfairness in the tax system. The parameter χ measures the equalizing effect of taxation and for simplicity Boghosian et al. take it to be proportional to wealth minus average wealth ("$w- W/N$"). But taxation, in the United States at least, is hardly like that. First, sales tax is regressive (poor people pay more), not progressive. Second rich people often pay *less* tax. They influence the tax laws and bend them so that they get off cheap. For example, hedge fund managers benefit from some sort of fiction and are currently paying 23.8% (US, federal) tax while ordinary taxpayers in the same income bracket pay 37%. Another example is depletion allowances. (But don't get me started.) Vox, a media outlet, claims that taxation affects almost all people the same; in other words, percentage of income is the same. The only exception (according to Vox) is billionaires: they pay less. Is Vox wrong? Is there some other factor? Whatever the reason, using ξ (the fitting parameter that reflects taxation) improves things. In any case, the quality of the fit is undeniable. And I've only shown you one fit. Boghosian does the same for other countries and other years.

9.2 Stock market bubbles and crashes

The market it going up, and up, and up. When will it crash? Should I buy now and take the chance that I'll sell before the crash? What to buy after it crashes? These are topics designed to raise your blood pressure, to say nothing of suicide after the big bubble bursts.

Sornette, in his many writings [215, 217], finds that bubbles exist in many societies and in many periods of history, and what they all have in common is humans. People. We herd. We imitate. We give feedback to each other which builds and builds—until— until it stops.

According to [214, 216] the main indicator of a bubble is an overly rapid increase in price. But it can be tricky to evaluate what "overly rapid" is. Inflation can (almost) always be counted on to cause increase, and then there is what people expect to earn from their investment. And the increase is not found to be linear, most likely because people respond to logarithmic signals, not linear ones. An example is our hearing. The loudness of sound is measured in decibels, which is the logarithm of power in the signal (to the base 10, multiplied by 10). The same is true of the Richter scale for earthquakes; it's logarithmic. Similarly when we evaluate something we tend to think in logarithms. Thus in buying a house you may not quibble about a hundred dollars up or down, while when going to a restaurant a $10 meal and a $110 meal are (or should be) very different things (USD).

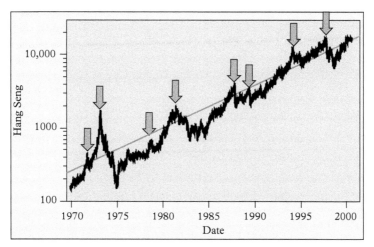

Fig. 9.2 The Hang Seng Index of the Hong Kong stock market. The "overly rapid" increase—faster than exponential—is indicated by arrows. (This is a semi-log plot. The normal increase is indicated by the line.) Adapted from [122]. Source: arXiv.

As for interest rates, which are measured in percent of the deposit or debt, the value of stocks is also expected to increase by fixed percentages. Thus the natural rate of price increase is a fixed fraction, that is, it is exponential. (Recall how a power 3, for example, can appear in the exponential: $1.04 \times 1.04 \times 1.04 = 1.04^3 = e^{3 \times \log 1.04} \approx e^{3 \times 0.392}$ (equaling approximately 1.1249).)

An "overly rapid" increase then means *faster* than exponential, or, in practice, faster than the usual exponential. Fig. 9.2 shows the behavior of the Hang Seng Index for stocks selling on the Hong Kong stock market. The plot is semi-logarithmic, so that exponential growth should look like a straight line (if price $= \exp(\gamma t)$ then $\log(\text{price}) = \gamma t$).

As found in [122], these "overly rapid" increases led to crashes. Of course there's a lot of judgement involved. For example, the paper cited defines a crash as a drop in the index of more than 15% in less than 3 weeks. If you defined a drop as a 10% decrease or were willing to wait 4 weeks, this criterion may overreact. Of course this is both the beauty and the danger of buying and selling in the stock market: judgement.[7] Nevertheless, if you see growth at a rate that is much greater than the historical value,

---

[7]I should say *alleged* judgement. As pointed out in the delightful book by Mlodinow [153] the stock market is largely random and out of thousands of fund managers some will be successful, some will not. But those who are often claim judgement! Mlodinow points out (in his Chapter 9) that one of the most successful of Wall Street managers, when all the years and all managers are allowed for, was expected (in the random model). Someone should do that well: there were three chances in four of that eventuality occurring *randomly*.

watch out! (In the case of the Hang Seng Index the (historical) rate was about 14% annually.)

Sornette et al. attribute this phenomenon to the "herding instinct." This is a positive feedback in which market strategists nervously try to get the most out of a price rise. Why nervous? Because among many of them there is an expectation of a crash, along with desire to get out before that happens. On the other hand, if they *don't* participate in the market during the period of rising values, then they'll miss out on the profits—another undesirable outcome.

Another view of market crashes is espoused by Harmon et al. [98]. They too blame the existence of the bubble/crash phenomenon on herding, although they prefer the words "mimicry" and "panic." As evidence they submit graphs of the number of stocks that have risen and the numbers that have fallen for the years 2000–2008. In each case they plot the number that have risen on each trading day. For the year 2000, the graph peaks at 50%, with a dropoff at larger and smaller values. In other words, it was rare for everyone to buy or sell and usually (on any given day) some equities increased while others decreased. But by 2008 the curves were much flatter: on a given day it would often happen that *everyone* would buy; on another day, they'd all sell. People were following what others did; there was mimicry, even panic.

Mathematically what Harmon et al. have is a non-linear positive feedback, with a different way of estimating the likelihood of a crash. As indicated, on any given day, some fraction of the stocks increase in price, some fraction decrease. Harmon et al. find that they can ignore the magnitude of the increase or decrease, only noting the sign. This conclusion is based on simulations, calculation and comparison with real data.

Here is the essence of their model. Suppose there are three kinds of equities: two kinds that *always* move in a certain direction, U for up, D for down, and a third kind that is influenced by a randomly selected security. There are N of this third kind, but the one that is randomly selected is chosen from the entire collection, all three types. Suppose then that there are $N + U + D$ equities (stocks, bonds, whatever) in total, with U of them *always* going up and D of them *always* going down. A given stock (from among the N) is chosen and may follow its "neighbors" or it may not, not following having a probability p. They simplify by having all the N changeable stocks be neighbors with one another. So this resembles a mean field version of the voter model (see Sec. 9.6).

With these assumptions the model is completely solvable. First, the state of the system is specified by a single number, m, which is the number of (changeable) stocks that increase in value (so they all correspond to +1). It is also assumed that with probability $(1-p)$ a stock changes, gets either a plus or minus one (recall that p is a probability, so that $0 \le p \le 1$). The number m can thus change to $m \pm 1$ or stay the same. The way the *given* stock decides is to pick an arbitrary *other* stock (to be known as the "selected" stock) and the *given* stock then moves up or down, mimicking the *other* stock. So first there is a probability p that m remains the same. All other options need to be multiplied by $(1-p)$. We go through the possibilities, one by one:

- Given stock is up. Probability is m/N.

  o Selected stock is up. Probability is $(U+m-1)/(N+U+D-1)$.

    The "selected" stock is not going to change; it's the one that influences the "given" stock. Probability of $m \to m$ is therefore $(1-p)\frac{m}{N} \cdot \frac{(U+m-1)}{N+U+D-1}$.

  o Selected stock is down. Probability is $(D+(N-m))/(N+U+D-1)$. Probability of $m \to m-1$ is therefore $(1-p)\frac{m}{N} \cdot \frac{D+(N-m)}{N+U+D-1}$.

- Given stock is down. Probability is $(N-m)/N$.

  o Selected stock is up. Probability is $(U+m)/(N+U+D-1)$. Probability of $m \to m+1$ is therefore $(1-p)\frac{N-m}{N} \cdot \frac{(U+m)}{N+U+D-1}$.

  o Selected stock is down. Probability is $(D+(N-m-1))/(N+U+D-1)$. Probability of $m \to m$ is therefore $(1-p)\frac{N-m}{N} \cdot \frac{D+(N-m-1)}{N+U+D-1}$.

There are thus three ways to stay the same and two ways to change. Since only a single number (m) characterizes the state, we have an evolving probability which can be described by a single function, $P(m,t)$. Note the capital P, which is the probability of each m at time-t—distinguishing it from (the lower case) p, a parameter of the problem. Summarizing the possibilities indicated, it satisfies

$$
\begin{aligned}
P(m,t+1) = & \left\{ (1-p)\frac{N-m}{N} \cdot \frac{(U+m)}{N+U+D-1} \right\} P(m-1,t) \\
& + \left\{ p + (1-p)\frac{m}{N} \cdot \frac{(U+m-1)}{N+U+D-1} \right. \\
& \left. + (1-p)\frac{N-m}{N} \cdot \frac{D+(N-m-1)}{N+U+D-1} \right\} P(m,t) \\
& + \left\{ (1-p)\frac{m}{N} \cdot \frac{D+(N-m)}{N+U+D-1} \right\} P(m+1,t).
\end{aligned} \tag{9.4}
$$

As a check, note that all the terms multiplying $(1-p)$ add to one, so that probability is conserved. Eq. (9.4) can be simplified to

$$P(m,t+1) = \left\{ \frac{1-p}{N(N+U+D-1)}(N-m)(U+m) \right\} P(m-1,t)$$

$$+ \left\{ p + \frac{1-p}{N(N+U+D-1)} [m(U+m-1) \right.$$

$$\left. + (N-m)(D+N-m-1)] \right\} P(m,t)$$

$$+ \left\{ \frac{1-p}{N(N+U+D-1)} m(D+N-m) \right\} P(m+1,t). \qquad (9.5)$$

Why did we go through this painful exercise? We'd like to know the stationary state of this process in order to compare it with annual records and deduce from them the effective value of U and D for the year in question. Now the term in p is a multiple of the identity, so the issue becomes finding the stationary state of the matrix R, where that matrix is the operator taking $P(\cdot, t)$ to $P(\cdot, t+1)$. The stationary state of that matrix is known[8] and is

$$P_0(m) = \frac{\binom{U+m-1}{m}\binom{N+D-m-1}{N-m}}{\binom{N+D+U-1}{N}}. \qquad (9.6)$$

The function P_0 is very different for different values of D and U. When these quantities are large the stationary state is nearly a Gaussian, a normal distribution. But for small values, for example $U = D = 1$, it is completely flat. This would mean that the likelihood of changes in stock prices varies all over the place: some days almost all stocks go up, others they all go down. According to [98] this wild behavior is a precursor of a crash. They assume $D = U$ and allow the values to be non-integers, using analytic continuation.[9] They fit these values to the years 2000–2008 and see an almost monotonic drop, from $(D = U =)$ 5.79 in 2000 to 1.24 in 2008—the year the market crashed.

The main thing to bear in mind, however, is that all these measures—Sornette et al.'s, Harmon et al.'s, and many others—don't tell you *when*. When will the debacle

---

[8]To verify this assertion it is sufficient to drop the denominator in Eq. (9.6), since that is independent of m and only normalizes P_0. Similarly, as remarked, the portion multiplying p is the identity matrix and also can be ignored. So finally what one must show is that

$$N(N+U+D-1)\overline{P}_0(m) = (N-m)(U+m)\overline{P}_0(m-1)$$
$$+ \{[m(U+m-1) + (N-m)(D+N-m-1)]\} \overline{P}_0(m)$$
$$+ m(D+N-m)\overline{P}_0(m+1),$$

where $\overline{P}_0$ is the un-normalized stationary state. This is an elaborate exercise in the manipulation of factorials—which I leave to the reader.

[9]It's a fact that n-factorial $= n! = \Gamma(n+1)$ with $\Gamma(z) = \int_0^\infty e^{-t}t^{z-1}dt$. This is inserted into $\binom{N}{m} = \frac{N!}{m!(N-m)!}$, allowing immediate analytic continuation. See Sec. D.7.

happen? You can know when things are getting dicey, but for those who are not risk-averse the temptation to pull in a bit more profit may lead to trouble.

9.3   Linguistics

9.3.1   Zipf's law

Take a big book, maybe Melville's *Moby Dick*. Count the number of times each word appears and, using this, give each word a rank. For example, if the definite article *the* appears most frequently, its rank is 1. Now plot the logarithm of the frequency against the logarithm of the rank. You get a straight line with a slope near (minus) one. This is true of essentially all big books in all languages, although obviously the most common words will be different (some languages don't even have articles). Getting a straight line in the log–log plot is equivalent to having a power law: $\log a = b \log c \Rightarrow a = c^b$. For 12 words and 30 languages[10] this is illustrated in Fig. 9.3.

What I just said is *almost* true. Clearly it can't be perfectly true, since in the tail of the distribution—really rare words—you get a lot of noise. But even when the statistics are good you can improve things by putting in an additional constant and taking

$$\text{frequency} \equiv f(r) \approx \frac{1}{(r + \beta)^\alpha}, \tag{9.7}$$

where r is the rank, α is a number close to 1 and β is a small constant. The relation Eq. (9.7) is known as Zipf's law because of the intensity of study by the eponymous George Kingsley Zipf (see his 1936 book [243]), not because of priority.[11] The constant β was added later by Mandelbrot[12] [146], yielding a better fit.

---

[10]The languages are Basque, Belorussian, Catalan, Croatian, Czech, Danish, Dutch, English, Esperanto, Finnish, French, Galician, German, Hebrew, Hungarian, Indonesian, Italian, Latin, Lithuanian, Malay, Polish, Portuguese, Romanian, Serbian, Slovak, Slovene, Spanish, Turkish, Ukrainian, and Uzbek.

[11]What is now called Zipf's law was discovered (separately) by Felix Auerbach and Jean-Baptiste Estoup. Auerbach was a professor of physics in Jena, who, together with his wife, committed suicide in 1933 because of the rise of Hitler, with its accompanying anti-Semitism. The context of Auerbach's 1913 work was city populations [13]. For the frequency of words, the law was found by Jean-Baptiste Estoup [56] at about the same time, in Paris. Zipf, a professor at Harvard, quotes both of these sources in his own work.

[12]Benoit Mandelbrot, 1924–2010, mathematician, scientist. The facts are available on Wikipedia, but I knew him as a supreme (if occasionally egocentric) raconteur. He knew the entire history of mathematics in terms of personalities, including those of spouses/partners. The one time I recall his being left speechless was in a conversation, at IBM Yorktown Heights, involving him, Phil Seiden (coauthor of [204] among other things) and myself. Mandelbrot bragged, "I lost 25 pounds in 3 months," and Seiden responded, "So did I, but I have cancer." Incidentally, Phil's wife, Lois, managed to have him enrolled in an early (1980s) experimental treatment of kidney cancer using interferon. Of the 10 people in the study only Phil was cured completely, and the technique was abandoned. Phil died many years later from the effects of prior radiation therapy. I understand that in recent years the body's immune system has again been harnessed in treating some cancers. See [173].

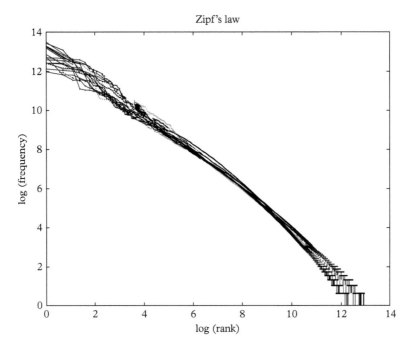

Fig. 9.3 A plot of the frequencies of words in 30 different languages. Adapted from [118].

It turns out that, even with Mandlebrot's enhancements, there's still a lot of controversy surrounding the Zipf "law." For starters, there are several natural questions to ask:

- Technical question: What's a word? Would you count both house and houses? Saw and seen? I presume those drawing up these lists have established a convention, but the whole thing seems a bit loose to me.
- Further to the previous item: What about homonyms? "Pen" is a writing instrument and a place for keeping animals; "wound" is an injury as well as a verb meaning wrapped around. Do these count as one or two words?
- What about other languages (... in some of which the articles "a," "an," and "the" don't exist)?
- What about other linguistic constructs (e.g., phrases)?
- When comparing languages, does the meaning of a word play a role in the ranking?
- Is all this power law stuff baloney?
- And the key question: **Why?**

All of these questions have been addressed by one or another researcher. The meaning or the concept behind a word is discussed by linguists and psychologists.

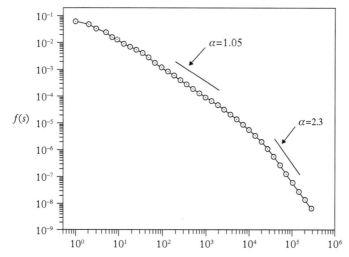

Fig. 9.4 A distribution of English words showing a break in the slope. The number of books used was 2006 and the number of words 183,403,300. Adapted from [154].

Both Piantadosi [176] and Calude and Pagel [36] go beyond the mere counting of words. There's a collection of 200 word concepts collected by Swadesh [221] that have essentially the same meaning in many languages, both extremely common words like 'mother' and less common ones like 'dirty' are on the list. (Following [36], I use a single quotation mark to indicate a concept and double quotes for the actual word form.) This allows one to see if the Zipf correspondence holds across languages and to find consequences. Of particular interest in [36] is the rate at which word forms change. Calude and Pagel find that changes are more common among less common words, which I suppose is not much of a surprise.

In [176] there's a graph (not reproduced here) in which the ranking is the universal one, that is, determined, for the Swadesh list, from uses in many languages.[13] The fits are not fits. It's clear that frequencies—and Zipf's law—are not universal. (To be fair: Piantadosi sees correlations and the graphs are marshalled as positive evidence. So—as usual—things are muddled.)

There are also other issues in connection with the Zipf law.

First it seems that for the tail of the distribution—really rare words—the slope changes. Montemurro [154] plotted the entire distribution and found a break in the

---

[13]The languages are Basque, Chilean, Chinese, Czech, English, Estonian, Finnish, French, German, Greek, Maori, Polish, Portuguese, Russian, Spanish, Swahili, Tok Pisin, and Turkish. Where there are different dialects or versions of the language, I don't know which was used. Clearly the authors were sensitive to this issue, since Spanish and Chilean are distinguished, although Chilean is a form of Spanish. Note that Piantadosi [176] (quoting Calude and Pagel [36]) does *not* consider these graphs as proof against Zipf: on the contrary.

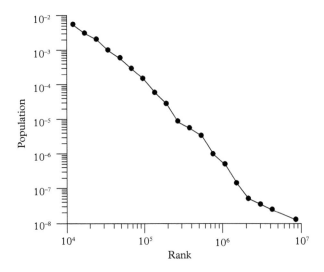

Fig. 9.5 Population of a city vs. rank of the city. Taken from [163].

slope. Somewhere in the rank ~5000 range the slope goes from close to 1 to between 2 and 3. A large corpus of data is needed to reach this conclusion and I would have been happier if the graph had error bars (say, between different books), but once you are suspicious of a "law" and once you find that it breaks down, other criticisms become more credible.

Which brings up the next question on my list above—the baloney factor. There are two aspects to this issue. First, is it really a power law? As indicated in Chapter 7, other distributions, such as the log-normal distribution, can mimic a power law over many decades. And second, do the data support any particular regularity? Some books do not follow the power law. We've also seen that a broader understanding of Zipf's law—involved with the meaning of words—does not obey even an approximate power law. There's a paper [238] that fits the curves with three dependencies. The curves are illustrated in their Fig. 1. To me, fitting these curves with three different analytic forms is a waste of time. It's proverbial that on a log–log plot everything looks like a straight line, but, in this case, there is no straight line. Maybe Zipf's law applies to other phenomena (see below), but it's in trouble for words. (For further negative views see [219] and [176].)

As indicated in Chapter 7, words are not the only objects that seem to be distributed according to a power law. If you plot city population vs. rank (Fig. 9.5)—or personal income or genes or a host of other quantities—you'll get power laws; they don't all have the same power, but they'll have the power law feature in common.

These phenomena will be considered separately since there is no reason for them to share the same underlying mathematical mechanism as the linguistic tendencies.

Finally there is the question of **Why?**

Piantadosi [176] has a long list of candidates addressing this question, some of which we've met before (Chapter 7). But even his list is not exhaustive [63, 139]. Some explanations do not use the fact that this is natural language: some minimize entropy, some point to economic reasons (for cities). But Piantadosi would like a linguistic explanation, and maybe there is none. He makes a reasonably convincing case that nothing offered provides an explanation. Is the power law just a coincidence? That's hard to believe, in view of the large number of other phenomena that allegedly follow the power law (cities, scientific quotations, etc.). Zipf himself attributed the relation to the "principle of least effort," that is, a communications convenience, a point of view adopted by other authors. I would say that there are several issues: Is there really a power law? If there is, the reason is disputed. And finally even if it is a power law, why is the power so close to 1?

9.4   Power laws for cities

Cities turn out to be better than words. The power laws look better and they had priority. In fact the first person to notice power laws in rank vs. population was Felix Auerbach in 1913 [13] (not Zipf, who came later). He looked at German cities (he was a professor in Jena, Germany[14]) and at other countries as well. He found a power that was roughly –1 but with different constants in other countries (the general rule is population = const · rank ^{power}). Auerbach opens his paper with a general comment:

> At first glance, the facts of human life, like the phenomena of nature, do not seem to be subordinate to certain general laws. Statistics, however, teaches us that this is a matter of degree due to the complexity of human relations. One only has to learn to read the statistics correctly in order to draw more general conclusions; and there are often interesting and strange laws.[15]

Many are the explanations for these "statistics," and in fact Krugman [131] laments the absence of a comprehensive solution.[16] A student of his, Xavier Gabaix [63], gave

---

[14]... where the optical company Zeiss was founded (by Carl Zeiss).

[15]The translation (from the German) is mostly due to Google translate, but I inserted some changes to bring Auerbach's declaration to more modern language.

[16]What is incontrovertible is that for four or five decades the log(population) vs. log(rank) curve is approximately a straight line and that its slope is close to –1. Krugman's (and his student Gabaix's) instead works with the integral: the probability that the population *exceeds* some given value (S in their notation). Since the eventual solution is an exponential these are the same (up to a constant), but in principle they can be different.

what many economists consider to be a definitive derivation but it has not been considered definitive by others—who gave further reasoning (e.g., [143] for an astrophysics related argument).

A simplified version of the Gabaix argument follows. First, you don't look at population but at relative population, that is, the variable of interest is

$$S = \frac{\text{population of a given city}}{\text{total } urban \text{ population}}. \tag{9.7}$$

Thus if there are N cities,[17] labeled by i, then $\sum_{i=1}^{N} S_t^i = 1$, where S_t^i is the relative size of city-i at time-t. Next is the central assumption that a city's growth or shrinkage is proportional to its attained size, that is, $S_{t+1}^i = \gamma_{t+1}^i S_t^i$ where the γs have some distribution $f(\gamma)$. This distribution has two immediate requirements. First, its sum (of positive quantities) must be 1. And second, you want the average normalized size to be constant. (Why? This is a convenient assumption, but I don't think it's always true.) Thus

$$\int_0^\infty \gamma^\beta f(\gamma) d\gamma = 1 \text{ for } \beta = 0, 1, \qquad \| f \text{ is the distribution of growth rates } (\gamma). \tag{9.9}$$

Let $G_t(S)$ be the probability that S_t for a specific city is greater than S. Then

$$G_{t+1}(S) = \Pr(S_{t+1} > S) = \Pr(\gamma_{t+1} S_t > S). \tag{9.10}$$

Here we have used the definition and have omitted an index (called earlier "i") indicating which city. But this last equality says that the expectation (i.e., the integral over $f(\gamma)$) of S_t must exceed S/γ_{t+1}. Next define χ to be the identity operator for a given "condition," that is, $\chi(\text{condition}) = 1$ if the "condition" is satisfied, 0 otherwise. (It's the same as the truth function, $\mathcal{T}$, of Sec. 2.1, but I wanted to stick to Gabaix's notation.) With this definition we have $G_{t+1}(S) = \int f(\gamma)\chi(S_t > S/\gamma) \, d\gamma$. (It would be pointless to place a label $t+1$ on γ unless f also carried a label. In practice we are only interested in steady-state solutions in which $f(\gamma)$ does not change.) But $\chi(S_t > S/\gamma)$ is just $G(S/\gamma)$ and therefore

$$G(S) = \int_0^\infty G(S/\gamma) f(\gamma) \, d\gamma. \tag{9.11}$$

There's an argument of Simon [212] that suggests that there should be a power law for cities (and other things too) but doesn't give any particular power. (Krugman comments [132] on Simon's argument, calling it "Simon's story (I [Krugman] call it a story because it is too nihilistic to call a model). ...") Simon's logic is similar to Yule's mentioned earlier (Sec. 7.3). As detailed there, you get a power law, but not a specific power.

---

[17]I presume Gabaix means total urban population of a given country. From Auerbach's writings I have the impression that his relation holds one country at a time. In particular, different countries have different parameters.

Once you've decided that the solution to Eq. (9.11) is to be a power law, there isn't much more to derive. Suppose $G(S) = A/S^\alpha$, then equation Eq. (9.9) for $\beta = 1$ demands $\alpha = 1$.[18,19]

Of course there are details that I'm neglecting and Gabaix's paper [63] is 29 pages long—so, even aside from footnotes, citations and more elaborate explanation, there are more features that I'm ignoring. Among other issues there's the problem of a minimum size for cities. The result is that there are both deviations from a slope of 1 and deviations from a straight line on a log–log plot.

I started this section saying that the power law behavior of cities was better than that of words. Unfortunately that doesn't make the case for a bona fide power law. One point that has been raised is, "What is a city?" Do you go by political boundaries, metropolitan regions or some other criterion. And then there are deviations. In the United States New York is #1, Los Angeles #2, but the ratio of population is not 2 to 1. There are also deviations at small cities; in fact, once again, but for different reasons, "What is a city?" There's an article [11] in which the various conclusions on "Zipf's law" are called into question, for a variety of reasons. As usual, the waters are muddied, but that's life.

9.5 Urban discrimination

Thomas Schelling was an expert in game theory. In 1971 he published a "game" [190] in which, with a small amount of preference by individuals, neighborhoods would become segregated. Let me give some background, since I lived through that period and saw these things happening at first hand. As a child, I delivered newspapers in neighborhoods of the Weequahic section of Newark, New Jersey. The population was changing—I could see it happening in real time. Schelling's model came in the wake of riots in Newark,[20] Detroit and Los Angeles. There were other places as well, but I mention these since each led the field in a particular way: the first in time was the riot in Los Angeles, the most financial damage was in Detroit, and the most lives lost was in Newark.

The model is simple! You take a large checkerboard-like region to be your city. So you can label locations in the city by a pair of integers (n_x, n_y) with each integer

---

[18]One can use a max-ent (maximum entropy) argument to get a specific form of $f(\gamma)$. Let $f(\gamma)$ be the fractional gain or loss of population for a particular city. Apply a variational principle in which $W = -\int f \log f \, d\gamma + \lambda_0 (\int f \, d\gamma - 1) + \lambda_1 (\int \gamma f \, d\gamma - 1)$ yields $f(\gamma) = \exp(-\gamma)$.

[19]There's a feature that troubles me about this argument. I had previously remarked that Krugman's argument sounded like "the rich get richer." But that line of reasoning always leads to powers (in the power law) equal to or greater than 2, whereas Gabaix is arguing for 1 as the power.

[20]All three cities mentioned are in the United States. Newark is in New Jersey, Detroit is in Michigan and Los Angeles is in California.

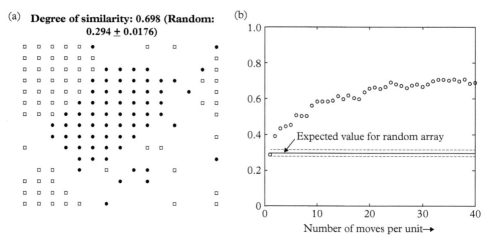

Fig. 9.6 (a) The result of a simulation implementing the Schelling game, with $p = 3/8 + 0.01$. As you can see, with this level of preference the population is almost entirely segregated. (The open squares are also segregated, recalling the periodic boundary conditions.) (b) The history of the segregation level as a function of time. ("Level of segregation" = average number of squares of the same type ÷8.) It takes quite a few steps for the 15-by-15 array to settle down. For lower p values this is a quicker process.

running from 1 to N, where N^2 is the number of locations. Each location is considered to have eight neighbors, the locations in the 3-by-3 square surrounding (n_x, n_y), not counting (n_x, n_y) itself. To avoid edge effects we identify $(0, n_y)$ with (N, n_y) (i.e., they are considered the same point) and similarly for $(n_x, 0)$ and (n_x, N) (so your city is a (two-dimensional) torus and all points have the same number of neighbors). Each location has one of three kinds of occupancy: A (which you can think of as type A), B (type B) and V (vacant). (A and B can be thought of as dark skinned and light skinned, speakers of Italian and English, immigrants from Spain or Afghanistan—pick your favorite prejudice.)

If a location is occupied by type A, the interpretation is that A lives there. Ditto for B. Each type, A and B, has a preference for the neighbors: they want at least a fraction p of their neighbors to be like themselves. If they look around and see that this is not true, they move to a vacant (V) location. Some details still need to be specified, but those are already the main features of the model. The details consist of specifying which vacancy they move to (in Schelling's original model it was the nearest), and whether the choices were made sequentially or simultaneously (with further details in case of conflicts in the latter case).

But pretty much whatever you do, if the number p is anything but zero, you get a degree of segregation. In Fig. 9.6 we show a typical outcome for the preference

p	Level of segregation
0	Random, 0.294 ± 0.019
0+0.01	0.332
1/8+0.01	0.416
2/8+0.01	0.584
3/8+0.01	0.812

Table 9.1 Array size is 15-by15, with 67 (out of $15^2 = 225$) occupied by each type, A and B. Except for $p = 0$ these are based on a single run, so that the average value may deviate slightly from the number quoted. For $p = 1/2 + 0.01$ and larger the degree of segregation drops back down again, presumably because changes become impossible. For larger lattice sizes, say 50-by-50 you again get segregation for higher p values, but there is a continuing tendency to have switches become impossible.

coefficient ("p") just over 3/8. As you can see there is near total segregation. The history is also shown, where time units are one-opportunity-to-move per individual. This history records the "level of segregation" defined as the average number of like individuals seen by each agent. Also shown is the level of segregation for a random array, with the given population inputs and with error bars reflecting the finite size of the array. In Table 9.1 I show the level of segregation for various preference values. In all cases the array is 15-by-15 and A and B are each 30% of the total number of locations (actually 67 each).

What's amusing is that the "degree of segregation" is not a monotonic function of p. As the city size (array size) varies, the number of iterations varies and the input fractions vary, so different results emerge. For example, with each person moving about 100 times in a city that is 100-by-100 and with a 30% fraction of both A and B, the degree of segregation is well over random, but still doesn't reach the levels of 15-by-15, with $p = 3/8 + .0\ 01$. Nevertheless, both visual inspection plus the "degree of segregation" tell you that this system is segregated. From a human perspective it seems to me that the most unrealistic aspect of the simulation is the number of times each person moves. Have you moved 40 times? (This is the number of times each "person" (unit) in the simulation has moved—cf. see Figure 9.1b.)

In the years since Schelling proposed his model many variants have been suggested, trying to get to the heart of urban segregation. Sociologists, philosophers and others of various persuasions have all created schemes in which other factors are considered, such as price and viewing range (not just the eight neighbors of the original model). There are also different degrees of preference for the two kinds of participant (which may be realistic). One amusing option is considered in [159] in which people *prefer* to live in integrated neighborhoods, but for some values of the viewing range this too leads to segregation. (This paper deals with three kinds, or types, of participants plus vacancies. There are particular rules for occupying a vacancy in case of dissatisfaction. They find that for intermediate ranges there can be segregation. I have not checked their results.)

From a mathematical point of view this model is an example of an *agent-based simulation*. One could also look at this as a cellular automaton with some randomness [201] with more than two states (three in the example given, although variants of the Schelling model consider more than just "A" and "B"—there could be "C," "D" and more). As such a certain amount of rigorous work on the model itself has been done, for example [21, 242].

Zhang [242] in particular generalized Young's one-dimensional analytic solution [237]. The techniques used, like those of Young, were game theory. Included were economic data as well as differences in preferences of type A and type B. In particular, Zhang focused on (so-called) white people vs. people of color. The "whites" had a preference for having other "whites" live nearby, while "blacks" had no such preference (this possibility was mentioned earlier). Still there was segregation, which had the outcome that "blacks" paid less for equivalent housing. What I found relevant was that there was a concept of "wrong" decisions, which in practice looked very much like a temperature. What I mean is that the probability of an agent's making a decision different from the "rational" was governed by a parameter that looked like inverse temperature. (Zhang even uses the letter β, which in physics is also used for inverse temperature.) See the rules articulated in Chapter 5. Presumably the result was a Boltzmann distribution, but I have not checked this.

Yet another approach (apparently also studied by Schelling) is to focus on a particular neighborhood. That way you just have a set of equations that are first order and only involve the total number of each "player," rather than a two-dimensional array, which is more difficult to solve. This is an approach taken by Montgomery [155].

Suppose there is a neighborhood in which N_A and N_B people of type A and B are trying to find homes. We allow individuals to have varying preferences. Among type A people they can tolerate up to n_A people of type B living in the same neighborhood, but their preference is "uniformly" distributed. This means that someone whose tolerance value was n_A would be happy if there were n_A type B people living in the same neighborhood, but not more. Or with a tolerance value of zero, any type B person would not be tolerated. Others are somewhere in between, with equal numbers having each preference. The probability density function for the distribution of preferences is

$$f_A(\theta) = \begin{cases} \frac{1}{n_A}, & 0 \le \theta \le n_a, \\ 0, & \text{otherwise,} \end{cases} \tag{9.12}$$

and similarly for B. A complete theory would use agents with values drawn from this distribution, but we'll be using a kind of mean field theory in which all possibilities enter the equations. That means we pretend that everyone moves (or doesn't move) at the same time.

Two further pieces of information are needed. First, A and B refer to the number of people living in the neighborhood, not just their type. And second, A and B may not be whole integers. The latter condition is mainly a convenience.

According to this rule the change in A is

$$\Delta A = \begin{cases} N_A\left(1 - \frac{1}{n_A}\frac{A}{B}\right) - A, & \text{if } \frac{A}{B} \le n_A, \\ -A, & \text{if } \frac{A}{B} > n_A. \end{cases} \tag{9.13}$$

The time for this change is not specified, since all we'll ultimately be interested in is the equilibrium, that is, when ΔA is zero. (We'll also take $\Delta B = 0$, once we've defined it.) Let me go through the logic behind Eq. (9.13). Say $\frac{A}{B} \le n_A$. Then the change in A is the number of people happy with the new value of A (which is $N_A\left(1 - \frac{1}{n_A}\frac{A}{B}\right)$) minus the old value of A. The number happy with the new arrangement is the total population (N_A) minus those who would be unhappy. The unhappy fraction is given by $\frac{1}{n_A}\frac{A}{B}$. On the other hand, if $\frac{A}{B} > n_A$ then no one is happy and all move out of the neighborhood. Similar considerations attach to B, so that

$$\Delta B = \begin{cases} N_B\left(1 - \frac{1}{n_B}\frac{B}{A}\right) - B, & \text{if } \frac{B}{A} \le n_B, \\ -B, & \text{if } \frac{B}{A} > n_B. \end{cases} \tag{9.14}$$

In the steady state $\Delta A = \Delta B = 0$, so that once again there are cases:

$$\begin{aligned} \Delta A = 0 &\implies \begin{cases} B = (N_A - A)\, A\frac{n_A}{N_A}, & \text{if } \frac{A}{B} \le n_A, \\ A = 0, & \text{if } \frac{A}{B} > n_A, \end{cases} \\ \Delta B = 0 &\implies \begin{cases} A = (N_B - B)\, B\frac{n_B}{N_B}, & \text{if } \frac{B}{A} \le n_B, \\ B = 0, & \text{if } \frac{B}{A} > n_B. \end{cases} \end{aligned} \tag{9.15}$$

Equilibrium is guaranteed by the conditions of Eq. (9.15), but it is not clear whether it is a stable equilibrium. You can analyze this analytically, but Montgomery uses the quiver function in MATLAB™ to see what's stable and what is not. First let me show you the plot (Fig. 9.7). The meaning of symbols is explained in the figure caption. The intersection of the lines represents a solution for when both $A/B \le n_A$ and $B/A \le n_B$. Unfortunately, as is evident from Figure 9.7b (a detail of Figure 9.7a), it is not a stable solution. So the neighborhood has *no* solution in which As and Bs live side by side.

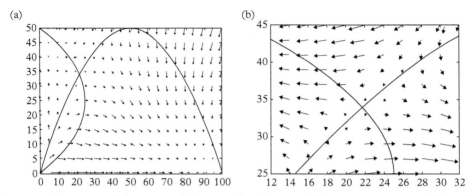

Fig. 9.7 Two views of the "quiver" diagram for $N_A = 100$, $N_B = 50$, $n_A = n_B = 2$. The solid lines are some of the points on the graph where ΔA or ΔB is zero. Other zero points are indicated in Eq. (9.15). The arrows point in the direction $(\Delta A, \Delta B)$ and are indicative of the direction to be taken in changes. Figure (b) is a detail of figure (a), showing the neighborhood of the point (22,33) (approximately).

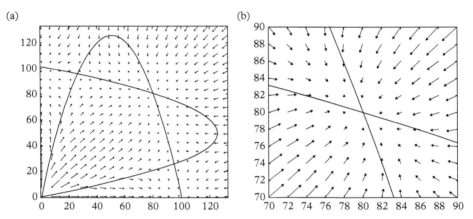

Fig. 9.8 Two views of the "quiver" diagram for $N_A = 100$, $N_B = 100$, $n_A = n_B = 5$. Same as Fig. 9.7, but with values of 5 for $n_A = n_B$ and $N_B = 100$ instead of 50.

However, for other parameter values you can get integrated neighborhoods (i.e., other than those in Fig. 9.7's caption). If people are more tolerant ($n_A = n_B = 5$), total numbers are more equal and, if the initial conditions are right, you can get integration. Here is a MATLAB™ program you can play with to get varying results. I also give a command line yielding Fig. 9.8.

```
function neighborhood(in);
if ~exist('in'), in=struct;end    % In case there's no input ("in")
% Values for relevant parameters:
[NA,in]=setdefault(in,'NA',100);  % # A's
```

```
[NB,in]=setdefault(in,'NB',50);    % # B's
[NAgrid,in]=setdefault(in,'NAgrid',NA); % Parameters affecting grid
[NBgrid,in]=setdefault(in,'NBgrid',NB); %   display
[nA,in]=setdefault(in,'nA',2);
[nB,in]=setdefault(in,'nB',2);
[grid_separation1,in]=setdefault(in,'grid_separation1',5);
[grid_separation2,in]=setdefault(in,'grid_separation2',2);
[ctr,in]=setdefault(in,'ctr',[22,35]);
% Global picture
[A,B] = meshgrid(0:grid_separation1:NAgrid,0:grid_separation1:NBgrid);
dA = (NA*(1-min((1/nA)*(B./A),1))-A);
dB = (NB*(1-min((1/nB)*(A./B),1))-B);
figure(100),hold off,  % The "hold" is in case figure 100 already exists
quiver(A,B,dA,dB,'k');
hold on;                     % Keeps the quiver diagram, but allows more
                             % input in the figure
A = 0:NA;
nullA = nA*A.*(NA-A)./NA;
B = 0:NB;
nullB = nB*B.*(NB-B)./NB;
plot(A,nullA,'k',nullB,B,'k'); % phase diagram with vectorfield
axis('tight'), hold off % figure is complets
% Detailed picture
dis_from_ctr=10; ctrx=ctr(1);ctry=ctr(2);
figure(102),hold off,
NA0=ctrx-dis_from_ctr;NB0=ctry-dis_from_ctr;
NA1=ctrx+dis_from_ctr;NB1=ctry+dis_from_ctr;
[A,B]=meshgrid(NA0:grid_separation2:NA1,NB0:grid_separation2:NB1);
dA = (NA*(1-min((1/nA)*(B./A),1))-A);
dB = (NB*(1-min((1/nB)*(A./B),1))-B);
quiver(A,B,dA,dB,'k');
hold on;
A = NA0:NA1;
nullA = nA*A.*(NA-A)./NA;
B = NB0:NB1;
nullB = nB*B.*(NB-B)./NB;
plot(A,nullA,'k',nullB,B,'k'); % phase diagram with vector field
```

```
hold off
axis([NA0,NA1,NB0,NB1])
%-------------------------------------
function [a,in]=setdefault(in,a_string,a_default);
% Example: [a,in]=setdefault(in,'a',5); creates in.a
%              as well as a, and sets both equal to 5.
% "in" MUST already exist as a structure.
if isfield(in,a_string),
    eval(['a=in.',a_string,';']);
else
    a=a_default;
    if nargout==2,
        eval(['in.',a_string,'=a;'])
    end
end
```

Command line for viewing Fig. 9.8:

```
clear in; in.NB=100; in.nA=5;in.nB=5;in.grid_separation1=7;in.NAgrid=137;...
          in.NBgrid=137; in.ctr=[80,80]; neighborhood(in)
```

For these values the point near $(83, 83)$ is stable.

9.6   Voter models and elections

There's an election coming up. Perhaps you're undecided; perhaps you tentatively made up your mind, but you're still open to influence.[21] Not to worry: there's a *voter model* to tell you how the election will turn out.

Not really. In its simplest (to me) form it doesn't match what people do and I would question the name. If you enhance it, it can approach reality. Let's start with the simplest form.

You have N people and just two choices: they're either Republicans or Democrats, Red or Blue, no third choice. In our model we describe them as 0 or 1. They live on a two-dimensional square lattice and can be influenced by any one of their (four) nearest

---

[21] There's a joke about the poor quality of candidates that's worth telling. ... A thief approaches a citizen and threatens: "Hands up or I'll shoot." Our obliging citizen raises his hands, shivering and shaking, and says, "OK, take it all, but don't shoot." But this thief is not after money—he only wants information. He says, "Hillary or Donald?" The victim is of course surprised, but he thinks and thinks, all the while shivering. Finally he says, "Shoot." (Hillary is Hillary Clinton, Donald is Donald Trump and I refer to the US presidential election of 2016. But you can change the principals and pick your most (un-)favorite election.)

neighbors. First you randomly pick the person who might change opinion or affiliation, from 0 to 1 or 1 to 0. Then you pick a random neighbor (from among the four). The neighbor is assumed to convert the first-selected individual to the neighbor's opinion (0 or 1). If the first agreed with the neighbor there is no change; otherwise the first individual follows the neighbor.

The result of this process is very much like the rubber band game (Chapter 3): very quickly everyone holds the same opinion, although *which* opinion depends on initial conditions and the randomness of the process. This is not the way politics works.

The enhancement that I would say makes the model most credible is the existence of "zealots," individuals who will not change opinion, no matter what. People belong to a party and "always" vote with their party, Republican or Democrat. The model then includes, among the N people, n_0 people who keep their 0 affiliation, no matter what, and n_1 people who are dyed-in-the-wool 1's. This may be an exaggeration, but it's much closer to reality than the simplest model.

Analyzing this can be a mess, but if we assume that the class of neighbors includes everyone (so we once again have a mean field theory) then the results are simple. Here's the formal mathematics. You would like a stochastic matrix that describes the transition probability from (say) k holding the opinion 1 to either $(k+1)$ or $(k-1)$ holding that opinion. Let $M \equiv N-n_0-n_1$ be the number of flexible people and just keep track of them.[22] Treating them all at once, you have $0 \le k \le M$ and R, the transition probability matrix, is $(M+1)$-by-$(M+1)$. The matrix elements are not difficult to work out. First, the transitions that increase the number of 1's:

$$R(k+1,k) = \frac{M-k}{M} \cdot \frac{k+n_1}{N}, \quad k = 0,\ldots,M \quad (k \text{ ones to } k+1 \text{ ones.}) \tag{9.16}$$

$$(\text{remember to read right to left.}),$$

which is the probability that the one influenced is a type 0 times the probability that the one doing the influencing is a type 1. For the value $k = M$, $R(M+1, M)$ is zero, so R does not need to increase dimension. The probability of a downward move is

$$R(k-1,k) = \frac{k}{M} \cdot \frac{M-k+n_0}{N}, \quad k = 0,\ldots,M, \tag{9.17}$$

and again for $k=0$ the value of $R(-1,0)$ is zero, so the index (-1) never comes into play. For a given number k of 1's, the sum of up and down steps is smaller than 1, because k may not change. This is reflected in the diagonal term, $R(k,k)$, which is just $1 - R(k+1,k) - R(k-1,k)$.

The results are predictable. From R one gets a stationary probability distribution (call it p_0)—the likelihood that after many time steps there are a certain number of

---

[22]The "M" of this subsection is the "N" of Sec. 9.2 [98].

1's and a corresponding number of 0's. The peak in p_0 follows in a simple way the number of zealots. Let's say $n_0 + n_1$ is fixed as some fraction of N and vary (say) n_1. Then the expectation of the number of 1's, $\sum_k k p_0(k)$, follows n_1—in fact it is a simple function of it.[23]

Exercise 30 Find the dependence of p_0 on the level of zealotry.

Immediate generalizations would be to other lattices, not just two dimensional-square, and other definitions of "neighbors," an example of which we've already given in our mean field equations (Eqs. (9.16) and (9.17)).

An important generalization is the inclusion of noise: sometimes the influence doesn't work. This makes the dynamics "ergodic," that is, any mixture of 1's and 0's can be reached from any other state [84].

Although I have described zealots as an important factor, there's an interesting article [60] that analyzes a recent US presidential election in detail and used both noise and a creative definition of "neighbor" to fit the election data. They posit two sets of neighbors for each person, around "home" and around "work." They use census data to estimate commuting destinations. Thus each person has two sources of influence and these are weighted by another parameter. They are also interested in correlation lengths: how does the voter share of each party vary as you move away from a particular district. Their data are microscopic, in that they go county by county throughout the United States. The agreement they obtain with actual election results is impressive. My own impression is that they underplay the role of "zealots," but you can't argue with their fits to the data.

The voter model is not only a model for voting, but covers other fields as well. In fact, it was introduced [41] to describe two antagonistic populations fighting over territory and has been used in genetics, cancer research, finance (cf. Sec. 9.2) and other fields. The descriptions in Wikipedia and some other sources tend to be rather abstract, but the simplest versions can be described more concretely. In principle this is a cellular automaton with probabilistic rules.[24]

---

[23] For another application of the mean field voter model, see Sec. 9.2. There we also give an analytic solution for the ground state, p_0.

[24] One of the first random cellular automata (possibly the first, but I'm not sure) was [201]. (On the other hand, there's the aphorism quoted to me by my older son: Who is a mathematical theorem named for? The first person after Euler to discover it.)

9.7 Crowd control

Don't yell "fire" in a crowded theater! And this application tells you why not (as if you didn't know). In addition, the article by Moussaid, Helbing, and Theraulaz [158] predicts that when moderately crowded you'll have two streams of people.[25] The motivation, however, is not so much the ordinary as the incidence of disaster. On several occasions—waiting for a store to open (Walmart, Black Friday 2008), stampedes (Mecca, 2006), even "parades" (Love Parade, Duisberg, Germany 2010)—there have been deaths. The problem is that people are confronted with a bottleneck and begin pushing. Nor does it help that some people fall and (1) are likely to die, and (2) get in the way, making others die as well. It's gruesome and while "control" may have negative associations, there are times when it would be helpful to find ways to avoid these incidents.

It also turns out that there are connections to flocking birds (Sec. 10.4), traffic (Sec. 9.8) and fluid dynamics. As for birds, there are a few rules and the overall behavior is emergent. For pedestrians the rules are simple, but there are a lot of parameters. Here are the two rules: (1) A person chooses the direction that allows the most direct path possible (avoiding obstacles—cf. the second rule) to the target point; and (2) a person maintains a distance from the first obstacle in the walking direction allowing a time of collision greater than a minimum value.

As I said, the rules are simple, but there are lots of parameters. First, how big is a person? Second, what is visible; in other words, what is the spread of angles in which a person sees obstacles? Then what is the minimum time of rule #2? And what speed does the person walk at? There are more, but in all cases reasonable values are taken and simulations done. (As far as I know, there's nothing analytic.) The simulations support the general philosophy of using limited rules and the parameters used. A cynic might say that with enough parameters you can fit anything, but in my experience that's not so. But see Appendix I.

What's new about the having of rules is that two-particle interactions are *not* used [53]. That is, earlier work, including works coauthored by Helbing [106], used interactions between people and between walls. This work only gives rules for one individual. Of course that individual must avoid obstacles, which might be other people, but the rules apply to each individual, very much like the flocking rules for birds (and other creatures).

---

[25]This must be a local phenomenon: my experience in Paris is that people do *not* move aside for easier (foot) traffic flow. As one (French) observer put it: "It's a game of chicken." I recall going with a lot of packages on a narrow Paris sidewalk. Two women were talking. Did they move aside? Did they acknowledge my presence in any way? No. But I'll take Paris as an anomaly.

These simple, "flocking," rules reproduce the behavior of both a single walker and streams of walkers. The latter, both in the model and in experiments with people, form streams when the density is higher. That is, people tend to group, one group or stream going (say) east and one west. (That's my experience also in New York City. Except that I wanted to walk more quickly—and I had to look for breaks and move from group to group.) The model also predicts correctly what people do when there are intersecting streams (they form stripes).

But this is not enough once people are *really* crowded. What really crowded means is that the density is so high that people cannot move freely and are pushed. In this case [158] finds the two rules insufficient—since they involve free walking—and need to supplement these rules with force laws from other people and walls. There are times that pushing and shoving leave one out of control. For example, the physical contact force for running into another person is $\vec{f}_g = kg(r_i - r_j - d_{ij})\hat{n}_{ij}$, where $g(x) = 0$ if pedestrians i and j do not touch each other. Otherwise it equals its argument. (r_k is the position of pedestrian k, $\hat{n}_{ij}$ is a normalized vector pointing from pedestrian j to i, and d_{ij} is the distance between the pedestrians' centers of mass, from j to i.) A similar equation holds for walls—when a person is pushed wall collisions also need to be taken into account.

With these enhanced rules—when things get really crowded—turmoil ensues. Of course we know that already, but according to the models there are ways to mitigate the turmoil. Mainly these improvements are architectural. There's a talk by Helbing that analyzes in detail the disaster in Mecca. In that 2014 talk [105] a number of suggestions are put forth. They are: (1) Lighting (if there's a choice, show which is better, e.g., by making the good choice green and the bad one red). (2) Correct placing of obstacles. Interestingly, although obstacles can be a problem, correctly placed they can slow things down and make escape easier (cf. ants, Sec. 10.2). (3) Exit width.

I'd say where the modeling kicks in is in designing obstacles. (The other ideas are dictated by common sense.) It's interesting that both in theory and in experiment a well-placed obstacle can prevent pileup. Moreover, by looking at different architecture in simulations good actual designs are possible. A lot of it is common sense (e.g., exit width), but the placement and curvature of walls can have unexpected effects.

And, by the way, there's even a power law. If you plot the distance covered between stops versus the frequency of occurrence there a power law of power about (minus) 2. Let me define terms. First, I am talking about bottlenecks. People can go, but there are all the problems of overcrowding. A "stop" however is when the speed is less than 0.05m/s. A graph is shown in Fig. 9.9.

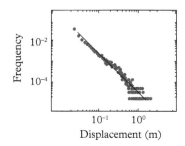

Fig. 9.9 Displacement between stops versus frequency of occurrence. (See the text for further explanation.) The curve covers about three decades in both directions and is a fairly good straight line. Simulations give a slope of -1.95 ± 0.09 while observation gives a slope of -2.01 ± 0.15, in good agreement. Adapted from [158]. Source: *Proceedings of the National Academy of Sciences of the United States of America.*

9.8 Traffic

This is a broad subject, of interest to everyone in "developed" countries, sometimes for reasons beyond saving time—for example the Dutch want to know how rapidly they can evacuate a given area if it is flooded [108]. (Evacuation should also have concerned planners in New Orleans before they were confronted with Hurricane Katrina in 2005—especially if they had taken the trouble to read the relevant 2001 *Scientific American* article [62].)

There are apparently three phases of motion that dominate: free flow, synchronized flow and traffic jams [125]. Quite a bit of work has been done to model traffic and competing ideas abound. In a moment I will go into this in detail, but first, just to be a bit shocking, I will describe a paradox that not only deals with traffic, but extends to other networks (e.g., the internet) as well. The paradox is due to Dietrich Braess and colleagues [31] and is based on a simple diagram (Fig. 9.10).

First look at the solid lines. The dashed line (B to C) is not a route (yet). "n" is the number of cars taking a particular route. So if a car, one of n, goes from "A" to "D" via "B" it will take $n/100$ minutes to cover the first leg of the journey and 45 minutes the rest. The same rules apply to trips via "C" but the order is different. The sensible thing for drivers to do is to divide their numbers on the two routes. Let's say there are altogether 4000 cars. Then it's best if 2000 take each route. The time to travel the entire distance for people going via B is $\frac{2000}{100} + 45 = 65$ minutes. Ditto for those going via C.

In the next scene the government has spent \$4 billion to build a new fast road from B to C. We can even idealize the trip from B to C as taking no time at all. Note that the B to C road is one way. So everybody decides they want to use this wonderful new expressway. How long does the trip now take? Now 4000 people go from the start to

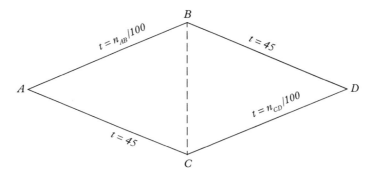

Fig. 9.10 Braess's paradox. A network with particular passage times. All times are given in minutes.

B, all of them take the fast route, and all go the final stretch from C to D. The time is $\frac{4000}{100} + 0 + \frac{4000}{100} = 80$ minutes; so with the new road it takes *longer* to go from start to end.

Exercise 31 This is not a complete proof. Show that the 80 minute trip is an attractor. If one driver takes the fast route there is a saving, but others will follow and finally all lose.

There is a related paradox, one that must be faced by every transportation authority in the developed world: building roads doesn't solve problems, it creates them. Well, sometimes yes, sometimes no. There's a phenomenon known as "induced driving" in which improved roads lead to more traffic, which in turn makes those "improved" roads be filled to capacity. For example, people are willing to take more distant commutes because the roads are better. There is improvement at first, but then it takes longer, and it's hard to find another job. It's a matter of controversy and uncertainty how important this is, but it's hard to ignore pressure to solve problems by building. I don't have an answer and can only draw attention to the problem.

9.8.1   Cellular automaton models of traffic

A traffic model due to Biham, Middleton and Levine [27] is based on a cellular automaton. It exhibits (what looks like) a phase transition between being completely jammed and being completely free. Drivers are going in two orthogonal directions, either (say) to the north or to the east. (They don't go south or west.) The model is constructed on an N-by-N square lattice with periodic boundary conditions. N is an integer and can vary; p, the occupation probability for both kinds of driver, is also varied. Initially

pN^2 easterly drivers and pN^2 northerly drivers are put in random positions, that is, corresponding numbers of cells are occupied. The remaining $(1-2p)N^2$ cells are vacant (so $p<1/2$ and some convention chosen in case pN^2 is not an integer). The rule for advancing one time step (say from t to $t+1$) is simple. Consider an easterly driver. If the square to the east of his present position is unoccupied *and* it doesn't become occupied (at time step $t+1$) by a northerly driver, then this driver can move to it. The same rule applies to northerly drivers, except that they are driving north. For the case that a given square is vacant but would be occupied by both an easterly and a northerly driver, there is again a simple rule: choose randomly who advances and who doesn't move. In other words, with probability $1/2$ the easterly car does advance (to the east) and the northerly car stays where it is. And with probability $1/2$ the opposite occurs. (In [27] other models are also considered, but are not discussed here.)

That's the model. The result is that at low p the cars move freely. And at high p they don't move at all: traffic is totally jammed. The transition is fairly sharp, but the actual p value at which the transition takes place depends on N. Near the transition there is a dependence on initial conditions, and sometimes it will be jammed, sometimes the cars will arrange themselves to allow flow. When there is jamming you get everyone stuck along a diagonal that runs from southwest to northeast. A typical result, at high density, is shown in Fig. 9.11.

Unfortunately it is difficult to do analytic work on this model and the information I've given is based on numerical simulation. In practice though it is clear what must happen and the model is far enough from reality that the actual numbers don't matter. There is a simple one-dimensional model for which analytic solutions do exist, but that takes us even further from reality. This model is known by various names, among them the "single-step," deterministic version. In it a ring is randomly occupied with probability p and a "car" can move to the right only if the site to its right is empty. The equilibrium of this problem is known analytically and the velocity is rather simple: it's 1 if $p \leq 1/2$ and $(1-p)p$ otherwise. (On the other hand, this model does—surprisingly— have something to do with surfaces. See [130].)

A second model, but this time of one-lane traffic, is due to Nagel and Schrekenberg [160]. In it there are no collisions because drivers reduce their speed to avoid them. Also, there's some randomness in that a driver may reduce speed arbitrarily. Let me be precise (following [40]). Sites are arranged along a line, and each site can be occupied by a single vehicle or none at all. Periodic boundary conditions are adopted. Time is measured in integer values, $t = 1, \ldots, T$. Each velocity is an integer ranging from zero to V_{max}. There are four stages or rules for the motion.

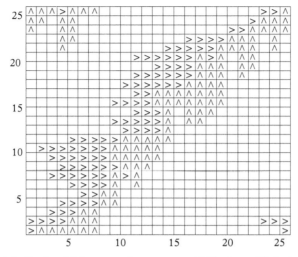

Fig. 9.11 Symbols indicate which way a given car *would like to go*. Unfortunately the density is high and all that happens is that a static configuration develops. (This is the *final* state, not the initial one, which was random.) This high-density phase is above the transition density. The system size is 25 by 25 and $p = 0.36$. This figure is similar to Fig. 2 in [27].

- Accelerate. Each velocity is increased by 1 (unless it is already at $V_{\max}$). Specifically, if the sites are labeled by integers, $i = 1, \ldots, N$, this step can be written

$$v(i,t) \rightarrow v^*(i, t+1) = \min(v(i,t) + 1, V_{\max}),\qquad (9.18)$$

with t the time. This is a tentative velocity (hence the asterisk *) and may subsequently be reduced by other factors.

- Decelerate to avoid collisions. Let $D(i,t)$ be the distance from $\#i$ to $\#(i+1)$ (remembering the periodicity in N). Then the $v^*(i, t+1)$ (the new, but not yet final) velocity is never bigger than $D(i,t)-1$ (remember: the time interval is 1). In particular

$$v^*(i, t+1) \rightarrow \min(v^*(i, t+1), D(i,t) - 1).\qquad (9.19)$$

- Random deceleration. There is a probability p such that a driver will slow down by 1 unit with that probability—but no one moves backwards. Thus

$$v^*(i, t+1) \rightarrow v(i, t+1) = \max(v^*(i, t+1) - 1, 0) \quad \text{with probability } p.\qquad (9.20)$$

- And, finally, *movement*,

$$i \rightarrow i + v(i, t+1).\qquad (9.21)$$

Each car advances according to its velocity.

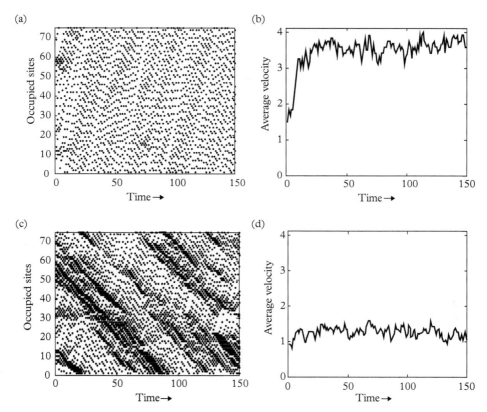

Fig. 9.12 Nagel and Schrekenberg's [160] one-dimensional model. Cars move upward, when they can (with periodicity in the boundary conditions). (a,b) Low density. (c,d) High density. Note that in the first row cars move at close to the maximum velocity, while an increase in density has them substantially slowed down. Parameters are: size of roadway, 75; time units, 100; maximum velocity, 4. Probability of spontaneous slowdown, 0.25. Low density, 12 cars (out of 75 positions). High density, 26 cars.

For practical applications the unit of time is about 1s and the unit of distance is about 7.5m. Speed (V_{max}) is an integer, generally 4m/s or 5m/s.

These rules give a realistic picture of traffic, although they do not (as far as I know) produce the *synchronized* phase. In Fig. 9.12 I show motion with different initial densities: low for which everyone can travel near to V_{max} and high for which there is jamming.

The rules just given are complicated and there's no exact solution. In [40] there is a mean field theory describing this situation, and for particular (but unrealistic) cases exact solutions can be found. But the above steps are easy to simulate on a computer and that is all that is reported here. Many variations of the rules above can be given. For example, the probability of spontaneous slowdown can be "personalized" to account for drivers having different dispositions. In other words there would be a different probability p_i that each driver would spontaneously slow down. Or one could have probability sensitive to the local velocity. It is also clear that my "universe" (of a 75 unit roadway) is too small, as is the time interval. But even with my clumsy programming these simulations took about 0.3 seconds to run for the denser state (and even less for the less dense situation). It should be noted though that running large two-dimensional simulations, as done in [27], runs up against time limitations.

It should also be noted that two significant types of events are missing: accidents and lane restrictions (due to construction, usually). I recall being on a four-lane interstate highway in Los Angeles where there had been an accident—sometime in the past—but there was still a delay due to "rubbernecking," those who wanted to stare at the unfortunates involved in the actual accident. Similar phenomena occur for lane restrictions. There is a significant delay typically when density is high and this is another density-dependent phenomenon.

More on cellular automata. There has been quite a bit of progress in the decade following the articles by Biham et al. [27] and Nagel and Schrekenberg [160]. By 2002 an article in *The New Yorker* [205] described cellular automata as a convenient way to model traffic, especially allowing for different temperaments of drivers. Apparently software developed by KLD Engineering (which can be found on the internet) can be useful for simulating traffic with a cellular automaton model. There is another article by Davis [48] on the role of physicists in traffic flow, basically applying statistical mechanics to this problem.

The true wave of the future. Maybe. There seems to be a lot of progress in self-driving cars. Automobile companies as well as computer giants are experimenting with cars that drive themselves. (And I'm not even talking about tanks, which have the additional feature that without human intervention they can kill people.) Yes, self-driving cars have caused fatalities, but hard-hearted traffic engineers deal with this issue all the time. Every improvement that gets rejected comes with a cost, and a human life—to these people—comes with a price tag. The self-driven cars now being tested need to blend in with human-driven vehicles. Two things may change that: first, in cities it may become forbidden to enter with anything but a self-driven vehicle, in which case cars will only have other computer-driven cars and pedestrians to contend with; second, it's not clear that our (distant?) future has any cars at all. The carbon footprint may be too large. But for now, traffic involves human drivers, with personalities that seem to be successfully handled by modeling. The approaches described here are just the tip of the iceberg and the subject is getting a lot of attention by the research community.

10
Biological sciences

Applications discussed in this chapter: Firefly synchronization, Flocking, Ecology, Biorobotics, Ants, Genes, Kuramoto model, Mobile phone localization, Neurology

---

Biological phenomena in which statistical mechanics can play a role are too numerous to list and surely I cannot do justice to the many applications. Books have been written about the subject [79] and there are articles for each of the many applications. However, I will give samples. Some topics are treated elsewhere in this book, for example flocking. I will omit subjects that are rather technical and focus on the opposite of "many." I have in mind the relation between work and free energy in *small* systems, a field that followed the appearance of the Jarzynski equalities [114, 115].

10.1 Firefly synchronization

Synchronization occurs in many biological phenomena. Notable is the action of the heart muscle. When not diseased, cells of the heart coordinate their action so as to drive blood through the system. Breakdown in this coordination signals disease, for example fibrillation in a heart attack. Mechanisms for achieving synchronization vary. Different models and equations may apply to cardiac muscle, to clocks in phase or to firefly synchronization. And even the last topic, to be taken up in the present section, varies in different species of fireflies. In this section a particular kind of firefly synchronization is discussed.

Remark: There is non-biological synchronization as well. A famous example is Huygens seeing two interacting pendulum clocks beating together. He saw them with opposite phases as well, but he did see synchrony. For details see [169]. Sec. 10.5 also deals with a non-biological application.

When Things Grow Many. Lawrence S. Schulman, Oxford University Press.
© Lawrence S. Schulman (2022). DOI: 10.1093/oso/9780198861881.003.0010

Fig. 10.1 Synchronized fireflies in Georgia, United States. Those are *not* stars in the sky! Source: By Jud McCranie - Own work, CC BY-SA 4.0, https://commons.wikimedia. org/w/index.php?curid=58229716.

The first remark about fireflies is that there are 2000 species of them. Fireflies are a kind of beetle (Lampyridae) and their variety and abundance are consistent with Haldane's assertion that "the Creator is inordinately fond of beetles" [96].[1] We are interested in their synchronization as an emergent behavior, but it will surprise no one to learn that some synchronize, some don't, and those that synchronize do it in different ways. Perhaps it *will* be surprising that no one is really sure of *why* they synchronize, although the consensus is that it has to do with mating behavior. In Fig. 10.1 is a lovely photo of what it looks like when many fireflies do their thing at the same time. This view though seems to be just the tip of the iceberg, and what you really want to do is travel to the big displays (Thailand, New Guinea, etc.) described in Buck's various publications [33, 34].

I won't go into the mechanism of lighting up. If you're interested (in luciferin and related substances) see for example [78], where some of the biochemistry is discussed. Another reference about fireflies is [141], where you'll learn among other things about "nuptial gifts." As far as I can determine for fireflies it's mostly spermatophores (packages of proteins that are raw materials for female production of eggs). Sexual cannibalism, as practiced by praying mantises (in some relatively small percentage of couplings), as far as I know, does not occur. But one article does mention the giving of body parts [141].

The exact mechanism of the synchrony varies with the species. One method is known as phase-advance synchronization, another as phase delay. For details of these and others, again see Buck's papers. I will discuss a simple model of entrainment due to Mirollo and Strogatz [152], which allows a bit of mathematics, although it goes only

---

[1]Haldane meant that there were a lot of beetle species.

partway in explaining the firefly exhibitions. On the other hand, entrainment comes up all over the place. Examples are women's menstrual cycles (for women spending a lot of time together),[2] mechanical clocks (all in one place, like a clock store) and pacemaker cells in the heart.

Actually the mathematics to be presented does not quite explain the phenomena just mentioned.[3] Instead I'll assume that there's some fixed signal coming in and the firefly tries to get in tune with it. It's as if all the other fireflies have managed to coordinate and this one bug needs to get in line. In the papers cited stronger results appear, but I'll follow Strogatz's text [218], which presents the simpler result.

Suppose there's a periodic signal (light pulse) coming in, with frequency Ω, so if the associated coordinate is Θ we have that

$$\dot{\Theta} = \Omega. \tag{10.1}$$

(The dot is d/dt, as usual.) We assume that each time Θ hits zero there's a signal—a burst of light—that reaches our firefly. Now our firefly has his own frequency, call it ω, and an associated coordinate θ. Each time *his* coordinate θ hits zero, he lights up. (We assume the duration of the light is such that only the moment it first flashes is significant. The subsequent dropoff, the tail of the light profile, isn't important. Also, I'm taking this firefly to be male. Females in some species also signal, but we focus on the males.) Thus if our firefly did not receive this outside signal, his equation of motion would be $\dot{\theta} = \omega$, and he'd light up every $2\pi/\omega$ (at $\theta = 0$, by convention).

But in fact our firefly *does* attend to this outside signal. There's a positive coupling constant, call it A, such that the undisturbed equation of motion is changed to

$$\dot{\theta} = \omega + A \sin(\Theta - \theta). \tag{10.2}$$

This additional term can advance or retard the firefly's signalling. Consider what happens: if the outside signal is a little ahead, so $\Theta > \theta$, then the positive coupling A speeds up the progress of θ ($\dot{\theta}$ is greater) and the firefly fires a bit sooner. On the other hand, if $\Theta < \theta$ then the effect of the outside signal is to reduce $\dot{\theta}$, and the firefly waits a bit longer, again (possibly) bringing him into sync with the outside signal.

How effective can this synchronization be? It turns out that both experiment and Eq. (10.2) provide limits to the firefly's abilities. We examine the limits imposed by the equation of motion. Note however that we will find one slightly embarrassing feature: there's a phase difference between the outside signal and the firefly. So this is not perfect entrainment.

---

[2]The coordination of women's menstrual cycles has been disputed. See [93].
[3]In later sections we do go into detail over other models of synchronization.

A way to see these features is to define new variables, $\phi \equiv \theta - \Theta$, $\tau = At$ and $\mu = \frac{\omega - \Omega}{A}$. A bit of arithmetic[4] shows that the equation for ϕ is

$$\phi' = \mu - \sin \phi, \tag{10.3}$$

where the prime (on ϕ) means $d/d\tau$. By the way, this "bit of arithmetic" is a fairly typical trick (you might call it). If you look at Eq. (10.2) you'd say that there are three parameters and the behavior as a function of these many parameters might be complicated. This change of variables shows that there is really only one (μ). This means that you don't have to explore a *three*-dimensional space to see the range of possibilities for solutions to the equation.

Let $\tau \to \infty$ and assume the system goes to a stable point. Then $\phi' = 0$ and

$$\phi = \arcsin \mu, \tag{10.4}$$

which for $\mu < 1$ has two real solutions, while for $\mu > 1$ it has none. We examine the case $\mu < 1$. One solution, which we'll call ϕ^*, lies in $[-\pi/2, \pi/2]$, while the other is $\phi^{**} = \pi - \phi^*$ (mod 2π). To find out how each behaves, consider the two cases in a small neighborhood of the stationary point. Suppose first that $\phi = \phi^* + \delta$ with small δ. Then $\phi' = \delta' = \mu - \sin(\phi^* + \delta) = \mu - \sin \phi^* \cos \delta - \cos \phi^* \sin \delta$; to first order in δ this is[5] $\delta' = -\sqrt{1 - \mu^2}\, \delta$. This can be written as $(\log \delta)' = -\sqrt{1 - \mu^2}$, implying—since $\sqrt{1 - \mu^2} > 0$—that $\delta(t)$ decreases exponentially. By contrast for the other root the cosine term is negative, so that $\delta' = +\sqrt{1 - \mu^2}\, \delta$. It follows that $|\delta|$ grows and ϕ^{**} repels points near it.[6]

These features are best illustrated by plotting ϕ' vs. ϕ and marking the points where ϕ' vanishes, also indicating whether those places where it vanishes are attractors or repellers. See Fig. 10.2.

As indicated, this demonstration has the virtue of simplicity, but it does not capture the full richness of entrainment. The major defects are that I've only dealt with an external signal, rather than showing how *many* signals come together, and I've only looked at a case with an imperfect match, where, if there's a slight deviation in the natural signal, there will be a phase shift in the firefly signal compared with the regulating one.[3]

---

[4]Here's the arithmetic. First you subtract Eq. (10.2) from Eq. (10.1) to get $\dot{\phi} = \Omega - \omega - A \sin \phi$; then you divide by A. Now use $d\phi/dt = (d\phi/d\tau)(d\tau/dt)$.

[5]If $\sin \mu = \phi^*$ and $-\pi/2 < \phi < \pi/2$ then $\cos \phi^* = \sqrt{1 - \mu^2}$.

[6]Note that the prime here means derivative, not the new value of δ. (In my notation, for percolitis, the logistic equation and other applications it means the new value.)

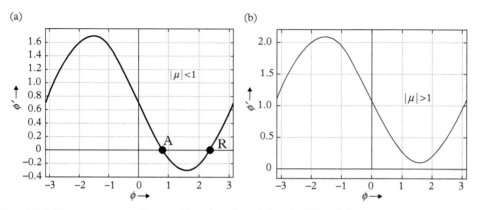

Fig. 10.2 Force versus angle—see Eqs. (10.2) and (10.3). When $|\mu| < 1$ there are two angles where the force vanishes, but only one of them is an attractor, marked "A" in the figure. The other zero of $\phi' = 0$ is a repeller and is marked 'R.' The condition that $|\mu| < 1$ translates to $|\Omega - \omega| < A$, which defines the range of frequencies that the firefly can match. At $\mu = 1$ repeller and attractor coalesce, and for larger μ the firefly cannot match the external signal.

10.1.1   Other models for synchronization

The demonstration of "synchronization" above has the advantage of simplicity, but assumes that there is an outside source of stimulation ($\Theta = \Omega t$). Mirollo and Strogatz [152] deal with more general equations, following a model of the heart developed by Peskin [175]. The Peskin model, known as "integrate and fire," has pulses. The scheme for coordination or synchronization is simple: start many oscillators at points between zero and one. Allow them to increase at some rate. When the first oscillator reaches one, the others get a boost in their phase also. Any oscillators whose phase then (after the aforementioned increase) equals or exceeds 1 is set to zero and the process begins again, with those that had not been set to 1 (and then 0) continuing—with the indicated increase—as they were. This gradually brings all oscillators into synchrony. Mirollo and Strogatz generalized this simple rule into more general behavior during the intermediate times. (Peskin assumed a linear first-order equation, a concept generalized in [152].) What [152] didn't do was to generalize the situation to oscillators having varying natural frequencies.

Integrate and fire.   We look more closely at "integrate and fire" models. Peskin's conjecture was that the cells could differ slightly in their underlying dynamics (e.g., if a linear model is assumed, have different frequencies) and that there could be many of them, and still they would synchronize. He only managed to give mathematical proof for two cells with the same dynamics. Peskin's model was

$$\frac{dx_i}{dt} = f_0 - \gamma x_i \,, \tag{10.5}$$

where f_0 and γ are constants and γ is small (and represents dissipation). This would be the dynamics until one cell hits 1 (or more hit, if they are really simultaneous). When that happens all would get an increment, ε. At this point any cell exceeding 1 (which would include the cell that originally hit, #1) would be reset to zero. And then all would obey Eq. (10.5)—until the next cell hit 1. As indicated, Mirollo and Strogatz generalized this in allowing a more general dynamics and extended the proof—with identical dynamics for all cells—to general populations.

I myself have done no analytic work on this model. However, I have simulated the linear version (Eq. (10.5)). My results are as follows. For systems having the same f_0 there is synchronization for sufficiently large ε (the distance from 1 for which one resets to zero, explained above). But if ε is too small and $N > 2$ you needn't have synchronization.

If the dynamics vary, that is, if f_0 is different for different oscillators, then you cannot have complete synchronization under Eq. (10.5), because, even if cells go to zero at the same time, once they have hit zero they head for 1 at different rates. However, they are reset at the same time, for sufficiently large ε. If they "fire" when reaching one and resetting, then all fire at the same time. It is reasonable to call this synchronization.

On the other hand, my experience is that if N becomes large synchronization is difficult to achieve. Personally I doubt if this is how the heart works. I can't speak for Peskin, but I get the feeling that these equations are not central to his thesis.

Let me mention though that there is also the matter of my programs. First, and this is a general affliction, one stops debugging a program when it gives reasonable answers. (I say "one," but what I mean is "I and probably most others." I don't doubt that there are more fastidious programmers.) Second, and this is a definite fault of my program, I have a fixed time step (called dt). So I must go slightly past the value 1 in order to reset. This is bounded by dt, which is small, but not zero. For this reason I cannot say whether I've found a loophole in [152]. Mirollo and Strogatz predict a number of properties that I haven't managed to recover. Is this my assumption on dt, or is this another of the phenomena mentioned in [152]? I don't know.

Thoughts about synchrony. It's a fact that synchrony is ubiquitous: from heart cells, to fireflies, to clocks in an old-fashioned clock store all showing the same—correct or incorrect—time. But what's not clear to me is that there's a common mechanism,

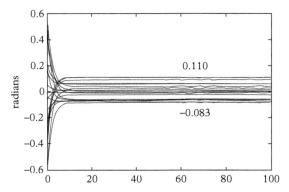

Fig. 10.3 Firefly angle as a function of time. There is a spread in individual firefly natural frequency, but eventually they do synchronize, with the extremes of angle corresponding to the extremes in natural frequency. The largest and smallest asymptotic angles are indicated (so there is a spread of about 0.2). Moreover, the average frequency for the oscillators is (very close to) the average of the natural frequencies. See the main text for details.

a common set of equations that unites all phenomena. For example, for a limited mass ratio range a pair of pendulum clocks will be coordinated, but at diametrically opposite phases of their cycles.[7] It would be nice to have a unified theory, but perhaps it doesn't exist.

Returning to the original model, I've explored a slight variation of Eq. (10.2),

$$\dot{\theta}_k = \omega_k + A \sin(\Theta - \theta_k), \quad k = 1, \ldots, N, \tag{10.6}$$

where now $\Theta = (1/N) \sum_i \theta_i$, an average of the nearby fireflies, so $N \approx 5\text{--}25$. (Actually N might be an interesting number to estimate, but the results of using Eq. (10.6) do not depend much on it—even for $N = 150$ the spread in asymptotic angles was about the same.) In the above equation I allow for variation of the fireflies themselves by positing a spread in their internal rates ($\{\omega_k\}$). In principle A could also vary but I consider that less essential. There may be a good analytic way of dealing with Eq. (10.6), but I've taken an easy way out by solving the equations numerically. So as not to get involved in the mod 2π properties of the angles I limit the spread of initial conditions, but it is nevertheless much wider than the values of the angles as $t \to \infty$. In other words, eventually the fireflies fire in unison, or nearly so. In Fig. 10.3 I show the results of solving Eq. (10.6) with parameter $A = 0.5$ and with an initial spread of about ± 0.015 around $\omega = 1.05$. The spread in asymptotic angles is about twice the spread in natural frequencies, so if the signal is short enough the fireflies will be seen to be pulsing together.

---

[7]This was discovered by Huygens, who is given credit for the invention of the pendulum clock. A lot of literature has sprung up to describe this one phenomenon, for example [24].

In this model there is nothing external, but the fireflies themselves come together (as can be seen from the disparate initial conditions in the figure). For $A = 0.2$ (so the insects are less sensitive to each other's signals) the spread in angles is about five times the spread of natural frequencies. Since this result is independent of N (my phenomenological observation), one might have a chance of estimating the coupling in Nature. This would involve learning the variation in natural rates and seeing to what degree there is sensitivity to details of the model (e.g., the sinusoidal coupling).

And now for a reality check. The behavior of fireflies is much richer than is here suggested. From Gould's book [83] I learned that their signaling also involves precise timing, thereby distinguishing different species from one another. After all you wouldn't want a *greenii* to mate with a *macdermotti* (two species); or, more important, they wouldn't want it. It turns out that species can be distinguished by the timing of their flashes. A male *greenii* will flash twice with a second between flashes and the female responds 1 second later, whereas male *macdermotti* allow a 2 second interval to elapse between flashes. The mathematical discussion above says nothing about these patterns and is presumed to be controlled by an instinctive oscillator circuit within the animal.

It is also plausible, based on the fine tuning, that Nature and evolution have brought natural frequencies of the firefly oscillators to within (roughly) 1% or 2% of one another as demanded above for synchronization.

For more on the subject of synchronization, see Sec. 10.5.

10.2 Biorobotics and glass

Following Solomon's advice [182], we look at (ראה) the ants. Actually there's quite a bit to be learned from them. For one thing they can crawl into tiny spaces—this touches on biorobotics, where you try to build a robot to do that [87]. But of more interest in the present context is the analogies—rather concrete ones as we'll see—that can be drawn to non-living systems.

I remark that Solomon's advice had to do with the work ethic and emergence. The quote[8] [182] is

> Go to the ant, thou sluggard;
> Consider her ways, and be wise;
> Which having no chief, overseer, or ruler,
> Provideth her bread in the summer,
> and gatherest her food in the harvest.

---

[8]In this translation what I earlier called "look at" (ראה) is rendered "consider."

You might think that the ant is the opposite of a sluggard, but—as Dan Goldman (of Georgia Tech) will tell you—there's idleness in ant behavior [4]. And this is for good reason. As we know from both pedestrian and vehicle traffic when too many people want to go to the same place, instead of increasing flow, you get immobility. Traffic jams. People die in a crush. So a good fraction of the ants stand around doing nothing, while other ants do the job! How they decide who works and who sits around, I don't know.

The context is ants in narrow confines [86]. When ants (in this case fire ants) build a nest, there are long, narrow tunnels and it often happens that two-way traffic is slowed. On top of that when two ants meet they "antennate," meaning they brush antennas and exchange chemical signals. Presumably this provides information about food sources, identities and other things the ant needs to know. The authors of [86] studied this phenomenon first by looking. They created an artificial nest and an artificial foraging area and connected them by tubes about 10 cm in length and of varying diameter. Then they "looked at" the ants. Usually the tunnels are underground and you can't see what the ants are doing. But instead artificial—and transparent—tunnels were built. Presumably, the ants behaved in them the same way they would behave in natural tunnels. The main issue was how it got *too* crowded to move. To study this it's useful to define a correlation function that relates aggregates of ants at different times.

First you need to define a density of ants. There's a quantity ρ_L that's 1 where an ant is, zero elsewhere. More interesting is ρ_n, which is the density of an *aggregate*, a bunch of n ants. They fill some linear dimension (position in the tunnels is a one-dimensional variable, x), from x_i to x_f (i is initial, f is final). ρ_L or ρ_n can exceed 1 because the ants can pile up and the various densities are functions of a single variable, the linear dimension along the tunnel. A defined density in this form resembles quantities defined in the study of grains, colloids and glasses, none of which is alive. Having defined ρ_n, it is possible to define a correlation function for an aggregate of n ants. As a function of temporal separation it is

$$Q_n(\tau) = \frac{\langle \rho_n(x,t_0)\rho_n(x,t_0+\tau)\rangle - \langle\rho_n(x,t_0)\rangle^2}{\langle\rho_n(x,t_0)\rho_n(x,t_0)\rangle - \langle\rho_n(x,t_0)\rangle^2}, \tag{10.7}$$

where the brackets represent integration over the transverse directions (to the tunnel). For ants, for colloids, for granular matter, and for other materials, $Q_n(\tau)$ represents persistence. (The density, ρ_n in the above, is predicated on the ants being part of an aggregate of size n.)

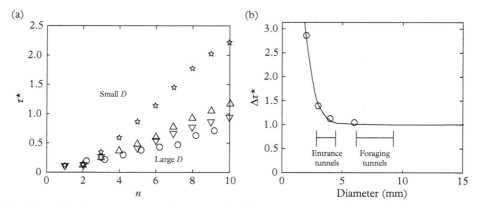

Fig. 10.4 (a) τ^* is shown as a function of n (the horizontal axis) for increasing Diameter (D). The largest diameter values are the lowest in the graph (and the D values are implicit). (b) The slopes of the functions in (a) ($\Delta\tau^*$) as a function of D (the horizonal axis, with D in millimeters). The curve shown in (b) is the function $1 + \frac{A}{(D-D_c)}$, involving two parameters (A, D_c) to fit four points. That doesn't seem much of a fit, but is justified by the simulation results, which follow. Adapted from [86].

Q measures the level of correlation in time. If there is a bunch of ants *now*, what will the scene look like in 1 second, in 2 and so on? Such correlations obviously decay in time, and it was found—from experiment—that the decay was roughly like a stretched exponential, with a τ^* (call it lifetime) such that $Q_n(\tau) \sim \exp\left(-\left(\frac{\tau}{\tau^*}\right)^\beta\right)$ with β a fitted parameter on the order of 1. What I've not written explicitly is that τ^* is a function of both n and the diameter (call it D) of the tunnel through which the ants crawl. For fixed D, as a function of n you get more or less a straight line—which I suppose is remarkable, but true. But you get different straight lines for different D. The bigger the diameter, the less time an aggregation would last. See Fig. 10.4.

In the spirit of what is seen for flocking (Sec. 10.4), in which a few rules were sufficient to model behavior of entire flocks, the researchers made a cellular automaton model of the ants. The model was two-dimensional, with lattice spacing corresponding to ant size: the lattice spacing is L, one body length (about 3.5 mm), and a typical model tunnel was 31 L long and 3 or more L in width. The rules were simple: one site could accommodate at most a single ant. Ants could enter the tunnel from either end and could advance either by moving in the original direction or by moving diagonally (with the component in the original direction). This meant that to pass each other ants would have to move sideways. The ants moved (what to each of them was) forward (if possible) on each iteration. However, when two ants met going in opposite directions there was introduced a probability p of moving sideways. This was to simulate the need to antennate and provided a delay of (expected) length which was a function of p: the smaller the p, the longer the interaction time (which could then be varied

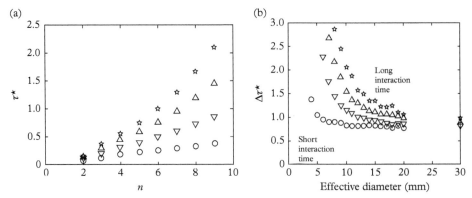

Fig. 10.5 Simulation results for the data of Fig. 10.4—same explanations of items graphed. (Not all values in one figure appear in the other.) In (b) more than one possible interaction time is plotted, with times increasing as one moves northeast (in the graph). Actual interaction time is about 1/2 second. Also in (b) slopes are plotted for various interaction times (related to "antennate"-time). Fits are not shown. The plots on the upper right (extreme northeast) do not have realistic time scales, but are included because of glass data. *If* the ants took correspondingly long to antennate they would correspond to "strong" glass transitions. (For the full plots see [86]. In the original plots, there are about a dozen plots in (a) and seven in (b).) Adapted from [86].

by varying p). The model was good at large diameters, less good at smaller ones; the results are shown in Fig. 10.5. In any case this allowed a better fit of slope vs. tunnel diameter and is some justification of the given formula.

In their article [86], Gravish et al. found that this behavior of τ^* looked a lot like the glass transition. I find this amazing. Ants, after all, were formed by evolution, while substances that undergo the glass transition just are. That is, they have certain immutable properties, and that's that. Nevertheless, the slowdown as things congealed or became crowded was similar.

First, a bit of information about glasses. Up to some density or temperature they flow smoothly and then they stop. But the stopping happens in two ways. In each case, and at each density (or temperature), there's a characteristic time, call it τ^* (the coincident notation is intentional), such that there is exponential (or stretched exponential) decay. But τ^*'s behavior can be used to distinguish *fragile* glasses from *strong* glasses. In strong glasses τ^* increases exponentially as a function of density. In fragile glasses, by contrast, there is smooth flow; in other words, there is a very slight dependence of τ^* up to some point, and then a sudden cessation of movement, τ^* increases super-exponentially and things slow down to the point of stoppage. Using simulations you can look at non-biological as well as biological ants. Biological ants have antennate-times of about half a second, well into the range of fragile transitions.

To see strong glass behavior you must go (in the simulations) to long antennate-times, well beyond the biological range. In other words, these (imaginary) ants can move well into the densities that might slow down objects at high densities—until they stopped.

This is in keeping with ideas about evolution, namely that a way was found to keep moving although crowded. However, I still think it remarkable that inanimate objects can also behave this way. (Well, while I'm marvelling, I could marvel at evolution—but that's another matter.)

10.3 Gene distributions

In this topic, there are power laws all over the place, more or less wherever you look [77, 99, 128, 129, 183, 208, 225]. The data are pretty good but less than perfect. Sometimes the lines on log–log plots are not straight and sometimes they don't cover enough orders of magnitude to make me happy (three or four seems to be enough, but it would also depend on the log–log plot).[9] The gene data are generally good for that number of

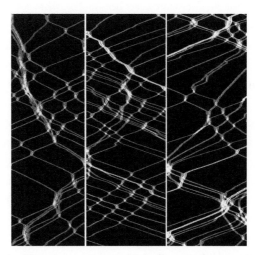

Fig. 10.6 Paths of ants. Thanks are due to Nick Gravish for permission to use this image from his Georgia Tech thesis.

---

[9]I have had unfortunate experience along these lines. There's something known as the "exponential disk" [52, 54] that comes up in astronomy. This refers to the light distribution from galaxies. The point is that the distribution in almost all cases is good (i.e., fits the data) for about 1 1/2 decades. That's not enough (although most card-carrying astrophysicists believe in it). An alternative was found, which also fit the data, in the form r^{-1}- const [207] and which had motivation (in my opinion) no less compelling than what appears in the citations given earlier. (Note that in the astrophysical case one deals with a semi-log plot, whereas the genetic data are log–log. But the point is that 1 1/2 decades of data can lead to an accidental fit.) See also [219] on power laws in general.

decades, although as pointed out in Chapter 7 the log-normal distribution can mimic a power law over several orders of magnitude. As I mentioned, power laws (or what looks like power laws) are everywhere. Ueda et al. [225] plot gene expression versus the rank of a particular gene and get power laws for several species, from humble bacteria (*Escherichia coli*) to people (*Homo sapiens*). Their study of the dynamics of genes is less compelling, extending over fewer decades and curving to some extent. In the book edited (and partly written) by Koonin, Wolf and Karev [129] quite a few examples of scale-free networks are given, including examples where the distribution of the number of nodes is $(\Pr(k) \sim k^{-\alpha}$, with k the number of edges possessed by a given node and α a number, typically near 2) also a power law. Note that in some of the graphs *more than* four decades of data are given and the appearance of a power law may be for as much as 10 decades. The slope typically lies between 1 and 3, but for most cases it's about 2. Of course it depends on what you're plotting. If you look at a cumulative distribution (e.g., $\int dx \, x^{-n}$) you get a higher power by 1 (so $\int dx \, x^{-n} \sim x^{-n+1}$). Gerstein and collaborators found power laws in folds (folding is the process by which a protein—manufactured in one dimension—assumes its "native" form, i.e., the three-dimensional form in which it acts). In other words, there are popular, less popular and rare folds, and they follow a power law. They also found power laws in other properties and found an overall resemblance to Zipf's law (Chapter 7). See [77, 99, 183]. Similarly the book edited by Koonin, Wolf and Karev has many examples of genes, proteins, protein networks and so on, following power laws.

It should be remarked that three or four decades may be as good as most data get. For example, the data for the coast of England, allegedly a fractal, originally [187] had data for only three or four orders of magnitude, and often not even that. A more recent calculation is that of Feder [57] and is good for slightly more than three decades. The data there are impressive in that they really follow a straight line (on a log–log plot); the genetic data are not as good.

The book by Koonin et al. [129] also contains an article that suggests skepticism. Wagner, in Chapter 4 of [129], attributes the various regularities observed to "mere chemistry." That is, other chapters find the power laws an indication of how life formed, deep properties [128], while Wagner suggests that non-living chemicals behave the same way. I am not one to judge, but the reader should keep in mind doubts about the thesis.

In Chapter 10 of [129] Gerstein and coauthors give an argument for the formation of a power law. It is very much like the "rich get richer" argument given earlier (see Sec. 7.3). The calculation seems acceptable, but the data (their Fig. 1) only extends over two orders of magnitude (and that's generous).

Finally, I want to report some of my own research [197]. It's primitive, but it attacks a particularly difficult problem and does not portend well for the human race. (Of course this only adds one to the list of things that do not portend well—take climate change, for instance.) It uses a model invented by H. J. Jensen and colleagues [97], which is interesting in its own right (it's a kind of evolution) and will now be described before I get to power laws.

A genome, according to this model, is a string of zeros and ones. Never mind that in DNA the smallest unit of information is 2 bits[10] (not 1), and—more importantly— even the shortest genomes are much longer than are studied in this model (a string of 20 0's and 1's already strains the computer).

Remark: Good news and bad news. The good news is that some of the properties of the genome, such as—as will be described below—degree of fecundity, are subsumed in genes not mentioned. So the complete genome can be thought of as larger, but we're focusing on a part that will distinguish species. The bad news is that the notion of species is muddled. Does each genome represent a species? After all, creatures of the same species have a variety of genomes. (Your fingerprints are different from mine, your blood type, eye color, skin color may be different—the list goes on.) Perhaps the various genes in a single species are like fecundity, part of the genome not among those studied. In this way the ones studied are particular to a species.

Leaving aside reservations, say each string is L units long. Then you have 2^L different genomes, and there is a 2^L-by-2^L matrix of preferences: which genome likes or dislikes having particular genomes around (this will become clear when I calculate the probability for reproduction). For convenience this matrix (call it J) is set between -1 and 1 and will be scaled later. A state of the system is $(n_1(t), n_2(t), \ldots, n_{2^L}(t))$; the subscript refers to a given genome and each n_k is an integer from 0 and up, indicating the number of that genome (k, with $k = 1, \ldots, 2^L$) alive at time-t.

The model then gives rules for death, reproduction and mutation. You randomly pick an individual (so genomes having a greater number of that genome alive are more likely to be picked). With some probability, p_{kill}, it is killed. If it is not killed, it can reproduce. Let the probability of this (assuming it is not killed) be $p_{reproduce}$. Now suppose it is one of those that reproduces. Then there is a chance, p_{mutate}, that there will be a copying error and there is a mutation, so the offspring will differ by one or more entries from the 0's and 1's of its ancestor.

[10]A bit is the smallest unit of information: yes or no, zero or one, etc.

The heart of the model is $p_{reproduce}$. This depends in an interesting way on who else is alive, and this is where the matrix of preferences, J, plays a role. You first calculate a quantity

$$H_\alpha \equiv \frac{1}{cN} \sum_\alpha J_{\alpha\beta} n_\beta - \mu N, \qquad (10.8)$$

where α and β are genome labels, $N \equiv \sum n_\beta$ and c and μ are new parameters. The quantity c is a normalization for J and μ is a kind of chemical potential. Note that expression (10.8) is neither extensive nor intensive (the first term stays the same if all is doubled (intensive) while the second term doubles). This may be typical of complex systems that have a definite size. In any case, having defined H_α,

$$p_{reproduce} = \frac{1}{1 + \exp(-H_\alpha)}. \qquad (10.9)$$

The quantity p_{mutate} is a given number, and, for convenience in allowing different L values, is taken as p_0/L, for some p_0. Given that the creature (bacterium or whatever) reproduces, this is the probability that its genes differ from its parent.[11]

There are two ways to produce J, one is simple and the other complicated (but saves computer time). However, they give qualitatively similar results. (The simple way is just to take random numbers between −1 and 1.) Then you randomly set three-quarters of the entries to zero (in both ways).

Remark: Although this model studies the action of a few genes (L or $L/2$ of them) the non-studied genes determine preferences, kill probabilities and so on.

That's it. The model is simple but naturally gives punctuated equilibrium, as seen in the fossil record, without an asteroid striking the planet.[12] A typical outcome has a few dominant genomes for a long period, followed by their downfall, a period of fibrillation, followed by another period of relative stability. It is unknown how many metastable states of the form I've described exist. Fig. 10.7 shows a typical output.

My own model adds a few bells and whistles to capture bacterial resistance to antibiotics. The assumption is that a special segment of the genome may carry a particular pattern. If the bacterium has this pattern it is destroyed by the antibiotic.

---

[11]One can have the mutations respect the phenotype to some extent by correlating J with the Hamming distance of a proposed change in genotype. This is described in [197] and other sources.

[12]This is not to say that an asteroid did *not* hit the planet. There's good evidence that it did, for example the presence of tektites. However, Jensen's model is an additional source of mortality. There are also those who claim that the dinosaurs were already (before the asteroid hit) on the way to extinction.

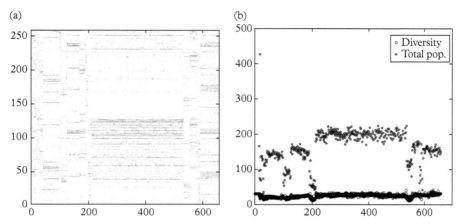

Fig. 10.7 Time (horizontal axis) is given in generations (about 15,000 time steps). (a) Occupation of each of 256 genomes. The actual state $(n_1, \ldots, n_{2^L})$ is not indicated, only whether $n_i = 0$ (no marker) or $n_i > 0$ (a dot) for $i = 1, \ldots, 2^L$. (b) Population and diversity. Both figures refer to the same run. (Diversity is the lower curve.) Values of parameters: $L = 8$, $\mu = 0.18$, $p_{\text{mutate}} = 0.05$, $c = 0.01$, $N_{\text{iterations}} = 1 \times 10^7$, $p_{\text{kill}} = 0.013$, generator of random numbers $= 10,756$. Note the reduction in population during periods of change.

The idea is that this part of the genome performs a vital function (say building an outer membrane) and this vital function is disrupted by the antibiotic. But the bacteria can fight back. They might have another portion of the genome that destroys the antibiotic. Or—and this is where things get interesting—they can engage in "HGT," horizontal gene transfer. In HGT bacteria acquire replacement parts that perform the vital function but are not attacked by the antibiotic or attack the antibiotic itself. In practice these pieces of DNA can be from other bacteria or from a soup of parts from dead bacteria. My model studies HGT where the length of the replacement DNA is the same at that lost, which is not realistic. Other features, such as the shortness of the genome, are also not realistic, but it's the best I can do; as far as I know, others have only dealt with this problem by mean field methods. An important variable is the timing of antibiotic administration, but this is too far from reality to be of clinical significance, except perhaps to stress the importance of a protocol for antibiotic administration.

Power laws enter when I study the variety of genomes available. You start with some population and let the system run for a while. That way you get a typical distribution of genomes in one of the metastable states. You do this many times (750 in my case). Then you take the logarithm of the numbers n_α for all non-zero n_α. You bin these in a histogram and plot against the logarithm of the histogram edges. The result is a straight line for maybe three decades (for the L's used in my simulations).

In other words, since the (absolute value of the) power (the slope of the line) is not high, *rare genomes* occur. Without the log–log plot of course they'd be buried in the dust, so the plot emphasizes them—but they are there. Why is this dangerous? It's rare genomes that provide "solutions" to the destructive power of antibiotics. In other words, there is often a way to provide an alternative to some vital function, and some bacterium will possess it. That bacterium will then prosper and pass on its resistance to other bacteria.

My take on this is not so much to convince me to worry about antibiotic resistance—this is already worrying—but as a confirmation that Jensen's model is realistic in this way also.

10.4   Flocking

This subject began (as far as I can tell) with an effort to fool people and reduce computational demands. The problem was to simulate the flight of a flock of birds on a computer screen, and Reynolds [186] was able to do this in a fully realistic fashion using minimal information. The objects doing the flying (or flocking, which includes fish, antelopes, etc.) are called "boids"; in computer-speak bird-oid became boid became a Brooklynite's pronunciation of bird (Brooklyn, New York, United States). Each object, each "boid," had only three rules:

- Stay separated. Steer to avoid crowding other boids.
- Stay aligned. Steer toward the average heading of nearby boids.
- Cohesion. Steer to the average position of nearby boids.

In this way Reynolds created a wonderful example of *emergent* behavior: you don't tell them to flock but they do! Or to put it as Grégoire et al. [90] did: "Moving and staying together without a leader."

Much of the work in this area relies on computer simulations, so I can tell you the results, but understanding is another matter,[13, 14] although as we'll see there exist several treatments of this problem, some of which use an intuition backed by equations.

---

[13]There's an (alleged?) quotation from Wigner, who was at a seminar during which computer results were presented. He commented, "It is nice to know that the computer understands the problem. But I would like to understand it too." Eugene Wigner, 1902–1995, Nobel Prize, group theory, nuclear, condensed matter, foundations. As a Hungarian–American he was fierce in his political views (e.g., he was instrumental in having Einstein write the letter to Roosevelt that started the atomic bomb project). At seminars he would make humble inquiries before sinking in the dagger.

[14]I have just been re-reading H. Bloom's *The Book of J* [28], only to learn that Wigner was *ironic*. According to Bloom, *ironic* comes from the Greek ειρνεια (eironia, dissembler) and is "Socratic: a feigned ignorance and humility designed to expose the inadequate assumptions of others." (Bloom, on page 25, also discusses other meanings of the term, but this meaning is suitable for the case at hand.)

With regard to Reynolds' "rules," at the moment they are incomplete, because you need to define "nearby" and other parameters in those instructions. I won't do this, since I'll give details for other models. (See `http://www.red3d.com/cwr/boids` for further information; accessed August 2020.)

This result (and others[15]) started an entire industry in the statistical physics community. An early model is due to Vicsek and collaborators [227] and keeps track of the position and velocity of each object—bird, fish or whatever. Each bird (let's call it that) has a position, x_i, and velocity, v_i, where i runs over the particle (bird) labels. The positions and velocities are simultaneously updated at intervals Δt according to the following rules. Position is simple:

$$x_i(t + \Delta t) = x_i(t) + v_i \Delta t. \tag{10.10}$$

The velocity has a more complicated rule. Its magnitude though is simple: it's the same for all birds, a quantity v_0, a parameter of the model. (This is justified by observation.) Its direction though is where details of the model kick in. You define the neighbors of a given bird, i, as all those birds within a distance R of i, where R is another parameter of the model. For all neighbors, including i, you calculate the average direction. First you evaluate the average velocity of the neighborhood, $\langle v_j(t) \rangle_{|x_j - x_i| \leq R}$, and then turn it into a unit vector (to be denoted $\widehat{\langle v_i(t) \rangle}$) by dividing by its magnitude. Next you perturb $\widehat{\langle v_i(t) \rangle}$. In two dimensions if the angle in which this unit vector points is $\theta_i(t)$[16] then you rotate it by a random amount:

$$\theta_i(t + \Delta t) = \theta_i(t) + \Delta_i(t), \tag{10.11}$$

where Δ_i is drawn from a uniform distribution on $[-\eta\pi, \eta\pi]$, with η another parameter of the model. (In a higher dimension you'd use a solid angle.) Once rotated, the new velocity is the rotated direction times the single magnitude of velocity, v_0. To summarize, the new velocity is $v_i(t + \Delta t) = v_0 \cdot \left(\text{rotated } \widehat{\langle v_i(t) \rangle} \right)$. One last parameter— an important one—that enters is the overall density of particles, call it ρ. Thus the parameters of the model are ρ (density), η (measure of randomness in velocity), v_0 (speed), R (radius of influence) and of course $\{x_{i0}\}$, the initial positions of all particles.

Is this what birds do? I don't know, but let's see the implications of the model. In the original 1995 paper, four images are shown (Fig. 10.8). Clearly for some parameter

---

[15]There is early literature of simulations like that of Reynolds and also journal articles that preceded or came out about the same time as [227]. The literature is enormous and I make no attempt at completeness or at establishing priorities. Here is a sample: [10, 90, 112, 223]. There were also observational studies of how birds actually flock. This is described in an article by Pomeroy and Heppner [179], which includes many references to observations that preceded Reynold's work.

[16]In two dimensions this would be the arctangent of the ratio of the y-component to the x-component of $\widehat{\langle v_i(t) \rangle}$.

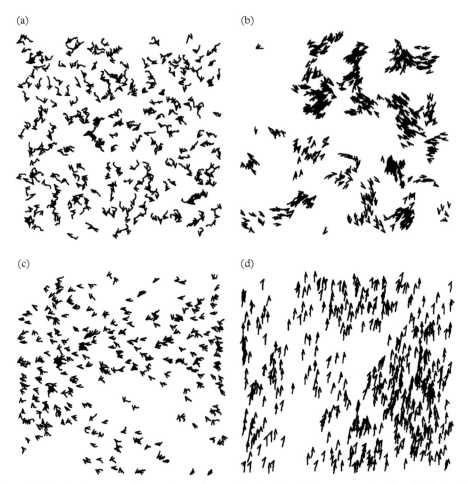

Fig. 10.8 This four-part figure shows the results of simulations at varying levels of density and noise. Since all velocity magnitudes are the same ($v_0 = 0.03$), only the direction need be shown. In addition for each particle the trajectory for the last 20 steps is shown as a continuous curve. The interaction distance is $R = 1$ and for all images the number of boids = $N = 300$. Varying the size of the two-dimensional square (area L^2) accomplishes the density changes. (a) shows the initial conditions for $L = 7$, $\eta = 2.0$. (b) uses $L = 25$, giving a much lower density, and $\eta = 0.1$, so there is also less noise. Under these conditions you get coherent groups moving in random directions. In (c) one again has higher density ($L = 7$) and higher noise ($\eta = 2.0$), but this is taken at a later time than in (a). There are still significant correlations among particles. The overall direction of motion is random. Yet higher density ($L = 5$) is shown in (d), but with lower noise, $\eta = 0.1$, leading to more ordered motion. From [227]. Source: *Physical Review*.

values there is flocking. In particular Fig. 10.8d most closely resembles a flock. There is relatively little noise and less variation of angle. Unfortunately it's not like flocks I have seen. Then again, I've never seen a two-dimensional flock, so this may be the best result possible. (But see below for a discussion of turning speed.)

Of course the scientific instincts of the investigators kicked in and besides showing the desired qualitative features they also studied the nature of the phase transition that occurs as you pass from the flocking state to the disordered one.

More recently a lot of attention was paid to what flocks actually do. It was found in particular that response to a change in direction (which is a way to avoid predators) travels linearly (in time) from the first bird to initiate the change to the furthest reaches of the flock. All this happens on the scale of milliseconds. The Vicsek et al. model predicts diffusive behavior, that is, the change would travel like the square root of the time.[17] If you've ever seen a cloud of fish suddenly change direction you would be convinced that the changes are not diffusive. To get linear behavior Attanasi et al. [12] introduce an inertial term into the equations of motion. Like Vicsek et al. they take the actual turning motion to be in a plane ; in other words, all the birds (or fish presumably) do their turning in a single plane. This allows each bird to be described by a single angle, ϕ_i, where the i refers to a particular bird. This assumption—that all birds turn in a certain plane—is backed up by observation.[18] There is also a question: *When* does a given bird turn? The authors have an answer to this (at maximum acceleration) and use this information to calculate how much time elapses between the turning of any pair of birds. Once this is known it is easy to realize that the turn begins locally—one bird decides to make a sudden turn—and the others follow, the message spreading rapidly and linearly (with distance) in time.

I would be curious if there is a sub-collection of "adventurous" birds, that is, whether any particular birds are likely to be the initiators of a turn. As far as I know, this has not been studied (and when you have thousands of starlings, how can you tell which is which?). Another point I'd be curious about has to do with ants. As

---

[17]The argument for diffusive behavior goes like this. If the noise is small then the flock is coordinated. This, plus the observations that the speed of each bird is about the same and when turning they all stay in a plane, allows one to write $v_i = v_0 \exp(i\phi_i)$, where the vector is now a point in the complex plane. (Noise is also neglected in this approximation.) This leads to a Hamiltonian of the form $H = -J \sum_{\langle ij \rangle} (\phi_i - \phi_j)^2$, where $\langle ij \rangle$ means sums over pairs with i and j "near" one another (which can be interpreted either as distance or as some other kind of neighborhood). When bird-to-bird distances are small this means the angle satisfies $\frac{\partial \phi}{\partial t} = -\frac{\delta H}{\delta \phi} = J' \nabla^2 \phi$. This is a diffusion equation, for which the change in angle travels as the square root of time elapsed, in contrast to the observed linear dependence on position.

[18]The authors of [12] take great pains to describe their equipment. And while scientific progress is often the result of improved technology, I'm not the one to describe the properties of their cameras and other machinery.

described in Sec. 10.2, some stand aside while others work. Which ones stand aside? Which work? Animals that flock presumably have more disparate genes than ants. Nevertheless in both cases some take initiative, some don't. Why?

The theory behind the linear dependence is counterintuitive and, as mentioned above, involves giving each bird inertia, a reluctance to turn. The numerical value is not high, but it suffices to turn a first-order equation (for ϕ_i) to a second-order equation, which supports a linear dependence of the turning time with distance. The equations are (see also [39])

$$\frac{d\vec{v}_i}{dt} = \frac{1}{\chi} \vec{s}_i \times \vec{v}_i \,,$$

$$\frac{d\vec{s}_i}{dt} = \vec{v}_i(t) \times \left[\frac{J}{v_0^2} \sum_j n_{ij} \vec{v}_j - \frac{\eta}{v_0^2} \frac{d\vec{v}_i}{dt} + \frac{\vec{\xi}_i}{v_0} \right] \,, \qquad (10.12)$$

$$\frac{d\vec{r}_i}{dt} = \vec{v}_i(t) \,,$$

where

- i labels the boid
- $\vec{s}_i$ and $\vec{v}_i$ are the three-dimensional spin and velocity of boid #i
- χ is a measure of the "inertia" for changes of the spin, how easy or difficult it is to change the spin value
- J is a coupling constant for the force between boids and n_{ij} ($=0$ or 1) which says whether there is a connection between boid i and boid j
- v_0 is the speed (|velocity|) of all boids
- η is a viscous coefficient and $\vec{\xi}_i$ a noise term. Each $\vec{\xi}_i$ is independent and identically distributed, and has variance

$$\langle \vec{\xi}_i(t) \cdot \vec{\xi}_j(t') \rangle = (2d)\eta T \delta_{ij} \delta(t - t') \,, \qquad (10.13)$$

with T a fictitious temperature and d the dimension, here 3.

n_{ij} has yet to be defined. It is the *connectivity* matrix. In simulations this was taken to be the six nearest neighbors of a given boid. There is a delicate balance in these equations. On the one hand, the constancy of the *magnitude* of velocity must come from non-conservative forces. (One assumes that ultimately Nature *does* conserve energy and the bird's food is the source of the delicate forces that keep speed constant. But, as far as the equations above are concerned, the velocity constraint is non-conservative.)

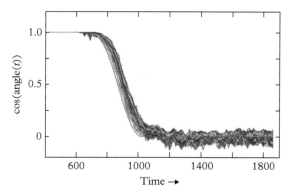

Fig. 10.9 The angle of a given bird relative to its initial angle. The first bird turns and the others follow. Simulations involve about 500 birds and each gets a line. (In the original image the lines are of various colors. Further details concerning parameters are also given although, as far as I can tell, the time is not. One can interpret the time scale as counting milliseconds. Two-tenths of a second for the turn seems about right.) Adapted from [39]. Source: *Journal of Statistical Physics.*

What promotes this from an (un-doable) exercise in classical mechanics is Noether's[19] theorem: rotation of a *spin* coordinate by a fixed amount leaves the Lagrangian unchanged. This implies that there is a constant of the motion. But the true measure of success comes from simulations and the authors are able to recover properties of bird flight. Their Fig. 4f is adapted here (Fig. 10.9). As you can see, the turn is rapid and, presumably, effective in turning away predators.

A decade after [227], equations had become more elaborate and the pictures were better too. If you look at a review article by Vicsek and Zafeiris [228] you can find beautiful pictures, both of natural flocking and of simulations (as well as equations and graphs). This review article cites over 300 articles and I'm sure that many have been omitted and that many more have appeared since 2012. The exact nature of the phase transition first found in [227] was challenged by Grégoire and Chaté [89].

Let me mention a related article that takes a previously mentioned approach, the maximum entropy method.

Bialek et al. [26] use observational data on starlings to deduce features of the flocking. For example, they conclude that, when a bird orients itself, it looks at a particular number of its neighbors rather than those within a particular distance. (This

---

[19]Emmy Noether, 1882–1935, an influential mathematician. In physics I would say her most important contribution is the theorem in the main text, namely that every symmetry of the action leads to a conservation law. As a woman she encountered severe prejudices. I was told an amusing story by (my thesis advisor) Arthur Wightman. When she was proposed as a professor at Göttingen some of the professors objected: What will happen to the professors' toilet? David Hilbert, her sponsor, responded: What are we running, a university or a locker room?

was adopted by [39] and is mentioned in Fig. 10.9.) But what I find interesting about what they do is their technique. They use the maximum entropy methods discussed in Sec. 6.2. Finally, in contrast to other work, they are not concerned with finding emergent behavior from simple rules (although in a sense this is recovered), but rather with a way of describing a rich data set. They take as the observational data the local pair correlation of bird velocity directions. Their situation is complicated by the fact that birds on the outer edges of the cluster behave a bit differently from those inside and also by the fact that they have more than one parameter to adjust. But finally the basic principle is that enunciated here.

My own view is that the analytical work is wonderful, but the exciting part is the emergence of collective behavior from local rules, only. That was already present in the work of Reynolds [186]. There is certainly a lot to learn from the nature of the transition (another topic discussed by [39] and other writers), and there are generalizations to hydrodynamic models, so this work has more general implications. I mention in passing that the flocking phenomenon shows that emergence and power laws are independent notions. As to universality, the phenomena studied by Cavagna et al. fall into classes of *dynamic* universality, that is, behavior that is both classifiable and dynamic. For example, the boid model of [39] turns out to be in the same universality class as the dynamics of liquid helium.[20] However, this topic is not pursued here and I refer the reader to [39] and references therein.

And finally an observation about geese. From my former home in northern New York, I watched flocks of them fly south in the fall and come back in the spring. They fly in loose formation, gigantic V's or occasionally W's, across the sky. There are not thousands, like starlings; sometimes there are two or three, sometimes 70 or more. They also make a lot of noise, honking, so that from far away you know they're there. Why? I don't know why they make noise but there is an explanation for the form. Apparently [180] there is an aerodynamic advantage to the V formation, the birds following have to expend less energy to fly in the wake of the leaders than do the leaders themselves. And they do take turns being the leader. But I have yet to hear a comprehensive explanation of their behavior.

10.5   Kuramoto model

There is a model of phase coherence that applies to lots of systems, so it's a bit arbitrary to include it in the section on biology. The model is based on work by Kuramoto [133] and is related to earlier work by Winfree [234]. It only addresses phases, but

---

[20]This would be a mapping of the hydrodynamic model onto the study of liquid helium. There the quantum phase corresponds to the flight direction and the superfluid component is the spin.

apparently lots of systems can be described this way, for example the flocking of birds, the synchronization of cardiac tissue and the flashing of fireflies [152].

In a way this is a take-off from earlier work on fireflies. The difference is that the earlier work was focused on Peskin's mechanism [175], known as "fire and integrate." When one of the components (cells, for example) reaches its threshold (say it went from 0 to 1, so 1 would be the threshold) it fired, and every other cell close to 1 (say greater than $1 - \varepsilon$) would also fire, and all those that fired would be set to 0 and start increasing again. This does lead to synchronization, although (as far as I know) this was only proved if all cells increased at the same rate, which is an artificial assumption.[21] In the following section we'll continue Sec. 10.1, where, near the end, we generalize to allow greater biological variety, as in reality.[22]

In the simplest form of the Kuramoto model you have N different oscillators, each described by a phase, $\theta_i(t)$, $i = 1, \ldots, N$. Each oscillator has its own natural frequency, ω_i, but they are coupled to one another. The coupling is where these models diverge. The easy form is mean field: all affect all, and the simplest of the simplest is the sine function, so that

$$\dot{\theta}_i = \omega_i + \frac{K}{N} \sum_j \sin(\theta_j - \theta_i), \qquad \text{for } i, j = 1 \text{ to } N. \tag{10.14}$$

K is a parameter and $\{\omega_i\}$ are chosen from some distribution, usually either the normal $(\Pr(\omega) \sim \exp(-(\omega - \omega_0)^2 / 2\sigma^2))$ or Cauchy $(\Pr(\omega) \sim 1/[(\omega - \omega_0)^2 + \sigma^2])$ distributions, with ω_0 and σ additional parameters. If K is large enough this leads to synchronization. That is, all frequencies become the same, at the mean of the various frequencies. Moreover, as K increases yet further, not only do all oscillators have the same frequency but they approach a common phase as well.

This is a generalization of Eq. (10.6), in that instead of putting the sum inside the sine function (recall, $\Theta = (1/N)\sum \theta_j$) you sum the sine of the various angles. (This makes more sense, as the previous definition led to ambiguities when θ went through multiples of 2π.)

Again for large enough K, once they have synchronized, recovery from a disturbance (a perturbation) is more rapid. For some values the initial synchronization takes 1 1/4 time units, while, once in synchrony, a perturbation of phases allowed recovery in about 1/4 of a time unit.

---

[21]I played with these equations numerically and did not find perfect synchronization. A lot depended on the fixed constant at which a cell swept into synchrony with others. The spread in natural frequencies was less important, although if it was too large one did not get synchrony.

[22]Peskin had other assumptions in his quest for rigorous results. However, my impression is that he really took into account the physiology of the heart and he had results beyond the conclusions I've so briefly reported.

For large N there's an analytic solution that can be useful. Define an order parameter $re^{i\psi}$, where r is non-negative and ψ real, between 0 and 2π. The definition is

$$re^{i\psi} \equiv \frac{1}{N} \sum_j e^{i\theta_j} . \tag{10.15}$$

Multiply by $e^{-\theta_i}$ to get

$$\text{Im}\left(re^{i(\psi - \theta_i)}\right) = r\sin(\psi - \theta_i) . \tag{10.16}$$

Using the order parameter and only the imaginary part of the equations of motion one obtains

$$\dot{\theta}_i = \omega_i + Kr\sin(\psi - \theta_i) . \tag{10.17}$$

This now looks very much like Eq. (10.6), with A taking the place of Kr. The difference though is essential: r is a consequence of the ordering. It takes the value 1 when the system is fully ordered, zero when not synchronized at all. It is also possible for r to take intermediate values. So Eq. (10.17) should be handled self-consistently. If $r = 0$ obviously there is no synchronization at all and every oscillator varies with its natural frequency. If $r = 1$ (or, in practice, is close to 1) then whether or not this is a solution clearly depends on K. Sufficiently large K leads to synchrony of all or, if $r < 1$, partial synchrony.

There is also a solution for $N \to \infty$ [2], but, frankly, I don't find it particularly revealing. You are free to read about it in the cited paper.

Here is a program that implements the Kuramoto model:

```
function out=KuramotoMeanField(in)
if ~exist('in'), in=struct; end % If the program is run without input
% Kuramoto model
% Command line for figures: (after clearing "in", if necessary)
% in.N=47;in.T=1;in.sig_om=1.5;in.ratio=.05;in.dt=.001;out=KuramotoMeanField(in);
global ratio oms N            % Makes these variables "global,'
                              % i.e., their value is known in a subprogram
                              % The subprogram also needs a global command

[ratio,in]=setdefault(in,'ratio',1);     % K over N
[N,in]=setdefault(in,'N',10);            % number of oscillators
[dth,in]=setdefault(in,'dth',.01);       %
[om0,in]=setdefault(in,'om0',1);         % mean oscillator frequency
[sig_om,in]=setdefault(in,'sig_om',pi/2); % std of frequency
[dt,in]=setdefault(in,'dt',.1);          %
[T,in]=setdefault(in,'T',30);            % total time
[type,in]=setdefault(in,'type',1);       % type of function in model
```

```
[seed,in]=setdefault(in,'seed',101);        % used to set random number generator
in.K=ratio*N;
out.input=in; clear in
% All parameters are now defined.
rng(seed)                                    % sets random number generator
oms=om0+sig_om*randn(N,1);                   % normal distribution of omegas
th0=2*pi*rand(N,1);                          % initial phases
Ts=0:dt:T;nT=length(Ts);
% The integration  (the "type"s correspond to various equations of motion)
if type==1,
    [t,ths]=ode45(@kur,Ts,th0);
elseif type==2,
    [t,ths]=ode45(@kur_cos,Ts,th0);
elseif type==3,
    [t,ths]=ode45(@kur_h,Ts,th0);
end
ths=mod(ths,2*pi);
% Viewing the results
msz=6;if N>5,msz=4;end,if N>9,msz=3;end      % size of markers
figure(102), plot(t*ones(1,N),ths,'o','markersize',msz),axis('tight')
u=(1/N)*sum(exp(i*ths),2);                   % order parameter and phase
[phi,r]=cart2pol(real(u),imag(u));
figure(108), plot(t,r,'k')
figure(110), plot(t,phi,'k')
out.results=[t,ths];
%---------------------------------------------
function thdot=kur(t,th)
global ratio oms N
th=mod(th,2*pi);
thdot=zeros(size(th));
for k1=1:N,
    thet1=th(k1);
    for k2=1:N,
        thet2=th(k2);
        thdot(k1)=thdot(k1)-ratio*sin(thet1-thet2);
    end
end
thdot=oms+thdot;
%---------------------------------------------
function thdot=kur_cos(t,th)
```

```
global ratio oms N
th=mod(th,2*pi);
thdot=zeros(size(th));
for k1=1:N,
    thet1=th(k1);
    for k2=1:N,
        thet2=th(k2);
        thdot(k1)=thdot(k1)-ratio*cos(thet1-thet2);
    end
end
thdot=oms+thdot;
%----------------------------------------
function thdot=kur_h(t,th)
global ratio oms N
% h is defined within this program
% In the present case it's a combination of cosines.
th=mod(th,2*pi);
thdot=zeros(size(th));
for k1=1:N,
    thet1=th(k1);
    for k2=1:N,
        thet2=th(k2);
        thet=thet1-thet2;
        thdot(k1)=thdot(k1)-ratio*[-.1*cos(thet)+2*cos(2*(thet))];
    end
end
thdot=oms+thdot;
%----------------------------------------
function [a,in]=setdefault(in,a_string,a_default);
% Example: [x,in]=setdefault(in,'x',5); creates x & in.x & sets them to 5.
% "in" MUST already exist as a structure.
if isfield(in,a_string),
    eval(['a=in.',a_string,';']);
else
    a=a_default;
    if nargout==2,
        eval(['in.',a_string,'=a;'])
    end
end
```

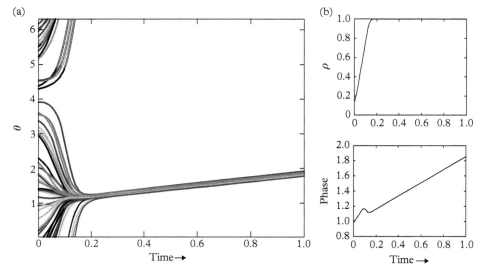

Fig. 10.10 Output of the program. There are $N = 47$ separate oscillators, which for the parameter values in the command line (given in the program) yield synchronization in 0.2 seconds (a). This is also reflected in the order parameters (b). However, as control parameters vary the system may erupt into chaos. For example, weakening K or increasing the range of frequencies (sufficiently) will lose synchronization.

The figures that result from this program are illustrated in Fig. 10.10. By varying input parameters one can sharpen synchronization or lose it completely. In particular, "type-2" uses the cosine instead of the sine and leads to chaos. For type-3, the function h is at the discretion of the programmer. (At present, it's a combination of cosines.)

There are many, many generalizations of the model. For example, you can posit some connectivity. You can put the system on a lattice and, say, only nearest neighbors influence each other. In two dimensions with a square lattice this would give four connections, or influencers, for each point (taking periodic boundary conditions). In that case, the usual practice is to divide K by the number of connections (four in the example), rather than the total number of spins. Another generalization is to replace the sine function by something else, call it a function $h(\theta)$. Then the equation of motion reads

$$\dot{\theta}_i = \omega_i + \frac{K}{n_i} \sum_j c_{ij} h\left(\theta_j - \theta_i\right), \tag{10.18}$$

with n_i the number of connections angle-i has. Thus each c_{ij} is either 0 or 1 and $n_i = \sum_j c_{ij}$. If h is an even function (e.g., cosine) there is no synchronization. (Apropos the program, KuramotoMeanField.m, just displayed, connectivity can easily be incorporated.)

However, for some applications it is appropriate to put in a delay, that is,

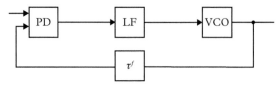

Fig. 10.11 The driving force is the current delivered by the voltage control oscillator (VCO). Current enters from the left, both the feedback associated with this particular device (to be known as $x_k(t)$ but delayed) and that of all the others (the collection $\{x_\ell(t)\}$). What is shown is a simplified diagram of one device (a phase locked loop, PLL), consisting of a phase detector (PD), a loop filter (LF), a VCO having a natural frequency ω_k, and a delayer (τ^f, the amount of feedback delay). This diagram is simplified in that a voltage frequency divider and phase shifter are absent. In particular the voltage divider is necessary since the frequencies associated with the VCO are about 24 megahertz, frequencies that the other components would have trouble handling. See the main text for a description of the operation of this device and its relation to the Kuramoto model. For more detail see Fig. 10.12. Figure due to L. Wetzel.

$$\dot{\theta}_i(t) = \omega_i + \frac{K}{n_i} \sum_j c_{ij} h \left(\theta_j(t - \tau_{ij}) - \theta_i(t) \right) , \tag{10.19}$$

with τ_{ij} the delay for a signal from j to reach i. In general this will give synchronization even for cosine (and for other even functions) since $\theta(t - \tau) \sim \theta(t) - \bar{\omega}\tau$ for some $\bar{\omega}$ and the cosine addition formula would contribute a sine function. It is also possible to add noise to the system, so there is a random spreading. The model often used is

$$\dot{\theta}_i(t) = \omega_i + \xi_i(t) + \frac{K}{n_i} \sum_j h \left(\theta_j(t - \tau_{ij}) - \theta_i(t) \right) , \tag{10.20}$$

with the expectation $\langle \xi_i(t) \rangle = 0$ and $\langle \xi_i(t)\xi_j(s) \rangle = D\delta(t - s)\delta_{ij}$ with D related to temperature.

Finally, higher order models can also synchronize. In the next example we find that a second derivative of the phase enters, and there is still synchronization.

The model describes electrical circuits. Some of the biological models have already been mentioned, so I will focus on a different sort. Lucas Wetzel of the Max Planck Institute for the Physics of Complex Systems (in Dresden, Germany) together with colleagues (see citations) has studied a way to get emergent coordination among disparate circuit elements without the help of a "reference clock." This—when all the details are put in, a feature that I will ignore—is a practical application of the Kuramoto equations. The features of his system [178, 232] are illustrated in Fig. 10.11. A table of acronyms used is given in Table 10.1.

As indicated in Fig. 10.11 the illustrated device, known as a phase-locked loop (PLL), is one of many (nine PLL's in the examples studied, but, in principle, many

Acronym	Stands for ...	Comment
PLL	Phase locked loop	Basic unit of oscillator
VCO	Voltage control oscillator	What drives the whole thing
LF	Loop filter	Kills wrong phase & transforms the signal
PD	Phase detector	Acts as multiplier
τ^f	Delay	Delay of the signal from this PLL In some applications there are other delays
INV	signal Inverter	Not used here

Table 10.1 Acronyms and short forms used in [178; 232].

more). The main signal is produced by the VCO, but other signals are involved (from other PLL's) so that all must be synchronous, without a reference clock.[23] The overall system might be a collection of cell phone towers that could be used to triangulate and obtain a precise location. (As of 2020 with towers synchronized only approximately, triangulation is only good to about 25 meters.) Current enters from the left and consists of $\{x_\ell\}$ as well as x_k. The component PD acts as a multiplier, forming the signals $x_\ell x_k$. We *assume* that each x_n can be written

$$x_n(t) = \sin\theta(t), \qquad (10.21)$$

with t an appropriate time. (The assumption is that an amplitude, A, as in $A\sin\theta$, is time-independent and can be taken to be unity.) Thus the signal coming out of PD is a sum of the form

$$\begin{aligned} x_k(t_k)x_\ell(t_\ell) &= \sin\theta(t_k)\sin\theta(t_\ell) \\ &= \frac{1}{2}\{\cos[\theta(t_k) - \theta(t_\ell)] - \cos[\theta(t_k) + \theta(t_\ell)]\}. \end{aligned} \qquad (10.22)$$

Next there is the action of LF, the loop filter. This is a device that to a good approximation kills the second term in Eq. (10.22) and gives the first term memory. To be precise, LF acts (approximately) on the output of PD as follows:

$$\begin{aligned} &\cos(\theta(t_k) - \theta(t_\ell)) - \cos(\theta(t_k) + \theta(t_\ell)) \longrightarrow \\ &\int_0^\infty du\, p(u)\cos(\theta(t_k - u) - \theta(t_\ell - u)), \end{aligned} \qquad (10.23)$$

where $\int du\, p(u) = 1$. What I have written as t_n $(n = \ell, k)$ includes an appropriate delay. Thus x_k is delayed by τ^f while x_ℓ has a transmission delay, in which for simplicity we do not include an ℓ dependence. Thus $t_k = t - \tau^f$ and $t_\ell = t - \tau$, for some τ.

---

[23]I'm giving you the easy version. Lucas pressed me to give more detail, which, while important in practice, might be distracting. First, in Fig. 10.11, the output signal from the VCO is sent *back* through the delayer and is fed into the PD, the same as external signals. (That much should have been clear, but it doesn't hurt to repeat it.) Also, I'm cavalier about the inverse Laplace transform. The fact is you need to use the initial condition $x_k^C(0)$, the initial state of the LF.

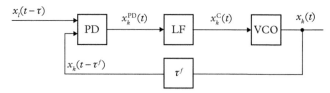

Fig. 10.12 More detail of a PLL (phase locked loop), described in the text. This is a single unit, of which there are, in general, many. Figure due to L. Wetzel.

At this point we make two observations. First, the function $p(u)$ can be taken to be of the form [38]

$$p(u) = \frac{u^{a-1}\exp(-u/b)}{b^a\Gamma(a)}. \tag{10.24}$$

And second, Eq. (10.23) is the Laplace transform of a product. And now for tractability, a is taken to be equal to 1, so that the Laplace transform, for $a = 1$, is

$$\hat{p}(\lambda) = \int_0^\infty du\,\exp(-\lambda u)\frac{\exp(-u/b)}{b\Gamma(1)} = \frac{1}{b\lambda + 1}. \tag{10.25}$$

Fig. 10.12 shows more detail than Fig. 10.11. In particular the output current, $x_k(t)$, is sent back to PD, with a delay, and combined with (multiplied by) all other currents (signified by $x_l(t-\tau)$). The output of PD is called $x_k^{PD}(t)$ and is then acted upon by the filter (as indicated earlier) to give $x_k^C(t)$, thus

$$x_k^C(t) = \int_0^\infty du\,p(u)x_k^{PD}(t-u), \tag{10.26}$$

which in turn leads to

$$\hat{x}_k^C(\lambda) = (1 + \lambda b)^{-1}\hat{x}_k^{PD}(\lambda). \tag{10.27}$$

Applying the inverse Laplace transform to this expression (and ignoring constants that appear, since they disappear anyway) we get

$$b\dot{x}_k^C(t) = -x_k^C(t) + x_k^{PD}(t). \tag{10.28}$$

Now we recall that

$$\dot{\theta}_k(t) = \omega_k + Kx_k^C(t), \tag{10.29}$$

and (for comparison with Kuramoto's equations) it is good to remember that although x_k^C carries a subscript k it has actually been combined with lots of other currents ($\{x_\ell\}$

for $\ell = 1, \ldots, N$). The time derivative of Eq. (10.29) is combined with Eq. (10.28) to yield

$$\frac{b}{K}\ddot{\theta}_k(t) = -\frac{\dot{\theta}_k(t) - \omega_k}{K} + x_k^{PD}(t). \tag{10.30}$$

Now $x_k^{PD}(t)$ contains cosine terms, but because they are delayed they are able to synchronize. To see this examine

$$\frac{1}{\xi}\ddot{\theta}_k(t) + \dot{\theta}_k(t) - \omega_k - \frac{K}{n_k}\sum_\ell c_{kl} \cos(\theta_\ell(t - \tau) - \theta_k(t - \tau^f)) = 0 \tag{10.31}$$

to find that it synchronizes! (As in Eq. (10.19), $n_k = \sum c_{k\ell}$ and $c_{k\ell} \in \{0,1\}$.) The quantity ξ is a kind of inverse mass (inertia), and arises because a was taken to be 1 in Eq. (10.24).

This is a second-order form of Kuramoto's equations, but still has the property of synchronization. Note, by the way, that a mass was also inserted into Eq. (10.12), in that case to allow a linear dependence on turning times for the case of flocking.

10.6 Ecology

No one doubts that the natural world—a forest, a tubeworm colony, a coral reef— is a complex system. There are so many factors that it's remarkable that sometimes there's a steady state. Until recently many ecologists were concerned with niches. They followed Darwinian rules. If two species occupied similar positions in the "trophic chain" then one would out-compete the other and there would be only one species that would survive. It was a corollary of Darwin's finches. (The trophic chain is based on who eats whom.) Then along came Stephen Hubbell, who said that it didn't matter [110]. He proposed a "neutral theory" in which similar species coexist and the relative proportion of each is random. This was heresy, but ... it worked![24] Well, it worked some of the time. What it did all of the time was create controversy. (Also I won't go into the history. I believe Hubbell had predecessors.)

What I'll present is not Hubbell's work but a variant due to John Harte. (Some references are [88, 100, 101, 102, 103, 104, 233].) What I like about Harte's work is that it is based on the maximum entropy principle. I've noted though that there are lots

---

[24]I exaggerate. What I'm referring to is the "competitive exclusion principle," also known as Gause's law. This is the assertion that if two species have the same demands (what eventually was called a niche) then one would out-compete the other and eliminate it. This was known not to hold well before Hubbell, or, to put it differently, was known to have exceptions.

Acronym	Stands for ...	Comment
SAR	Species–area relationship	log-log plot, slope $< 1/4$?
EAR	Endemics–area relationship	
SAD	Species abundance distribution	
AER	Abundance–area relationship	
NBD	Negative binomial distribution	See Footnote 25
dbh	Diameter at breast height	Apparently general usage (including by a tree surgeon in Atlanta!)
RPM	Random placement model	
HEAP	Hypothesis of equal allocation probabilities	
MaxEnt	Maximum entropy	Proposed by Jaynes [116, 117]
BCI	Barro Colorado Island	
METE	Maximum entropy theory of ecology	A theory
ASNE	State variables A, S, N & E	Special case of METE
AGSNE	... also with state variable G	Special case of METE

Table 10.2 Acronyms and short forms used in [102, 104].

of acronyms,[25] familiar to the ecologist, but not to me—in any case, I've provided a table (Table 10.2). (And considering Sec. 10.5 I shouldn't complain about acronyms.) We won't use them all, but, just in case you decide to read the paper, the table should serve as a key.

The philosophy is that of MaxEnt (see Chapter 6), which is to say, the philosophy of Jaynes: aside from what you know, you maximize ignorance of everything else. In practice what you know must be embodied in constraints (which translates to Lagrange multipliers) and to say you maximize ignorance means the probability distribution maximizes entropy, $-\sum_\alpha p_\alpha \log p_\alpha$, or its continuum version.

Harte et al. [104] start from a number of state variables: A_0 = area, S_0 = number of species (restricted to some taxonomic group, such as trees), N_0 = number of individuals, E_0 = the total metabolic rate, and sometimes G = the number of genera. Then they define two probability distributions. One they call R depends on ε, the level of metabolic activity, and n, the number of individuals of that species. There are then three constraints on R:

---

[25] In the table, the negative binomial distribution is defined as follows. Let independent Bernoulli trials have probability p. Let X be the trial at which the rth success occurs. Then $P(x = X|r, p) = \binom{x-1}{r-1} p^r (1-p)^{x-r}$.

$$\sum_n \int d\epsilon\, R(n,\epsilon) = 1, \qquad (10.32)$$

$$\sum_n \int d\epsilon\, n R(n,\epsilon) = \frac{N_0}{S_0}, \qquad (10.33)$$

$$\sum_n \int d\epsilon\, n\epsilon R(n,\epsilon) = \frac{E_0}{S_0}. \qquad (10.34)$$

The first is normalization of probability; it had better add to 1. The second is the average number of individuals per species. And the third is the metabolic expenditure per species. A second family of probability distributions relates to area. For this you need to know the total abundance of each *species* (or to have their distribution function, which we will later obtain). Call this n_{0j}, the j labeling the species. You then impose (in the absence of other information on the subject) a uniformity with respect to area:

$$\sum_{n=0}^{n_{0j}} n P_A^{n_{0j}}(n) = n_{0j}\frac{A}{A_0}. \qquad (10.35)$$

From this information, with the help of Lagrange multipliers (as in Sec. 6.2 and Appendix G) we find

$$R(n,\epsilon) = \frac{\exp(\lambda_1 n - \lambda_2 n\epsilon)}{Z}, \qquad (10.36)$$

with

$$-\lambda_1 \log(\lambda_1) = \frac{S_0}{N_0}, \qquad (10.37)$$

$$\lambda_2 = \frac{S_0}{E_0} \qquad \text{with the neglect of } \exp(-\lambda_2 n E_0) \qquad (10.38)$$
$$\text{(justified in [104], Supplement),}$$

$$Z = \sum_n \int d\epsilon\, \exp(-\lambda_1 n - \lambda_2 n\epsilon), \qquad (10.39)$$

and

$$P_A^{n_{0j}}(n) = \frac{\exp(-\lambda^{(j)}n)}{Z_j}, \qquad (10.40)$$

with

$$x = \exp(-\lambda_P^{(j)}) \text{ and } x \text{ satisfies } \frac{mA}{A_0} = \frac{\sum_{n=1}^m n x^n}{\sum_{n=1}^m x^n} \text{ where } m = n_0^{(j)}, \qquad (10.41)$$

$$Z^{(n_0)} = \sum_{k=1}^{n_0} \exp(-\lambda_P k).$$

The foregoing equations are a straightforward application of the maximum entropy principle.[26]

[26]Eq. (10.37) is only good to $O(\lambda^2\log\lambda)$. But it is stated that the exact equations were used for the fits.

From these quantities a host of other—relevant—quantities follow. Let's say you know R. Then you'll know the distribution of species size by integrating:

$$\Phi(n) = \int d\epsilon \, R(n, \epsilon). \tag{10.42}$$

From Φ you know the distribution of species values, the n_{0j} that was mentioned above. So you know whether a given species is present—that's just $[1 - P_A^n(0)]$. With this information you know SAR (the species–area relation—see Table 10.2):

$$S(A) = S_0 \sum_{n=1}^{N_0} [1 - P_A^n(0)] \, \Phi(n). \tag{10.43}$$

According to Gotelli ([82], Ch. 7) this "SAR" (see the table) is one of the few genuine "laws" of ecology. However, it should be noted[27] that this does *not* imply a power law. Although the best fitting power law is given in his Fig. 7b this does not mean that the data provide a power law. What is true is the trend: the more area, the more species.

Among other quantities, Harte et al. [104] calculate the SAR. Several quantities, or steps, are needed. First there's the probability that there are n of a certain species. That's given by $\Phi(n)$ above, and when the form of R is inserted one gets

$$\Phi(n) = \frac{1}{\log\left(\lambda_1^{-1}\right)} \frac{e^{n\lambda_1}}{n}, \tag{10.44}$$

with λ_1 given by Eq. (10.37) above. Then, there is the probability that something is there, namely $1 - P(0)$. It follows that the SAR is given by

$$
\begin{aligned}
S(A) &= S_0 \sum_{n_0=1}^{N_0} \left[1 - P_A^{(n_0)}\right] \Phi(n_0) \\
&= S_0 \sum_{n_0=1}^{N_0} \left[1 - \frac{1}{\sum_{n=0}^{n_0} \exp(-n\lambda_P)}\right] \frac{1}{\log\left(\lambda_1^{-1}\right)} \frac{e^{n_0\lambda_1}}{n_0},
\end{aligned} \tag{10.45}
$$

These formulas give good agreement with SARs in the regions studied. In Fig. 10.13 I show the SAD relation. There is excellent agreement with the predictions of the MaxEnt theory. I have some reservations about the extent of the range of species, not quite 1 1/2 decades. Of course that may not be under the control of the experimenter/observer: it will depend on the number of species. Also of interest is the SAD, measuring abundance. There the range is larger (3.7 decades) and in good agreement with the predictions of MaxEnt.

---

[27] N. Gotelli, personal communication (email of September 25, 2020).

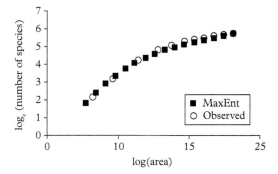

Fig. 10.13 Species-area relationship (SAR) for Barro Colorado Island (BCI). Note that the logarithm is to the base e so that the range of species is log 2.1 to log 5.8, which when translated to $\log_{10}$ is slightly under 1 1/2 decades ($\log_{10} = \log e \times \log$). Adapted from [104]. Source: Ecological Society of America.

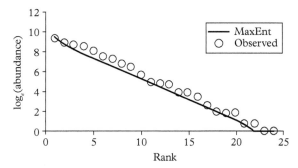

Fig. 10.14 Species-abundance relationship (SAD) for serpentine grassland. As in the previous figure the logarithm is to the base e so that the range of species is log 0.9 to log 9.4, which when translated to $\log_{10}$ is about 3.7 decades ($\log_{10} = \log e \times \log$). Adapted from [104]. Source: Ecological Society of America.

Am I happy? A little, but I have reservations. First, the data in favor of a power law for the SAD are weak. Gotelli backs off from this assertion and Harte goes for predictions of MaxEnt, which look a little like a power law but aren't really—recall the old maxim that *everything* looks like a straight line in a logarithmic plot. Of course that's not true, and Fig. 10.13 can stand as a counterexample. Fig. 10.14 looks better and Harte prefers to fit it with MaxEnt. Note though that it is *not* a power law since it is rank, not its logarithm, that is plotted (making the relation an exponential).

What's different about ecology? I think that this is a difficult problem. There's the complexity of the subject itself. What is an area? Are there any natural areas left? Do experimental plots with controlled environments count? There are many kinds of plants and animals, and each has its own dynamic. I don't want to detract from the many things that seem right. It's just that as you read, for example, Gotelli [82], with variations on the Lotka–Volterra model,[28] you realize that the subject matter does not always give definite answers. One problem though is that ecologists are often called upon to decide public policy—What happens if I fragment this area? What to do about some invasive species?—and they are forced to give answers when they know great uncertainty lurks.

Let me give some balance though to my treatment. There's an article [165] that traces the history of thermodynamics and ecology. For history I can recommend it. Unfortunately, despite the fact that it's from 2020, it doesn't mention Hubbell and as far as Harte is concerned he is cited but as one of a bunch of papers that present a controversy that "still awaits further discussion and resolution." The article, despite its historical qualities, omits those developments that I consider most promising.

10.7   Neurology

The subject (for our purposes) is built on the assumption that neuronal signals are on or off, that there are spikes when a neuron emits a signal, no spike when it doesn't. There is some judgement in this—you need to decide how long a spike lasts. If it's 100 ms then a lot of spikes will be counted as one. On the other hand, 0.5 ms will mean a single spike is counted multiple times. In practice, for the visual system about 5–20 ms is taken [213]. (But one of the sources for [213] seems to use 1 second [95], although that is for other purposes.) Bear in mind also that spikes can be both excitatory and inhibitory. What this means is that in some cases a spike will cause (or be part of the cause) another neuron to fire (excitatory) and in other cases the spike will act to prevent firing.

There is more to bear in mind: the power laws (and other possibly critical phenomena) are *without external stimulation*. That is, there is no external source for the excitation; the tissue emits a spontaneous signal. This is in the title to some articles, but after that it's understood.

---

[28]This is a model for the competition of two species. In its simplest form it is $\frac{dN_i}{dt} = r_i \left(\frac{K_i - N_i - \alpha N_j}{K_i} \right)$ with subscripts (to r, K, α and N) i and j taking the values 1 and 2 and the reverse. N is the total number of individuals, r is the rate of reproduction and K is the carrying capacity for the species in question. More complicated forms than αN_j can be used to express the competition. For example if i is some kind of cat and j its prey, then several of the prey species would be required for each cat. That would mean (at the very least) that α would need a subscript (i.e., would be different for the two species). One of the predictions of this model is a 3 year cycle of increase and decrease in lynxes (predator) and rabbits (prey). (See Gotelli [82] for examples.) I have a personal communication from Mark Kac to the effect that this variation could not be distinguished from randomness. (It's possible that this is published somewhere; I don't know.)

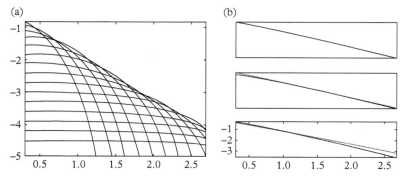

Fig. 10.15 All plots are log–log to the base 10. (a) Sum of exponentials. The exponents are the negative square roots of the numbers from 1 to 15, while their coefficients are 1 over twice the exponentials of 1 to 15 (in the same order). As is clear they simulate a power law from an argument of 2 through 500. (b) Three images. The top one is a sum of contributions in (a). The middle one is a fit (dotted line) of the logs (shifted upward a bit); you can't do better. The bottom one is a fit of the original curves, ax^{-b} and the sum of exponentials. This not only illustrates the dangers of assuming that something is a power law, but also shows that the fit depends on what's fitted! It's also clear that although this artificial example only covers approximately 2 1/2 decades (2–500), further exponential terms could extend the "good" fit.

And finally, to add another dash of reality, there's the fact that a sum of exponentials can look very much like a power law, as observed in [213]. See Fig. 10.15.

In practice there was a lot of speculation about the brain. In the 1980s it was a job to get signals from a few neurons. I recall Roger Traub (at IBM, Yorktown) made an electrical model of a neuron and then went on (the computer) to do entire networks of them [224] but as far as experiment was concerned he was dependent on careful placement of individual electrodes. More recently (circa 2010) there are chips that connect to an entire array of neurons. Still, one only looks at a small sample of active neurons, but that sample size has grown. We'll see that there are interesting statistics—power laws—involved. I should also mention that statistical physics has also played a role. Amit, Gutfreund and Sompolinsky [7, 8] proposed that the many metastable states of a spin glass provide a model for memory storage.

Here we are concerned with power laws. Beggs and Plenz [22, 23] looked at signals from the cortex of rats. (The details of dissection are not for the faint hearted.) Each spontaneous "avalanche" was counted. What was meant by an "avalanche" varied. Beggs ([22], Fig. 2) is detailed in his description: you take pictures every 4 ms and count the number of neurons that exceed a "suprathreshold" (high by 3 standard deviations) in their activity. An avalanche is then a sequence of positive results that begin and end with a blank. The number of pulses, the number of excited neurons that exceed the suprathreshold, counted from start to finish, is the size of an avalanche. *This*

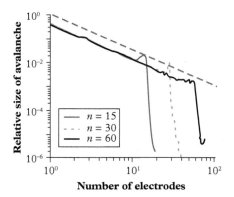

Fig. 10.16 Number of participants in the avalanche vs. size of the chip (sizes 15, 30 and 60). The slope is close to −1.5. Adapted from Fig. 4F, [23].

size follows a power law. See Fig. 10.16. True, in the smaller dimension there are not many decades of data. But the line is really straight and as you increase the number of electrodes it continues that way. Thus, the power law is limited by the size of the chip on which these excitations take place (although occasionally the count exceeds the size of the chip, since neurons are excited by outside neurons and some neurons fire more than once). Moreover, as you vary the time interval (4 ms, in Beggs's example) the power changes, but that was to be expected. After all, if you take a long interval for each picture you'll get a different pattern from a short interval. This doesn't by itself guarantee a different exponent, except that's what happens in practice. The ideal is when intervals are short enough to match the length of an excitation and that gives an exponent of (approximately) 3/2. However, I would not call this number sacred, since it *also* depends on the spacing between electrodes.

Let's suppose then that there really is a power law. What would be its significance?

First there is the business of SOC (self-organized criticality) [19; 172]. For some reason (to be elaborated) the system is primed. The avalanches occur and probably without a mean. What this means is that if the probability of an avalanche of size x is $p(x)$ then the mean of x is $\bar{x} = \int_0^\infty p(x)x\,dx$ (where "$\int$" can also be a discrete sum). But if $p(x) \sim x^{-\alpha}$ and $\alpha < 2$ that integral doesn't exist. (It's borderline if $\alpha = 2$ and there could be a cutoff.) In any case, it seems that $\alpha = 3/2$ so the integral is divergent. This implies that occasionally there will be large bursts. So for some reason, if indeed SOC is at work, it pays for neural systems to be poised at criticality.

So it comes to asking what is the advantage of criticality? The short answer is, I don't know. But there are hypotheses, all of which seem reasonable. They are enumerated by Beggs [22]. First there is the possibility of exciting many neurons. There are occasions when a large number of them need to be excited. Although the studies mentioned deal with spontaneous excitation it may be expected that similar, controlled, events would occur. Second there is the repetition of particular patterns. This may be a way to store information. Third it can be argued that computation depends on a delicate balance between order and disorder, and that being near a critical point is what's needed. And finally the tuning implies that (when excited) similar causes have similar effects in the brain, a kind of stability. See also [210] for elaboration, for additional material, and many citations.

How much of these assertions have been verified experimentally? As far as I know, nothing definitive, although [210] reports some work. (But I may be out of date.) What I can say is that the idea of criticality in neural processes has been an attractive one so that there are entire communities devoted to this idea.

I don't plan to go into artificial neural networks, although these are interesting in themselves. In 2013 I had papers in Italian and German to translate. The Italian went smoothly on Google translate, but the German was poorly handled. Too many compound words and the verb at the end was too difficult to figure out. Within a few years Google had improved tremendously, and German was smoothly translated. What happened? I believe artificial intelligence got much better. If you look at articles (e.g., Wikipedia) on deep learning and related topics you'll see that there have been significant developments. I'm having trouble accepting that a computer can do well in the game of Go (which requires significant pattern recognition) using deep learning. In fact, I believe that captchas will soon be out of business.

11
Physical sciences

Applications discussed in this chapter: Luminescence, Faking power laws, Large scale structure (galaxies are points!), Galactic morphology

This book isn't really about traditional physics, only the techniques of physics. However, some digressions are in order, in particular how a power law can be faked.

We also touch on topics in astrophysics and cosmology that may be controversial.

11.1 Power laws for luminescence

One of the problems facing people who make detectors is that the signal from the detector may be delayed [166–168]. Many detectors are scintillating materials and have what are called color centers that send out a flash of light when excited, a typical source of excitation being the passage of the particle to be detected. What's going on microscopically is that an intentional impurity, a color center atom, becomes excited when hit by the passing object (which is to be detected). From its excited state it *usually* drops back to its ground state in a matter of nanoseconds, which is what people want. But every once in a while it tunnels to a trap and can be passed from trap to trap until finally it locates an available color center and drops down emitting a photon. So you get what is called *delayed recombination*.

The amount of delayed light you see, as a function of time, is a power law. I say this glibly, but practically speaking you must give the range of values over which this is true. After a long time you have few photons, the signal is barely there, and all you get is background. For early times it may be difficult to distinguish the delayed signal from the tail of the normal signal. What you say therefore is "It's a power law for n decades," where n is typically about 4 or 5 but may be a large as 9. A general explanation was found by Huntley [111] and is more complicated than the clean integral in Eq. (7.21). In fact the point is that it's *not* a power law, but manages to appear as one over all measurable ranges of the variable.

When Things Grow Many. Lawrence S. Schulman, Oxford University Press.
© Lawrence S. Schulman (2022). DOI: 10.1093/oso/9780198861881.003.0011

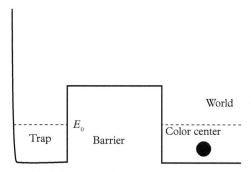

Fig. 11.1 Potential energy (as a function of radial coordinate—position) seen by an electron in a trap. The trap is on the left and the electron energy is E_0, below the barrier height. But if the electron tunnels *through* the barrier, it can then move freely (in the "world") to find a recombination center (a color center).

Huntley assumes there are defects throughout the host (crystal), and that those defects are able to trap electrons. Moreover, these are randomly distributed so that if an electron is a distance r from the nearest place it can decay into, the lifetime for this tunneling is

$$\tau = \frac{1}{s} e^{\alpha r}.$$
(11.1)

To explain s and α you need to be initiated into the world of quantum tunneling. The potential energy seen by the electron is something like that pictured in Fig. 11.1. The electron is within the potential well (marked "trap" in the figure) and, having energy E_0 below the barrier maximum, classically would stay there forever. But under the rules of quantum mechanics it can tunnel through the barrier, with a lifetime that would (in the semi-classical approximation) be $C \exp\left(\int dr \sqrt{(2m/\hbar^2)(E - V(r))}\right)$ with the "constant" C also dependent on the potential. (By the way it's the exponential in the foregoing formula that allows the wide variation in lifetimes that's known to exist for nuclei.) In practice this is often not correct but the general idea is that the coefficient (variously C or $1/s$) is roughly constant. The potential between the well and the world, V, is also assumed relatively constant, so that the lifetime increases exponentially with distance tunneled. The quantity s is to be thought of as an attempt-to-escape frequency (having dimensions of inverse time). The actual value of s is not important in our discussion but for the record it's about $3 \times 10^{15}\,\text{s}^{-1}$. Given the approximate constancy of the potential, α also has an interpretation, namely as a kind of constant times the square root of the barrier height. We do not need a numerical value for α, except to say that it is large enough so that one can ignore the possibility that two electrons are in range of a target color center.

Let the number density of traps be ρ. Then the probability that the nearest trap to a given color center is at a distance r is

$p(r)dr$ = [volume of shell of radius r] × [density] × [prob. that no electron is closer]

$$= 4\pi r^2 dr \times \rho \times \exp(-4\pi\rho r^3/3),\tag{11.2}$$

where the product indicators (×) occupy parallel positions in the verbal and symbolic versions of the equation.[1] What we want to know is the number of trapped electrons at a time-t after the disturbance, because its time derivative will reflect the level of light emerging after the disturbance has passed. This will be the initial number of electrons times the probability that the nearest trap is at distance r (just calculated) times the the probability that no decay has taken place, which is $\exp(-t/\tau)$ with τ given above (Eq. (11.1)). Dividing by the initial number of electrons we have

$$\frac{n(t)}{n_{\text{initial}}} = \int_0^\infty 4\pi r^2 dr\, \rho \exp(-4\pi\rho r^3/3)\exp\left(-\frac{t}{s^{-1}e^{\alpha r}}\right).\tag{11.3}$$

This integral is truly a mess. It can improve though with some changes of variable. First, let $R \equiv \left(\frac{4}{3}\pi\rho r^3\right)^{1/3}$. Also measure time in units of st, that is, $\tilde{t} \equiv st$. Finally we introduce a parameter $\gamma = \left(\frac{3}{4\pi\rho}\right)^{1/3}\alpha$. With a bit of algebra, this yields

$$\frac{n(t)}{n_{\text{initial}}} = \int_0^\infty R^2 dR\, \exp(-R^3)\exp\left(-\tilde{t}e^{-\gamma R}\right).\tag{11.4}$$

The illumination at any time is the negative of the time derivative of this quantity,

$$\begin{aligned}I(t) &= -\frac{\partial}{\partial t}\frac{n(t)}{n_{\text{initial}}} = -\frac{\partial\tilde{t}}{\partial t}\frac{\partial}{\partial\tilde{t}}\int_0^\infty R^2 dR\, \exp(-R^3)\exp\left(-\tilde{t}e^{-\gamma R}\right)\\ &= s\int_0^\infty R^2 dR\, \exp(-R^3)e^{-\gamma R}\exp\left(-\tilde{t}e^{-\gamma R}\right).\end{aligned}\tag{11.5}$$

The good news is that this depends only on two parameters, γ (which is essentially $\alpha/\rho^{1/3}$) and $\tilde{t}$.

I will show the graphs in a moment, but first let's consider the relation Eq. (11.1), which can be written

$$r_c = \frac{\log st}{\alpha},\tag{11.6}$$

where t is now τ the time of decay. We make the approximation that for times greater than t $(=\tau)$ all electrons have finished tunneling, while for times earlier than t nothing

---

[1] Since the traps are random and uniformly distributed the probability that n traps are in a given volume, $V = \frac{4}{3}\pi r^3$, is the Poisson distribution $P(n) = \frac{\rho^n}{n!}\exp(-\rho V)$. Therefore the probability that a volume $\frac{4}{3}\pi r^3$ is empty is the value given in Eq. (11.2). Finally, the "electron" of Eq. (11.2) is in a trap, hence the other factors.

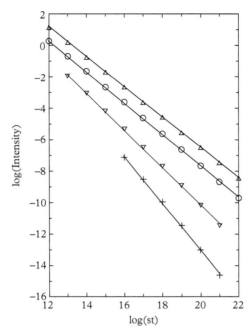

Fig. 11.2 The figure shows in descending order various reduced densities of traps. (Reduced density = $\rho' = \frac{4\pi}{3} \frac{\rho}{\alpha^3}$, with ρ usual density and α an effective tunneling rate (see Eq. (11.1).) The reduced densities and slopes of the various curves are: $\triangle$: $\rho' = 2 \times 10^{-6}$ & $I \propto t^{-0.958}$; $\bigcirc$: $\rho' = 10^{-5}$ & $I \propto t^{-0.997}$; $\triangledown$: $\rho' = 5 \times 10^{-5}$ & $I \propto t^{-1.185}$; $+$: $\rho' = 10^{-4}$ & $I \propto t^{-1.50}$. Adapted from [111]. Source: *Journal of Physics: Condensed Matter.*

has tunneled. With this assumption, let's examine the integral at time-t. It is maximum when the derivative (with respect to R) vanishes, that is,

$$n(t) = n_{\text{initial}} \exp\left(\frac{4\pi r_c^3}{3}\right) \tag{11.7}$$

have tunneled. The intensity ($\propto$ number tunneling) at any particular time is

$$I \propto -\frac{1}{n_{\text{initial}}} \frac{dn}{dt}. \tag{11.8}$$

This can be calculated using the integral Eq. (11.4) or Eq. (11.7), which differ from one another by less than 5% (for the ranges illustrated). In Fig. 11.2, Eq. (11.4) is used.

As one can readily see, there is an apparent power law for none of the reasons given in Chapter 7. Even for (effective) densities of 10^{-4} the power law form holds for half a dozen orders of magnitude. It is even true that the slope can be close to minus 1. I remark that in plotting the recombination luminescence from experimental data power law behavior is only exhibited for four to nine decades, and my guess is that some of the assumptions break down. In particular, I would be suspicious of the constancy of s and α. The bottom line though is that to a good approximation Huntley's calculation gives power laws. (See also Sec. 10.7 and especially Fig. 10.15.)

11.2   Large scale structure

At one point you could make an astrophysicist or cosmologist see red, just by mentioning endless fractal structure.[2] But the idea is natural, given that galaxies are members of clusters, clusters are part of larger clusters, and so on. There are also voids, lending further validity to the claim—for "nearby" structure. The condensed matter physicist L. Pietronero, who had calculated the fractal dimension of several condensed matter substances and had studied self-organized criticality [177], turned his attention skyward and found (together with collaborators Sylos Labini, Montuori and several others—see the citations [42, 64, 134]) that "large scale structure" is fractal! (In "large scale structure" studies, galaxies are thought of as points.)

One should also bear in mind an aphorism due to Landau[3] in his evaluation of cosmologists: "often in error, never in doubt." This is true today, of some of them, anyway.

The original claim of Pietronero et al. was that as far as they could tell the distribution of galaxies was fractal all the way—meaning as far as one could see, all catalogues [134]. They also noticed an inconsistency in the astrophysicist's use of "correlation function." (I had also noticed this, and it certainly confirmed Landau's opinion of cosmologists.) In the astrophysics literature they would define a correlation by dividing by the average density, even though that average was dependent on the sample looked at, and, in fact, their equations told them that average density for sufficiently large radius tended to zero. I got interested in this subject because I had been impressed by *Scientific American* articles on voids [91, 211] and felt that Phil Seiden and I could contribute [203] by identifying the initial configuration of galaxies as a percolation cluster, which near its critical point would have voids (and excess densities, which is also observed).

---

[2]I do not speak theoretically. (Well maybe they don't see red, but they become angry.) Perhaps I shouldn't mention names, but when Y (X is already used [80]) was asked shortly after he'd won the Harvey Prize about Pietronero's fractal universe ideas, he did see red—at least figuratively. To be fair, cosmologists do admit that "nearby" matter is fractally distributed.

[3]I will give a short biography of Lev Landau, but it doesn't do him justice. He had studied in Europe (and considered himself a student of Niels Bohr), but returned to Russia. There he became a towering figure in physics and founded a school of theoretical physics (of which the Landau–Lifshitz books on physics are a remnant). The number of effects and (physical) phenomena named after him is phenomenal. For details see Wikipedia, or even *Physics Today* (I recall articles about him). And besides his accomplishments in physics there was also his humor, of which I give a small sample in the text.

On the other hand, at age 380,000 years (minus epsilon, just before "recombination") the universe was quite uniform. This is known because just after this—that is, after recombination—atoms became neutral, and the resulting radiation is uniform to one part in 10^5. (The "recombination" refers to the fact that the universe had cooled enough so that atomic hydrogen could form and the dominant force was no longer the (short-range shielded) electromagnetism that dominated an uncombined plasma.[4] The resulting transparent universe gave rise to what is now known as CMB, cosmic microwave background radiation.[5]) So there is reason to believe that eventually things become uniform with a finite density—just not the density used in that fake (or variant?) definition of correlation length.[6] There is still a way to reconcile this with zero density, if the universe is infinite, while the total amount of matter is finite, it just seems unlikely (but who knows?).

The short answer is, I don't know whom to believe. Or to put it differently, not all the facts are in. The early work [134] demonstrates fractal structure, but most graphs only cover 1 1/2 or 2 orders of magnitude, although within that range they are pretty good straight lines. An exception is the information on radio galaxies, but that is a broken line. Given my caution (cf. Footnote 9, Chapter 10) I hesitate to draw firm conclusions. In later work [119] they seem to have yielded that things eventually become homogeneous; they look for a crossover distance (at which things are homogeneous), but don't really fix on anything. Finally in a recent paper [43] they take the idea of no crossover seriously and attribute "accelerated expansion" to the fractal nature of the matter distribution rather than to dark energy.

Now in case you're not tuned in to this, a big question these days is the nature of dark energy. Is it a non-zero cosmological constant (off by *many* orders of mag-

---

[4]This also accounts for the arrow of time since then. The dominant force went from being electromagnetism to being gravity. In gravity things prefer to clump, so that the universe—which was relatively uniform—was in an extremely unlikely state. As a result of gravitational clumping stars formed and our low entropy lives are "fueled" by that of our primary star, the Sun. See [196]. So if your little sister or brother asks, "What do we get from the Sun?" the answer is *not* energy, which in an era of global warming we do all we can to stop an energy accumulation. As pointed out by Schrödinger [193] it is negative entropy.

[5]This radiation earned its discoverers, Arno Penzias and Robert Wilson, a Nobel Prize. I had thought the prize should go to Robert Dicke, who at the time was building an antenna to view this radiation. But shortly before the award, Jim Peebles had given a seminar in Maryland outlining Dicke's plans. This was picked up by someone from Bell labs who told Penzias and Wilson what they had been viewing (they had been cleaning pigeon droppings off their antenna, thinking that was the source of their static). (I know about the seminar in Maryland from Peebles.) However, I later learned that the effect had actually been predicted, years earlier, by Alpher and Herman [6]. According to an interview with Alpher, Peebles had submitted an article on the subject to *Physical Review* (surely with the knowledge and approval of Dicke) but had refused to put the relevant citation in a *Physical Review* article, refereed by Alpher, instead preferring to go to a different journal.

[6]It is ironic (*not* in the sense of Footnote 14, Chapter 10) that one of the chief users of this "variant" correlation length, Jim Peebles, won a Nobel Prize in 2019.

nitude), is it "quintessence," or what? (It's also possible that it's a big mistake: the "accelerated expansion" may have been due to an error in estimating distances or to the mass distribution of the larger universe.) The question has been around since 1998 when accelerated expansion was discovered [174]. So Pietronero et al. are taking on a difficult question. To their disadvantage there isn't any exact solution for a fractal distribution, while there most definitely is for a uniform matter distribution. That's also part of the reason that these ideas are rejected by most cosmologists: what would one do without the FRW (Friedmann–Robinson–Walker) metric (based on homogeneity)? Poor cosmologists (or some of them, anyway)! They expect Nature to follow their solutions, and it should be the other way around! It may be that after all they're right, but if so, it's not because they've found an exact solution. Another problem for the fractal distribution people is that their arguments are based on a fractal dimension of 2.9 ± 0.02.[7] To be brutally honest that's not all that far from 3 and frankly, given the quality of the graphs I've seen, I'm not sure whether to believe the error bars. And I'm sympathetic.

My true opinion, for what it's worth, is that at large scales things do become uniform and that clumping (under gravity) is a gradual process. At first there was some clumping (due to dark matter, not what gave rise to the CMB[8]), but as the universe grew older larger and larger regions clumped under normal gravitational interaction. This "clumping" gives rise to fractal structure. At the present stage in the universe's history the fractal nature extends to some distance (500 million light years?), but at larger scales you have uniformity. One question not answered (and which came up in [203]) is how much "participation in the Hubble flow" obtains at any given distance. Yes, at a scale that there is uniformity the galaxies are departing from us at a velocity proportional to their distance (as predicted), but what about closer? This was a problem that troubled Einstein and was taken up by a number of individuals and summarized by Bonner [30], but which has not been solved to my knowledge.

Nevertheless, it has been dealt with in a phenomenological way in~[203].

This subsection, although it deals with fractals, is a bit far from the main theme of this book, but it's a chance to get some of my annoyance off my chest. And to show that physics, often thought of as an example of an exact science, has gaps, errors, personal issues and prejudices! And I won't deny that some of them may be mine. (The gaps, etc. might make it more of a human endeavor.)

---

[7] Reference [43] gives 2.9 ± 0.02 as the value in the abstract. However. in the article itself 2.87 is the figure given, without error bars.

[8] ... but deduced from it.

11.3   Galactic morphology

Here is a topic in which studies of cellular automata, statistical mechanics, and astrophysics all have contributions. It is a model of star formation and gives rise to spiral structure.[9] But it didn't start out that way. The model I'll talk about started playfully: Martin Gardner's columns in *Scientific American* [67] described John Conway's Game of Life. That "game" began as a way to have reproduction in fewer steps than a corresponding scheme by John von Neumann. But it took on a "life" of its own as many aficionados used those rules and saw all sorts of systematic forms of behavior. At that time Phil Seiden was spending a sabbatical at the Technion and he and I worked together, adding randomness to the cellular automaton. This destroyed (most of) the nice patterns seen by others, but led to a phase transition (as a function of the strength of the randomness). At the end of his sabbatical Phil returned to IBM in Yorktown Heights, New York. In those years IBM indulged in a certain amount of pure science. In particular, Humberto Gerola was an astrophysicist in IBM's employ. When Humberto heard Phil's description of our efforts he said: "Hey, that's the way stars are born in galaxies"" Here is the common thread of stars and cells (in cellular automata): An episode of star formation involves a cloud of gas in the form of molecules. So the gas is relatively cool and consists of mostly H_2. That gas turns into many stars. It can be thought of as taking place roughly in an area 600 light years on a side and involves a few big stars, maybe 30 or even 50 times the mass of our Sun. But it also creates many smaller stars and probably tiny objects, say the size of Jupiter, that never shine by their own power at all. (Well, "tiny" is relative.) What causes a molecular cloud to form stars? Lots of things, the collisions of galaxies, collisions of clouds and, significantly, the shock wave from a nearby supernova. And where did the supernova come from? Typically from a big star that's not too far away. The star that exploded itself came from an earlier episode of star formation. (In some cases this can be seen explicitly, something that looks like a string of firecrackers in the sky, in which several generations of star formation are evident.) And how about cellular automata? For the probabilistic sort of cellular automaton there will be several attempts by "living" nearest neighbors to make the cell that we're interested in be alive. This is what supernovas do. If a molecular cloud has enough neighbors and if some of them have given rise to supernovas, there's a fair chance that one of them will ignite—cause to be alive—a cloud in an adjacent cell. And this works. I do need to give rough dimensions of the grid, the cells of the cellular automaton. First, a spiral galaxy is flat, a ratio of diameter to thickness being around 100 (or more) to 1. So two dimensions is adequate.

---

[9]This model has come up before, in particular in connection with percolitis and self-organized criticality.

(a)   (b)

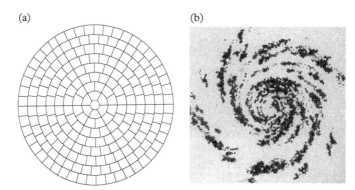

Fig. 11.3 (a) shows the grid, replacing the simple square lattice of the original Conway game. (b) is the result of a simple simulation, simple in the sense that there is no memory. Illustrations from [204]. Source: *Science*.

Second, each "cell" was approximately 600 light years on a side; I say "approximately" because they were not perfect squares and did not line up in a rectangular grid. Rather the centers of the cells formed an approximate triangular lattice and slid past one another, in keeping with the differential rotation of the galaxy.[10] Finally, I need to tell you what the time interval is, since we are approximating a real physical phenomenon. The duratation of each time step is governed by how long it takes for a star to go supernova and for the shock wave to reach nearby regions. You may think of stars as living for billions (10^9) of years. But the bigger a star is, the brighter it burns—more than the increase in its mass. So the really big stars are short lived; 10 million years is longer than it takes for them to become supernovas, so that time period is reasonable.

In Fig. 11.3 I show both the grid and the results of a simulation using this rule. In practice more elaborate rules were used. We kept track of the gas, meaning that if it was too hot star formation was prevented; it cooled at a certain rate, meaning that (to use an analogy to percolitis) immunity was present. A cell that fired (had star formation) was unlikely to do so for several time steps.[11]

The results of simulations were impressive, to me at least, although the astrophysical community preferred more elaborate explanations. Nevertheless, this approach was able to explain the luminescence of dwarf galaxies as well as predict phenomena. There was a logic also to the beautiful spiral arms that emerged from simulations. First, as

---

[10]To a good approximation the radial velocity is nearly constant as you move away from the center of the galaxy. This means that the *angular* velocity is highest in the center ($\omega = v/r$). As a consequence, cells near the center rotated more rapidly than those further out. This constancy of the radial velocity, by the way, is one of the reasons that people believe there's dark matter. But that is a long story (which you can read in many texts on mechanics, in particular [20], if you can get a copy).

[11]The "percolitis" model was invented to help explain the star formation process [204]. It had previously [202] simply been called "($\infty+1$)-percolation"(or "($N+1$)-percolation"), reflecting its mean field nature.

pointed out in the sections on self-organized criticality (Secs. 4.6 and 4.6.1), with immunity the system tends to the critical point. Near the critical point correlations become large so that the correlation associated with long arms becomes significant. Moreover, there is enhanced fertility near where shear has caused an exchange of neighbors, because it's less likely that an adjacent site has had star formation (recently).

The analytic work on this problem—of interest in the present context, since computer results aren't easily checked—gave the wrong critical exponent. This was expected, even when higher order correlations were included, as did real space renormalization methods. The bright spot was a specialized calculation, which we motivate as follows. The problem calculating critical exponents, and so on, was identified as the vacuum, large regions as you approached criticality. These were poorly treated because of the high-order correlations needed to obtain big empty regions. So define a new quantity, ρ_L, the living, no vacuum, density, where vacuum is defined as an empty cell, surrounded by (six) empty cells. Thus

$$\rho_L \equiv \frac{L}{M-V},\qquad(11.9)$$

where L is the number of "living" cells, V the number of vacuum cells and M the total number of cells, including vacuum cells. In terms of the total ρ $(= L/M)$ this is

$$\rho_L \equiv \frac{\rho}{1 - \langle \prod_{\alpha=0}^{6}(1-\sigma_\alpha)\rangle},\qquad(11.10)$$

where the σ's are either 0 or 1 depending on whether the cell is dead or alive. We use cumulants, defined in Appendix D.3, which show a relation between different random variables. Thus

$$\frac{V}{M} = \left\langle \prod_{\alpha=0}^{6}(1-\sigma_\alpha)\right\rangle = (1-\rho^7) + (1-\rho^5)\sum_{\text{pairs}}\langle\sigma_\gamma\sigma_\delta\rangle_{\text{cumulant}},\qquad(11.11)$$

with yet higher order cumulants ignored. There are two kinds of cumulants to consider. There are correlations between sites that share one neighbor (to be called "S") and those which have two neighbors in common ("D"). To see how many of each is involved one can look at a diagram of a triangular lattice, Fig. 11.4. In that diagram, sites 7, 8 and 9 contribute nothing. There are only three sharing one neighbor: 1–4, 2–5 and 3–6. However, there are many involving two. Altogether there are 18 of them, so that

$$\frac{V}{M} = \left\langle \prod_{\alpha=0}^{6}(1-\sigma_\alpha)\right\rangle = (1-\rho^7) + (1-\rho^5)(18D+3S).\qquad(11.12)$$

The cumulants can in turn be related to the density: The evaluation of these two sorts of cumulants assumes that things have settled down and $\rho(t-1) = \rho(t) = \rho$. For them we use the basic formula (see Sec. 4.6.1, Eq. (4.19))

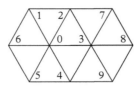

Fig. 11.4 Diagram for evaluating cumulants.

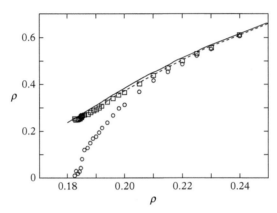

Fig. 11.5 Data on the local density are the square boxes, while (for reference) the total data for total density are the circles. The fits (including cumulants) are the dashed line while the mean field values are the solid line, which doesn't do badly either.

$$S(n, t+1) = 1 - \prod_{\ell} [1 - A(\ell, n, t) S(\ell, t)] \,, \tag{11.13}$$

with each A an independent random variable taking values 0 (with probability $1-p$) and 1 (with probability p), and $S(n, t)$ is the probability that individual-n is sick on day t (using the terminology of that section). To evaluate cumulants we take a product of two such terms. The correlation arises from the fact that in products of σ's, when two labels take the same value (say δ) the product is the same as σ_δ alone. Thus (for example)

$$S = p^2 \rho(1 - \rho)(1 - \rho p)^{10} \,, \tag{11.14}$$

and similarly for D. When all is plugged in, this gives an accurate account of the local density. See Fig. 11.5. Even the mean field (neglecting cumulants) doesn't do badly. For further details see [202]. N.B. The work just reported was performed on a (static) triangular lattice, not the moving version needed to describe real galaxies.

12
Putting it all together

Many techniques have been used in this book and it may seem like a hodge-podge of methods. To some extent this impression is true, but the same could be said of statistical mechanics in general. If there's any unifying principle it's the use of mathematics. There's also what I've called the baloney factor. I won't review the times I've used that expression or its equivalents; let's just say that not everything fits smoothly into the framework I've outlined. Sometimes the subject is *too* complicated, sometimes you just need to scratch your head. But you should be careful before you pronounce something a power law or a complex system.

Here are some of the methods (and buzzwords): maximum entropy, probability theory, power laws, universality, renormalization, emergence, agent-based modeling, scaling, fractals, synchronization, self-organized criticality, on and on. Let me review some of them. There won't result a unified picture, rather approaches you can use as you consider a problem.

The first lesson is the size of fluctuations. If you have N participants (random variables or something else, like people) it's $\sqrt{N}$, unless it's not. There are two things to consider: when it's true and when it's not. It's true when the variables are independent of one another and themselves have a finite variance; that's the central limit theorem (Appendix D.3), and there's even a cartoon proof (Sec. 2.1). The more interesting case is when it's not. One possibility is that the random variables themselves have infinite variance, such as $P(x) = \frac{a/\pi}{x^2+a^2}$, the Cauchy distribution. (The quantity a is a parameter. $\int_{-\infty}^{\infty} x^\alpha P(x)dx$ does not exist for $\alpha > 1$ and is conditionally convergent for $\alpha = 1$.) The other possibility is that there are correlations among the variables (participants?) and some kind of criticality, whether self-organized or not. There was an explicit example given in the Curie–Weiss model of a ferromagnet (Sec. 5.3), but it's far more general.

Another example has to do with powers, not power laws. You can ask how many appearances do the left-hand endpoints in the Cantor set make up to level n? Recall (Appendix C) that the elements of the Cantor set are numbers of the form $\sum_1^\infty \frac{a_k}{3^k}$ with

When Things Grow Many. Lawrence S. Schulman, Oxford University Press.
© Lawrence S. Schulman (2022). DOI: 10.1093/oso/9780198861881.003.0012

$a_k = 0$ or 2—never 1. (The construction in Appendix C is poetic, but this is the prose version.) The number of left-hand points is the number of a_k's in which the last entry is a 0. There is one of these at the first level, two at the second, four at the third, and so on. So that number is $1 + 2 + \cdots + 2^n = 2^{n+1} - 1$. Taking the logarithm of one over that number and neglecting the 1 gives

$$\log(\text{one over the number of left-hand endpoints}) = -(n+1)\log 2.$$

Writing this as # left endpoints $= 2^{n+1}$ we get a value of $1/\log 2$ as the length scale. (The quantity n is the level number of the Cantor set and is like the x that often appears.)

As to power laws, I could give real life examples, like words in a book or cities—except that it's not clear they are indeed powers. But mathematically examples abound and I have an entire chapter of examples: diffusion, exponential of exponential, and so on—and significantly, SOC, self-organized criticality.

An example of self-organized criticality was given in Sec. 4.6.2. Here I'll give the chemical version: you have three species, S, I and R (*not* the SIR model of epidemiology) and they have the following reactions and rates:

$$
\begin{array}{lll}
S & \rightarrow R, & \text{with a rate of 1 (fixing the time scale)} \\
R + S \rightarrow 2S, & \text{with a rate } x \\
3S & \rightarrow 2S + I, & \text{creation of an inert atom (or individual) at rate } y \\
N & \equiv R + S + R, & \text{a conserved quantity.}
\end{array}
\qquad (12.1)
$$

If you want the epidemiological version, you can think of S as sick, R as recovered (or not sick[1]) and I as inert, out of the picture, but we'll stick to chemistry here. The system studied earlier had $2S \rightarrow S + I$ and gave criticality in the mean field model, but was killed by fluctuations. Now we take $3S \rightarrow 2S + I$. The rules of 12.1 are implemented by taking the following probabilities of transition: For $S \rightarrow R$ the probability is proportional to S alone.[2] For $R + S \rightarrow 2S$ the rate is proportional to $xSR/(N-1)$ and finally for $3S \rightarrow 2S + I$ the rate is $yS(S-1)(S-2)/((N-1)(N-2))$. (They are normalized to total 1 and the process stops when S reaches zero.) In the article [69] we use the transition matrix approach to prove that criticality is reached (and not destroyed by fluctuations). Here we simulate the stochastic process.

This system decays by a power law. As "proof" I exhibit Fig. 12.1. Of course it's numerics, but the approximate constancy of time$^{1/4}$ times S is an indication that for

---

[1] As in Footnote 24, Chapter 4, R replaces H in the notation. For S and I, both italic and roman forms are used.

[2] The "rates" I give in the following are the transition probabilities. They involve—besides the raw rates $(1, x$ and $y)$—the state of the system, as described by the various quantities of each substance, involving S, R and N.

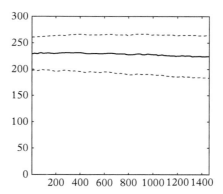

Fig. 12.1 The mean value of S for 4×10^3 runs of the program described in the main text. What is shown is $\text{time}^{1/4}$ times the mean of S as well as the standard deviation (also times the fourth root of time). The product $\text{time}^{1/4} \times S$ is roughly constant, indicating that the decline is $\text{time}^{-1/4}$, a power law. Under the rules of Eq. (12.1) there is therefore self-organized criticality.

the real system the decay goes like $S \approx \text{time}^{-1/4}$, a power law decay, indicative of self-organized criticality.

The point of the previous discussion is that a certain classes of phenomena are interrelated. Scaling, power laws, large fluctuations (like $\sqrt[4]{N}$), critical phenomena are related to one another. That's a strong statement, because the coastline of Norway and carbon resistors (which exhibit $1/f$ noise) appear to have something to do with natural complex phenomena, like language, city population and genes, products of intelligence or evolution. Why? Instead of saying "I don't know," I'll say, "Surely There's A Research Topic," acronym START (and I'll use this acronym frequently in this summary).

Another kind of natural ordering has to do with synchronization. The heart must pump as a unit and fireflies must flash in particular patterns to retain the properties of a species. Do these operate by the same mechanisms as clocks in a(n old fashioned) clock store or like a planned collection of "PLL"s (phase locked loops), synchronized for the purpose of localization?

Then there's the business of maximum entropy, or "MaxEnt" as it's come to be known. This was proposed as a general method by Jaynes [116, 117] and is simply a way of saying, what you don't know, you don't know, and maximize your lack of information (information being the opposite of entropy). This has been applied in many fields; examples here include votes of the US Supreme Court, ecology, and the retina of a frog. But there are many other applications. For example (and this is something missing here) MaxEnt has been applied to pictures. In fact there was a conference right under my nose (at Clarkson University in 2015 organized by Adom Giffen) on this subject (except that I was in Dresden at the time). In any case, it's a powerful method, not—as far as I can tell—related to complexity. True, it's been used to study flocking birds [26]. There I would say that the miracle is how few local rules are needed to attain flocking, without global rules. This is emergence. Is it the same as complexity? (START)

... which leads to the subject of emergence. This topic is a triumph of "bottom up" as opposed to "top down." Let me explain. Take an airport. Everyone has a job. This is true for the people who resolve flight problems, the people who wash the floors, gate agents, those who check that there's nothing "dangerous" in your luggage, and many others. They report to a boss who in turn reports to someone else, and finally to whatever authority runs the airport.[3] This is top down. It's typical of most human organizations. Bottom up is like ants or flocks of birds. A few local rules provide order. As to why there are lazy ants (cf. Sec. 10.2) and which ones choose to be lazy, START. Crowd control seems to be in between these extremes. On the one hand, given a few rules you can recover streams of people (Sec. 9.7), but when they get really close to each other, start pushing—for survival (or sometimes for bargains)—then no one waits for the others. This too can be modeled, but there's a departure from the simplest of rules.

Renormalization seems entirely to be a physics concept, but it shouldn't be. What it says is that details don't matter when you have strong collective behavior. It divides the (physical) universe into universality classes. At criticality everything in one universality class behaves in the same way (has the same critical exponents). Well, the world does not divide itself into universality classes, but maybe it should. Why are city populations, words in books and genes so similar, in that they are dominated by similar power laws? And maybe insurance companies should use this technique. (START)

And then there is the computer. An important tool is "agent-based modeling." You have a complicated system, maybe lots of rules, and often you have probabilities. This can be a study of urban segregation, of traffic or even of neutrons in a reactor (not dealt with here). You then simulate the behavior of one or several individuals on the computer, according to the rules and probabilities. This is repeated many times— as is possible with the computer. Such programs can take 5 seconds or 5 days, depending

---

[3]This is an idealized version. The TSA people (who check your luggage) report to a TSA manager who may coordinate with the head of airport, but doesn't take orders from the airport. Similarly in a store there may be representatives of suppliers who stock the shelves but answer to whatever company they represent. Of course their boss will try to coordinate with the store, but—in principle, at least—is not obliged to. So even in "top down" situations there's a bit of "bottom up."

on what's being modeled (or simulated). It's a good technique for finding answers, although it often leaves you without an explanation for why.

Don't underestimate this feature of computers. It's often that you have rules for some stochastic process, but the general solution is either evasive or not useful. One tribute is the number of programs written in this book: simulations of games with rubber bands, percolitis, urban discrimination, on and on; there may be less understanding, but there are also more answers.

The computer is useful (and dangerous) for other activities also. There is the issue of artificial intelligence that has morphed into study of neural networks and machines that can think. For example, and this is truly mind boggling, the game of Go demands a high level of pattern recognition,[4] more so than even checkers and chess. And computers have won against world champions in all three cases. Learning algorithms were not taken up in this book but they are becoming significant. This may lead to self-driving cars, but also to devices that kill "enemies" without human intervention. And perhaps most troubling, their intelligence (whatever that is) may rise to the point that humans become obsolete. START.

And finally there's complexity itself. Does it have something to do with currents? With scaling? I have speculations in Appendix H, but so do others, and none gets the whole picture. This indeed is a topic of START, but that is well known.

---

[4]Go is played on a 19 by 19 board with lines drawn, so that it's the intersection of lines (including the ends which may have three or two lines) that are where pieces (known as "stones") are put. The game begins with an empty board. The two players have an unlimited supply of stones, one taking black and the other taking white stones. The object is to use stones to form territories by surrounding vacant areas of the board. It is also possible to capture your opponent's stones by completely surrounding them. Players take turns, placing a stone on a vacant point at each turn. Black plays first. Stones are placed on the intersections of the lines rather than in the squares and once played stones are not moved. However they may be captured, in which case they are removed from the board, and kept by the capturing player as prisoners. At the end of the game, the players count one point for each vacant point inside their own territory, and one point for every stone they have captured. The player with the larger total of territory plus prisoners is the winner.

Appendix A
Notation

Mostly I'll employ standard physics notation, although occasionally there will be usages more commonly associated with mathematics. The list below is not comprehensive.

Pr	Probability, may appear with varying arguments. Can be the probability of some logical assertion, for example $\Pr(x \leftarrow y)$, or the assertion may be understood, as in $\Pr(x) =$ probability that the random variable X takes the value x.
$p_X(x)$	Probability that a random variable X takes the value x. Same as $\Pr(X = x)$.
$\langle X \rangle$	Expectation of the random variable, X. Can be written $\sum_x x \Pr (X = x)$ or $\int xp(x)dx$.
$\overline{X}$	Same as $\langle X \rangle$.
$E(X)$	Same as $\langle X \rangle$.
$\dot{x}$	Overdot, a dot over a quantity (x is the quantity in this example). $\dot{x}$ is its derivative, usually with respect to time.
y'	Prime, the derivative, usually with respect to a variable different from time. $y' = \frac{dy}{dx}$. Occasionally it will be a new value of the variable "y" (and *not* a derivative).
$\times$	Times (used for emphasis—usually omitted).
$\cdot$	Multiple usage. Sometimes it means "times"; sometimes it is an unidentified argument, for example $f(\cdot)$ or $\langle \cdot \rangle$.
$\boldsymbol{r}, \boldsymbol{v}$, etc.	*Vectors* (boldface). Can also be written, for example, $\vec{r}$, $\vec{v}$. The corresponding scalars are written r, v, etc., that is, *not* boldface. Boldface is sometimes used in other contexts.
$\forall$	For all.

const   A constant. Can take different values within a single equation; thus $(d/dx)(\text{const} \times x^2) = \text{const} \times x$.

log   log or logarithm, *to the base e*, unless otherwise explicitly indicated. ($\log_e$ is the *natural* logarithm. It is sometimes written "ln," but not here.)

| or

&   and

t (On a matrix, as in M^t with M a matrix.) Meaning: transpose.

$\equiv$ Can mean either "is defined as" or "is identically equal to."

β $= 1/(k_B T)$, where k_B is the Boltzmann constant ($\approx 1.38\,\text{J/K}$) and T is the absolute temperature (usually measured in kelvin). The dimension of β is inverse energy. It's also the name of a critical exponent, but I'll be explicit when used for that purpose. Usually this is written $\beta = 1/k_B T$ but here I wanted no ambiguity.

a e n   Where a is a number and n an integer of either sign. A way of writing a$\times 10^n$. (e is simply the letter "e."). Used by (among others) MATLAB™.

Tr   Trace of a matrix, $\text{Tr }A = \sum_i A_{ii}$ where A is a matrix.

sgn   Sign function (a.k.a. signum): x real & $\text{sgn }x = \begin{cases} 1 \text{ for } x > 0 \\ 0 \text{ for } x = 0 \\ -1 \text{ for } x < 0 \end{cases}$

*   Multiplication in MATLAB™

∧   To the power of in MATLAB™

Appendix B
Background in statistical physics

This is not a comprehensive treatment of statistical mechanics or thermodynamics by any means and mainly focuses on ideas used in the main text.

Statistical physics grew out of thermodynamics as an effort to learn how macroscopic "laws" followed from the detailed interactions of atoms. The pioneer in this field was Ludwig Boltzmann,[1] whose name is associated with the H-theorem (showing that entropy can only be constant or increase, and was subject to controversy even by those who believed in atoms), the Boltzmann distribution, the Maxwell–Boltzmann distribution and on and on, and finally with a constant of nature, the Boltzmann constant. All these matters will be discussed below.

There are four laws of thermodynamics, numbered 0 to 3. The zeroth law states that if two systems (say A and B) are in equilibrium with each other, and B is in equilibrium with a third system (say C), then A and C are in equilibrium with each other. This "law" presupposes a lot. First, you're supposed to know what equilibrium is. Of course thermodynamics is a macroscopic theory, and the meaning of equilibrium is that nothing macroscopic is happening. Second, it presupposes ergodicity, since that's also required for its validity. So I have to tell you what ergodicity means.[2] It's the statement that phase space averages are the same as time averages. Let's talk classical mechanics. There will be some path the system takes in a possibly high dimensional space: you need to specify the position (q) of all system particles and

---

[1]Ludwig Boltzmann, 1844–1906, born in Vienna, but moved often, mostly to German-speaking places, but also gave lectures in the United States. I won't summarize all his achievements: basically he founded an entire subject, statistical mechanics. His belief in the atomic theory, at the basis of his work, was challenged by the savants at the time, notably Ernst Mach, who, as professor in Vienna, would not accept the existence of atoms. Einstein's work on Brownian motion which allowed Perrin to measure Avogadro's number demonstrating the existence of microscopic atoms (see [161]) came too late to save Boltzmann from himself (he seems to have suffered from bipolar disorder and committed suicide in 1906). On his tomb is inscribed a formula for the entropy of a macroscopic state: $S = k \log W$, with S the entropy, k Boltzmann's constant ($\approx 1.38 \times 10^{-23}$ joules/kelvin) and W (in the modern language of quantum mechanics) the number of microscopic states corresponding to the macroscopic state.

[2]This was, I believe, the original meaning given by Boltzmann. Since then mathematicians have "refined" the notion. We here stick to the original meaning.

you need to give their momenta (p), so if I have n particles in three-dimensional space, $(q_1(t), \ldots, q_{3n}(t), p_1(t), \ldots, p_{3n}(t))$ would be the path in phase space, assuming $(q_1(t), \ldots)$ obey the equations of motion. It is also assumed that the system is isolated. Let f be some function on the entire phase space then ergodicity states

$$\int d^{3n}q \, d^{3n}p \, f(q_1, \ldots, q_{3n}, p_1, \ldots, p_{3n})$$

$$= \lim_{T \to \infty} \frac{1}{T} \int_0^T dt \, f(q_1(t), \ldots, q_{3n}(t), p_1(t), \ldots, p_{3n}(t)). \qquad \text{(B.1)}$$

The rationale for this hypothesis is that the system explores the entire phase space. Let $\gamma(t) = (q_1(t), \ldots, q_{3n}(t), p_1(t), \ldots, p_{3n}(t))$. Ergodicity then means that integrating f over the entire space is the same as letting $\gamma(t)$ explore everywhere.[3] (There are measure theoretic points to be made, but we ignore them here.) For a single particle this makes sense, but for macroscopic systems it is total nonsense. You can estimate that since the Big Bang for a mole of gas in a cubic meter, the system has explored $10^{-(10^{-25})}$ of the states available (see [74], although the fact that the assumptions break down was already known [157]). What happens if you don't have ergodicity? You don't have the zeroth law. See [73] for a counterexample.

The first law of thermodynamics is conservation of energy. There are some bells and whistles associated with this law, concerning work and heat. If a system is in contact with another system, the change in the (first) system's energy is $\Delta E = -W + Q$, where W is the work performed *by* the system and Q is the heat absorbed by the system. What's the difference between work and heat? Macroscopically one knows, but when you deal with smaller systems the distinction seems to depend on technology: work is energy you can recover, heat is not.[4]

The second law involves entropy, and here is where the mysteries truly occur. First I need to define entropy. A good definition involves both technology and atomic physics. Technology enters since you need to know what a macroscopic state is. Can you apply thermodynamics to a single particle caught in a trap? What about small biological systems? Leaving these questions aside, suppose you know what a macroscopic state is. Then the entropy of that macroscopic state is k_B (Boltzmann's constant) times

---

[3]There is also an assumption of energy conservation, so that exploring *all* of phase space means exploring an entire energy surface. In fact some formulations of ergodicity state the requirement as "phase space has a single constant of the motion, energy." (For bodies in free space there is also conservation of total momentum and total angular momentum.)

[4]The work–heat distinction is fraught with controversy. There are many learned discussions, including a formulation in terms of change in parameters, as in $\Delta \lambda V$, where V is a potential energy and $\Delta \lambda$ is the change in some parameter λ (where there's a Hamiltonian, $H = $ kinetic energy$+\lambda V$). The distinction becomes particularly acute when dealing with non-equilibrium thermodynamics. In this book we choose a simple route (cf. the definition given above) but bear in mind (especially if you need to calculate anything delicate) that there is an issue on this point.

the logarithm of the number of microscopic states consistent with the macroscopic one. That is, $S = k_B \log \mathcal{M}$, with $\mathcal{M}$ the number of microscopic states. But how to count microscopic states? To answer this you need to know about $\hbar$ (which Boltzmann didn't[5]). If you would talk about a one-dimensional system it has two degrees of freedom, call them q (position) and p (momentum). Then in the q–p plane (which is phase space for a one-space-dimensional system) each area of size $\hbar$ can accommodate a single state. In $3n$ dimensions, you take the volume of phase space and divide it by $\hbar^{3n}$. (Yes, I've carried the world of atomic physics into a classical statement. All this can be said quantum mechanically, but it would make things more complicated.)

On the other hand, a chemist would say that entropy is simple, and *measurable*. Take 100 grams of ice and convert it to water, at the same temperature (0°C). The change in entropy is $\frac{\text{heat of fusion}}{\text{Temperature}}$, and will be about $100 \, \text{gm} \times 79.7 \frac{\text{J}}{\text{gm}} / 273 \, \text{K} \approx 29.2 \frac{\text{J}}{\text{K}}$.[6] There's a relation between information entropy (see below) and the chemist's entropy (they are the same up to a multiple), but it only applies to the equilibrium situation.

The second law then states that in a closed system the entropy, S, either increases or stays the same. This is the content of the "H-theorem" proved by Boltzmann, in which H is a quantity closely related to S. (If the system is not closed and heat is exchanged, the second law becomes $S \geq Q/T$, with Q the amount of heat absorbed by the system and T the temperature.)

As I said, the second law is where the mysteries are. It provides an arrow of time: *S increases in the "forward" direction of time.* There has been endless discussion on this point (including by me) and I will not recall it here. There are other formulations of the second law: heat at a constant temperature cannot be converted to work. There have been serious proposals to the contrary, but, as far as I know, none has been (or could be?) implemented. Another feature of entropy is that it is often equivalent (up to constants) to missing information about a system. Making the relation precise would take us further afield in this appendix (but see Sec. 6.1), in particular you'd need a definition of "information." But for the appropriate definition for *closed* systems you only lose information as time goes on, so the same arrow of time is involved.

The third law is included here for the sake of completeness but has no role in the present volume. It is the statement that as temperature approaches zero (*absolute*

---

[5]There is a famous equation on Boltzmann's tomb in Vienna: $S = k \log W$, the entropy is the logarithm of the number of microstates consistent with the macrostates (that's W) times Boltzmann's constant ($k = k_B$; there's no "B" on the tomb). The funny thing is, this is not due to Boltzmann, but rather is a consequence of the work of Max Planck, who found that states were quantized and you could count them. Even the constant, k_B, was named after Boltzmann's death, so this equation would have seemed, well, ahead of its time, to Boltzmann.

[6]The final answer is given in SI units. T is the temperature, *measured from absolute zero.*

temperature, e.g., kelvins) so does entropy. This too has been contradicted in some artificial cases [14], but in our experience remains true in the real world.

There are three "ensembles" that are most used in physics. The simplest is the microcanonical. It refers to a completely isolated system—it does not exchange anything with anything else. The system has fixed energy. The assumption is that any state consistent with conservation of energy is equally likely. What justifies that assumption? Nothing. Nothing but experience. Conclusions drawn from this assumption do not contradict experience. (As you can see, physics has a lot of faith-based assumptions.)

Then there is the canonical ensemble. These systems are in contact with a "reservoir" of temperature. A reservoir is a system large enough that the exchanges of energy with the (supposedly smaller) systems do not affect its temperature (although work and heat can be exchanged, the latter leading to entropy changes). The most important conclusion from these assumptions is the probability of finding a system in a given state. This is known as the Boltzmann distribution and is

$$\Pr(\text{system is in state } \alpha) = \frac{\exp\left(-E_\alpha/k_B T\right)}{Z}, \tag{B.2}$$

where E_α is the energy of state-α and T is the temperature of the reservoir. (The k_B converts temperature to energy units, and would be superfluous if temperature were measured in energy units.) The quantity Z makes its appearance here as a normalization constant, $Z = \sum_\alpha \exp\left(-E_\alpha/k_B T\right)$, so probabilities add to 1.[7] (N.B. The sum for Z is a sum over *states*; if two states have the same energy they appear twice in the sum.[8])

Although I am mostly giving results in this appendix, let me give a heuristic proof of Eq. (B.2). What I need to add before giving the proof is that temperature is given by $1/T = dS/dE$; as temperature and energy increase the rate of entropy increase decreases. That this temperature agrees with the "T" in the ideal gas relation ($PV = Nk_B T$) is a small miracle, commented on by Baierlein ([15], Sec. 4.3), among others.

Our system is in contact with another, larger one, a "reservoir" at temperature T. It can exchange energy but the total is conserved. Thus ("system" is subscript s and "reservoir" subscript r)

$$0 = \Delta E_{\text{total}} = \Delta E_s + \Delta E_r \tag{B.3}$$

---

[7] Z looks innocent enough as a normalization factor, but it has considerable thermodynamic significance. In fact $Z = \exp(-F/k_B T)$, where F is the free energy. The letter Z is the first letter of *Zustandssumme*, which in German means "sum over states."

[8] OK, it's time for my cannibal story. A traveler in the deepest jungle is captured by cannibals. They are about to throw him into a large pot of boiling water. Does he worry? No. He's sure the most likely state is that of lowest energy, in other words, ice. After all the Boltzmann distribution tells him that. Should he nevertheless worry?

(where Δ means "the change in"). The change in entropy is given by a similar equation, and our assumption that the system is in equilibrium also implies constancy of the total:

$$0 = \Delta S_{\text{total}} = \Delta S_s + \Delta S_r . \tag{B.4}$$

Letting the Δ's refer to infinitesimal changes and taking E_{system} as the independent variable implies

$$0 = \frac{\partial S_s(E_s)}{\partial E_s} + \frac{\partial S_r(E_{\text{total}} - E_s)}{\partial F_s} = \frac{\partial S_s(E_s)}{\partial E_s} + \frac{\partial S_r(E_r)}{\partial E_r}\frac{\partial(E_{\text{total}} - E_s)}{\partial E_s} . \tag{B.5}$$

Using both $\frac{\partial(E_{\text{total}}-E_s)}{\partial E_s} = -1$ and $1/T = dS/dE$ implies that the system has the same temperature as the reservoir. Now imagine that the system is in some state α with energy E_α. By the conservation of total energy the reservoir must possess energy $E_{\text{total}} - E_\alpha$. The multiplicity of microscopic states in the reservoir is therefore $\mathcal{M} = \exp\left(\frac{S_r(E_{\text{total}}-E_\alpha)}{k_B}\right)$ (by the definition of S). We truncate[9] the power series expansion of S_r at the first non-trivial term, $S_r(E_{\text{total}} - E_\alpha) = S_r(E_{\text{total}}) - E_\alpha\frac{S_r(E_{\text{total}})}{\partial E_{\text{total}}}$, and ignore the proportionality factor $S_r(E_{\text{total}})$ common to all states, to get

$$\Pr(\alpha) \propto \exp\left(\frac{-E_\alpha}{k_B T}\right), \tag{B.6}$$

since $1/T = \frac{S_r(E_{\text{total}})}{\partial E_{\text{total}}}$. Inserting the proportionality factor $1/Z$, one gets the Boltzmann distribution, Eq. (B.2).

The third ensemble popular among physicists is the grand canonical ensemble. Not only can the reservoir exchange energy, but it can exchange particles as well. Often this makes proving things easier, but it does not play a role in the present volume. As for the canonical ensemble a new parameter is introduced, the chemical potential, usually designated by μ. This is the analog of T for the canonical ensemble.

Change of ensemble can also be phrased as a Legendre transformation. Thus in the microcanonical ensemble energy is a state variable, while in the canonical ensemble the variable that replaces energy is "free energy," F. Thus E and T are related by

$$\frac{1}{T} = \frac{\partial S}{\partial E} \tag{B.7}$$

(with S the entropy, E energy and T temperature). The Legendre transformation replacing energy by temperature is the replacement of a function by its derivative

---

[9]The truncation relies on the "reservoir" being much larger than the "system." Subsequent terms in the expansion involve the ratio of system to reservoir and in this idealization can be made arbitrarily small.

$$F = E - TS.$$
(B.8)

Similarly for the other ensembles, in each case defining a new thermodynamic potential. This transformation also occurs in classical mechanics, where the Lagrangian is replaced by the Hamiltonian, using the momentum, a derivative of the Lagrangian.

There are other ensembles. For example, there is also something known as enthalpy, but that too makes no appearance in this book, and we will not address its ensemble nor define it.

Appendix C
Fractals

If you have a straight line that's 1 meter in length and you use rulers whose length is $1/N$ (in meters), you'll find that you need N rulers to cover the line. For the record, N is N^1, that is, N to the power 1. So a straight line has dimension 1. Now take the coastline of Norway. If you use rulers of length $\ell = 100$ km, you'll get a certain length for the coast and you'll need $N(\ell)$ rulers to cover it, with $\ell = 100$ km. As you make the ruler shorter though the number needed to cover the coastline increases more rapidly than 1/[rulerlength]. This is because as you use a smaller ruler you are able to follow landscape features, like fjords and streams. In fact as Jens Feder[1] has shown [57], for ℓ ranging over a factor of about three orders of magnitude, you'll find $N(\ell) \sim \text{const} \times \ell^{-1.52}$ for the coast of Norway. That number, 1.52, is the *fractal dimension* of the coastline. In this case you're dealing with a physical phenomenon, so the range of values of ℓ is bounded.

For mathematical idealizations you can really go to the limit and define the fractal dimension as

$$d \equiv \lim_{\ell \to 0} \frac{\log(N(\ell))}{\log(1/\ell)} . \tag{C.1}$$

This formula holds for other figures embedded in some dimension (and you can use "rulers" that are hypercubes of side ℓ), for example the Sierpiński gasket or the Cantor set.

For the Cantor set I don't need pictures, although the first few steps in the construction (reading down) are illustrated in Fig. C.1. You take a line of length 1, and cut out the middle third, so instead of $[0, 1]$, you have $[0, 1/3]$ and $[2/3, 1]$ (note that I remove the *open set*, leaving the endpoints[2]). Next, remove the middle third from each of the two remaining pieces. Keep going—removing more and more, tinier and tinier, "thirds." What's left is the Cantor set.

---

[1] As a physicist Jens Feder was both a theorist and experimentalist. I knew him personally and was unhappy to learn that he had died in 2019, but such is the fate of all of us. (His son lent me a bicycle for my second participation in the Store Styrkeprøven.) Jens wrote a well-known book on fractals [57], translated into many languages, from which I take this information on the coast of Norway.

[2] A square bracket indicates inclusion of the endpoint in the set and an ordinary parenthesis indicates the exclusion of the endpoint. Thus $[0, 1) = \{x \mid x \geq 0 \text{ and } x < 1\}$.

Fig. C.1 First few steps in the construction of the Cantor set.

Now let's calculate its dimension. For some integer k, take a ruler of length 3^{-k}. You'll need 2^k of them to cover the set (convince yourself of this). So $d = \lim_{\ell \to 0} \frac{\log(2^k)}{\log(3^k)} = \frac{\log 2}{\log 3} \approx 0.6309 < 1$. This is a measure of the evanescence of the Cantor set. A typical feature for fractals is hierarchical structure, a feature observed (over broad ranges) for many physical phenomena. For idealized mathematical fractals (such as the Cantor set) the self-similarity can go on forever, so that the pattern looks the same at all sizes. In this sense the fractal is scale-free.

Exercise 32   Show that the cardinality of the Cantor set is the same as that of the line of all real numbers.

Appendix D
Review of probability

D.1 Basics

You have a set of objects, S, with elements $s \in S$. The set can be finite, countably infinite or uncountably infinite, the three options are epitomized by the following three examples:

- finite: the pips on a die
- countably infinite: the positive integers, $\{1, 2, \dots\}$
- uncountably infinite: all real numbers greater than 0 and less than 1.

In the first two cases you assign a probability to each element, meaning a likelihood that the element would be the result of a trial, a selection of a single element. These probabilities would be numbers p_s, $s \in S$, where $p_s \geq 0$ and $\sum p_s = 1$. In the third case you would take intervals, say ds such that $p(s) ds$ is the probability that the trial would yield a result between s and $s + ds$. We still require $p(s) \geq 0$, but the summation takes a different form: $\int_S p(s) ds = 1$. For this third situation the function $p(s)$ is known as the *probability density function*, often abbreviated PDF. Another abbreviation that you'll often find is CDF, the cumulative density function. This is the total probability up to some value, and would be $\int_{-\infty}^{x} p(s) ds$. One can also mix these or take a limit of intervals, so that (for example) there is probability 0.1 of getting (say) 0.95 and $\frac{1}{1-0.1}$ probability of getting values between (say) 0 and 0.5. This can involve Stieltjes or Lebesgue integration, but for our applications we don't need to get involved in this possibility.

Remark: Examples:

- For (two fair, distinguishable) dice you have 1/6 chance of each one hitting a number of pips between 1 and 6. So the chance of the total being 9 is

$$2\left(\frac{1}{6}\right)^2 (\text{3 and 6, and 6 and 3}) \ + 2\left(\frac{1}{6}\right)^2 (\text{4 and 5, and 5 and 4}) \ = \frac{1}{9}.$$

- The Poisson distribution is $\Pr(n) = \frac{r^n}{n!} \exp(-r)$, $\Pr(n)$ is the probability of n events and the rate of occurrence is r. Thus if you have one event every minute—on the average—then the probability of having three events in 1 minute is

$$\frac{1^3}{3!}\exp(-1) \approx 0.0613.$$

- The uniform distribution of the unit interval gives equal weight to all numbers between 0 and 1. Thus $\Pr(0.5 \le x < 0.51) = 1 \cdot (dx) = 0.01$, where $dx = 0.51 - 0.50$.

One can formalize the notion of trial by means of a *random variable*. Call the random variable X. It is a function from S, typically to a real number. Thus there is probability p_s that a trial will give X the numerical value $X(s)$. For example, suppose S is the set {heads, tails}, the result of a coin flip. Suppose further that if the result is heads you win \$10 and if it's tails you must pay \$9; moreover the coin is not fair, and has 60% probability of landing heads and (therefore) 40% probability of landing tails. (Landing on an edge is not an option.) This would mean X would take the value \$10 with probability 0.6 and the value minus \$9 with probability 0.4.

A natural question for the example just given is should you play? Well, it's not my place to tell anyone whether or not to gamble, but if you're going to you would like to know what to expect. This leads to the notion of *expectation value*.

The *expectation value* of a random variable is defined as

$$\langle X \rangle = \sum_S p_s X(s) \qquad \text{or} \qquad \int_S p(s)X(s). \tag{D.1}$$

(In the future I won't bother writing the integral form.) The expectation of the coin flip described above (omitting the dollar signs) is thus $\langle X \rangle = 0.6 \times 10 + 0.4 \times (-9) = +2.4$. If you only play once and if you've only got \$5, maybe you shouldn't play at all. But if there would be many identical flips and you can risk a bit, it's extremely likely that you'll come out ahead. This is the strategy of casinos, where the odds are generally not quite so unfair, but the games are played many, many times. (In the language of stochastic processes there is a *drift*, *not* zero expectation, and it's in favor of the casino.)

A word about notation: the brackets I used in Eq. (D.1) is physics notation. Another physics notation is the bar; that is, $\overline{X}$ is the same as $\langle X \rangle$. Mathematicians generally prefer the letter "E" (standing for **E**xpectation), that is, $E(X) = \langle X \rangle$.

You can also define functions of a random variable. Suppose you have a random variable X and a function of X (e.g., $f(X) = X^2$) then the expected value of $f(X)$ is $\langle f(X) \rangle = \sum_S f(X(s))p_s$.

Another shortcut that is used is to suppress the set S. One would write

$$\langle X \rangle = \sum_x \Pr(x)x. \tag{D.2}$$

In this language lower case x is the value taken by X and the sum is over all those values. The abbreviation "Pr" means probability. Sometimes for additional clarity I'll write $\Pr(X = x)$ (the probability that the random variable X equals x) and sometimes I'll simply write $\Pr(x)$. This may seem careless of me, but I am being no less careless than most physics writers. So please get used to these various notations.

Now suppose you have two random variables, X and Y, and let the values they take be designated x and y. They might or might not be independent: let me explain. In principle, the value taken say by Y might depend on the value of X.

Here is an example. Take a much simplified deck of cards: ace ($=1$), 2 and 3 for the values of the cards and only two suites, say spades (having the value 1) and diamonds (with value 2). Let x be the card values, 1, 2 or 3, and y the suites, 1 and 2. Then the x and y probabilities are independent of one another, and are equal to $1/6$ for each card. However, suppose you remove one card, say the ace of spades (leaving a deck of five cards, again with equal probabilities, i.e., $1/5$ for each). Then for random selection among the remaining cards the value of the probability of a particular x value is unchanged for diamonds, but is changed for spades. The probabilities are no longer independent. We'll shortly see how that can be made precise.

In general the probability distribution is a joint function, $f(x, y) = \Pr(x, y)$. Again you require that every term in the summand be non-negative and $\sum_{x,y}\Pr(x, y) = 1$. From this it follows that the *marginals*, for example $\Pr(x) = g(x) \equiv \sum_y \Pr(x, y)$, also sum to 1, that is, $\sum_x g(x) = 1$. This is also clear because $g(x)$ is the probability distribution for x, and *some* value of x must occur; similarly for Y.

Consider the (new) random variable $X + Y$. Since the expectation value is a linear function, $\langle X + Y \rangle$ is just $\langle X \rangle + \langle Y \rangle$, in equations

$$\langle X + Y \rangle = \sum_{x,y} \Pr(x, y)(x + y) = \sum_{x,y} \Pr(x, y)x + \sum_{x,y} \Pr(x, y)y$$

$$= \sum_x x \sum_y \Pr(x, y) + \sum_y y \sum_x \Pr(x, y)$$

$$= \sum_x x \Pr(x) + \sum_y y \Pr(y) = \langle X \rangle + \langle Y \rangle. \tag{D.3}$$

Where independence (or lack thereof) matters is in products. Consider X *times* Y:

$$\langle XY \rangle = \sum_{x,y} \Pr(x, y)xy. \tag{D.4}$$

If $\Pr(X = x \ \& \ Y = y) = \Pr(X = x) \cdot \Pr(Y = y)$, then Eq. (D.4) simplifies:

$$\langle XY \rangle = \sum_{x,y} \Pr(x)\Pr(y)xy = \sum_x \Pr(x)x \cdot \sum_y \Pr(y)y = \langle X \rangle \langle Y \rangle. \tag{D.5}$$

If this is *not* the case, then the product is what it is, and there is additional richness. In fact this product is a measure of the degree to which the two random variables are not independent, or, rather, are *correlated*. The *correlation coefficient* is a function of this, but I need to define a few other quantities first. These—in contrast to the correlation coefficient—are only functions of the individual random variables. The mean of X is simply its expectation. It is often called μ_X (= $\langle X \rangle$). Then there is the standard deviation, σ_X, defined as

$$\sigma_X \equiv \sqrt{\langle (X - \mu_X)^2 \rangle}. \tag{D.6}$$

Obviously we make similar definitions for the random variable Y. With these definitions, the (Pearson) correlation function is

$$\rho_{XY} \equiv \frac{\langle (X - \mu_X)(Y - \mu_Y) \rangle}{\sigma_X \sigma_Y}. \tag{D.7}$$

A little algebra shows that $\rho_{XY} = (\langle XY \rangle - \mu_X \mu_Y)/(\sigma_X \sigma_Y)$. Thus the only term in the correlation coefficient that hasn't been defined through one-variable definitions is $\langle XY \rangle$.

We check independence and lack thereof on our simplified deck of cards. If you have not drawn anything, each card has probability $1/6$ of appearing. The various sums are

$$\langle XY \rangle = \frac{1}{6}(1 + 2 + 3 + 2 + 4 + 6) = 3,$$
$$\langle X \rangle = \sum_x x \sum_y P(x,y) = \frac{1}{3}(1 + 2 + 3) = 2$$
$$\text{and} \quad \langle Y \rangle = \sum_y y \sum_x P(x,y) = \frac{1}{2}(1 + 2) = \frac{3}{2}, \tag{D.8}$$
$$\langle XY \rangle = \langle X \rangle \langle Y \rangle,$$

so that before picking one card the variables are independent. However, after picking the ace of spades...

$$\langle XY \rangle = \frac{1}{5}(2 + 3 + 2 + 4 + 6) = \frac{17}{5},$$
$$\langle X \rangle = \sum_x x \sum_y P(x,y) = \frac{1}{5}(1 + 2 \cdot 2 + 3 \cdot 2) = \frac{11}{5}$$
$$\text{and} \quad \langle Y \rangle = \sum_y y \sum_x P(x,y) = \frac{1}{5}(2 + 3 \cdot 2) = \frac{8}{5}, \tag{D.9}$$
$$\langle XY \rangle = \frac{17}{5} \neq \langle X \rangle \langle Y \rangle = \frac{11}{5} \cdot \frac{8}{5}.$$

The variables are no longer independent. (One is $\frac{85}{25}$, the other $\frac{88}{25}$.)

Exercise 33 Show that $\langle X \rangle \langle Y \rangle = \langle XY \rangle$ implies that $\Pr(X=x \,\&\, Y=y) = \Pr(x)\Pr(y)$, the converse of Eq. (D.4).

D.2   Counting

A lot of probability theory involves counting. This is because one assumes many things are equally likely and then it's a matter of counting all those that lead to a given outcome, giving the probability of that outcome. The "things" just mentioned could be microscopic states of a system, letters of an alphabet, terms in an expansion, and many other examples—which is why I used the vague term "things."

The most famous example of counting involves the combinatorial coefficient,

$$C_k^N \equiv \binom{N}{k} \equiv \frac{N!}{k!(N-k)!} , \tag{D.10}$$

where $k!$ is the factorial, $k! \equiv k(k-1)\ldots 1$, and $0!$ is defined to be 1. C_k^N is the number of ways to select k objects out of a total of N. The objects could be spins that are "up" (cf. Sec. 5.1) or steps to the right in a random walk or terms in a binomial product. The main requirement is that it doesn't matter in what order they are selected. For example, in the random walk a step to the right on the third or fifth step is still a step to the right, and the position of the walker at a later time does not depend on which is which.[1] For approximations to the factorial, see Sec. D.7.

Justification of Eq. (D.10): You have N objects of type A or type B. There are k A's and $(N-k)$ B's. How many different ways are there to order them? Imagine that they had labels, $A_1, A_2, \ldots A_k$, $B_{k+1}, \ldots, B_N$. Then there would be $N!$ ways to list them, $A_{17}B_{42}\ldots A_{11}$ (using all N numbers), and all permutations. However, you've over-counted by numbering them. The A's can appear in any order—that's over-counting by $k!$ and the same for the $(N-k)$ B's.

The binomial expansion $(x+y)^N$ is another example of the use of $\binom{N}{k}$. We have the following relation:

$$(x+y)^N = \sum_{k=0}^{N} \binom{N}{k} x^k y^{N-k} . \tag{D.11}$$

Incidentally, by setting $x = y = 1$ you have the useful identity $2^N = \sum_0^N \binom{N}{k}$.

Eq. (D.10) can be generalized to situations where you have three or more (m in the formula) kinds of objects:

$$\binom{N}{n_1 n_2 \ldots n_m} = \frac{N!}{n_1! n_2! \ldots n_m!} , \tag{D.12}$$

with the proviso that $\sum_{\ell=1}^{m} n_\ell = N$. This is the number of ways of choosing n_1 objects of type 1, n_2 objects of type 2, and so on, without regard to their order of selection. In Sec. 6.1 this coefficient makes an appearance.

---

[1] Of course the *total* number of steps to the right must remain the same, so that if the fifth rather than the thrid step is to the right, then the third step (or some other step that was formerly to the right) should be to the left.

The combinatorial coefficient appears for walks on a "Manhattan" grid. Consider the number of ways to walk on a square two-dimensional grid from the point $(0,0)$ to the point $(k, N-k)$ (where $0 \le k \le N$ and both are positive integers). (This could be Manhattan (New York) and you could be walking from 11th Ave. corner 43rd St. to 8th Ave. and 56th St.) The number of ways to walk: $\binom{N}{k}$ with $N = 16$ and $k = 3$ or 13, since $13 + 3 = 16$. (I've picked streets where Broadway does not mess things up.)

D.3 The central limit theorem

Square-root-of-N fluctuations are closely related to the central limit theorem. I'll prove a simple version of the theorem.

For a probability distribution[2] $p_X(x)$ of a random variable X, I define its characteristic function

$$\phi_X(s) \equiv \langle e^{isX} \rangle = \int_{-\infty}^{\infty} p_X(x) e^{isx}\, dx. \tag{D.13}$$

To go from ϕ to p you use the inverse Fourier transform:

$$p_X(x) = \frac{1}{2\pi} \int_{-\infty}^{\infty} e^{-isx} \phi_X(s)\, ds. \tag{D.14}$$

The characteristic function can also be used to find the moments of X (those that exist, anyway) and ϕ is known as the moment generating function. This is based on the fact that

$$\frac{1}{i^n} \frac{d^n}{ds^n} \phi(s) = \frac{1}{i^n} \langle i^n X^n e^{isX} \rangle \underset{\text{for } s \to 0}{\longrightarrow} \langle X^n \rangle \equiv \mu_n. \tag{D.15}$$

In particular

$$\phi(0) = 1 \qquad \text{probabilities sum to 1,}$$
$$-i\phi'(0) = \mu_1 \qquad \text{first moment,}$$
$$-\phi''(0) = \mu_2 \qquad \text{second moment.}$$

Sometimes it is useful to put the power series in the exponent and one writes

$$\log \phi(s) \equiv \sum_{n=1}^{\infty} \kappa_n \frac{i^n s^n}{n!}. \tag{D.16}$$

The κ's are called *cumulants*. The relations between the first few μ's and κ's are

$$\kappa_1 = \mu_1,$$
$$\kappa_2 = \mu_2 - \mu_1^2 \qquad (= \sigma^2),$$
$$\kappa_3 = \mu_3 - 3\mu_2\mu_1 + 2\mu_1^2,$$
$$\kappa_4 = \mu_4 - 4\mu_3\mu_1 - 3\mu_2^2 + 12\mu_2\mu_1^2 - 6\mu_1^4.$$

[2] ... a.k.a. $\Pr(X = x)$.

Note that κ_2 is the same as σ^2, where σ is the standard deviation. (σ^2 is known as the variance.) The moments μ_3 and μ_4 are given by Eq. (D.15).

An important property of the characteristic function is that for *independent* random variables X and Y the characteristic function of their sum is the product of their characteristic functions. This is because the expectation of a product is the product of its expectations—cf. Eq. (D.5):

$$\phi_{X+Y}(s) = \left\langle e^{is(X+Y)} \right\rangle = \left\langle e^{isX} e^{isY} \right\rangle = \left\langle e^{isX} \right\rangle \left\langle e^{isY} \right\rangle = \phi_X(s)\phi_Y(s). \tag{D.17}$$

N.B. This depends on $\Pr(x, y) = \Pr(x)\Pr(y)$, that is, X and Y are independent. By the definition of logarithm, it is also clear that the cumulants of $X + Y$ are simply the sums of the cumulants of X and Y separately.

Consider the sum of N independent, identically distributed (i.i.d.) random variables:

$$S_N = \sum_{\ell=1}^{N} X_\ell. \tag{D.18}$$

By Eq. (D.17) its characteristic function is

$$\phi_{S_N}(s) = [\phi_X(s)]^N, \tag{D.19}$$

where the function $\phi_X(s)$ is $\phi_{X_\ell}(s)$ for any ℓ, since all the X_ℓ's have the same characteristic function (by assumption).

It's clear that the expectation of S_N is the sum of the expectations of the X's so this part of S's probability distribution is fairly trivial and for simplicity we assume $\langle X \rangle = \mu_1 = 0$. (If you want to keep track of the mean you can replace X by $Y \equiv X - \langle X \rangle$ and undo this substitution at the end.) It follows that

$$\log \phi_{S_N}(s) = N \cdot \left(\sigma^2 \frac{i^2 s^2}{2} + \kappa_3 \frac{i^3 s^3}{6} + \dots \right) \tag{D.20}$$

(recall, $\sigma^2 = \kappa_2$).

We finally come to the central limit theorem. It's a statement not about S_N but about $\tilde{S}_N \equiv \frac{S_N}{\sigma\sqrt{N}}$, with σ (the square root of the second cumulant) corresponding to that of the random variable X. First a fact that is obvious from the definition, Eq. (D.13), of the characteristic function. If α is a real number, then

$$\phi_{\alpha X}(s) = \left\langle e^{is(\alpha X)} \right\rangle = \left\langle e^{i(s\alpha)X} \right\rangle = \phi_X(\alpha s). \tag{D.21}$$

(αX is a random variable—if it should happen that X takes the value $1/2$, then αX would take the value $\alpha/2$.) Letting "α"$= 1/\sigma\sqrt{N}$, we have

$$\log \phi_{\tilde{S}_N}(s) = \log \phi_{S_N}(s/\sigma\sqrt{N}) = -\frac{s^2}{2} + \frac{\kappa_3}{\sigma^3} \frac{1}{\sqrt{N}} \frac{i^3 s^3}{6} + \dots. \tag{D.22}$$

To find the probability distribution function for $\tilde{S}_N$ we Fourier transform, as in Eq. (D.14),

$$p_{\tilde{S}_N}(x) = \frac{1}{2\pi} \int_{-\infty}^{\infty} ds \exp\left(-isx - \frac{s^2}{2} + \frac{s^3}{\sqrt{N}} \cdot \text{something}\right). \tag{D.23}$$

The next step is to say that as $N \to \infty$, the term with "something" disappears because of the $\frac{1}{\sqrt{N}}$. The integral is then simply $\exp(-x^2/2)/\sqrt{2\pi}$, which is the normal (or Gaussian) distribution with variance 1. This leads to the statement:

The random variable $\dfrac{\sum_{\ell=1}^{N} (X_\ell - \langle X \rangle)}{\sigma\sqrt{N}}$ approaches the normal distribution with variance 1, as $N \to \infty$.

That's the central limit theorem. A particular consequence is that the original variable, S_N, has standard deviation $\sigma\sqrt{N}$, which is the (famous) square root of N that we've been encountering all along.

Of course the mathematically interesting part is getting rid of the "something" in Eq. (D.23), which can fail for a variety of reasons. The one that is most significant physically is when "something" is infinite or even if σ itself doesn't exist. When σ does not exist the attractors are the Lévy distributions, for example, the Cauchy distribution, $C_a(x) = \frac{a/\pi}{a^2+x^2}$. However, in the most general proof, you don't need much more than the existence of σ, nor are things restricted to the summands being identically distributed. For the systems we'll consider the breakdown in the $\sqrt{N}$ measure of fluctuations arises because the physical variables are *not* independent. The world is not an ideal gas!

D.4 Markov processes

A Markov process is something without memory. Every change of state depends only on the current state, not on previous history.[3] The archetype is the random walk. Let X_i, $i = 1,\dots,N$, be "i.i.d." (independent, identically distributed). So all these (random) variables have the same distribution and are independent of one another. For a random walk one thinks of each X_i as a step to the left or the right, and without loss of generality one can take $X_i = \pm 1$. The main interest is in $S_N = \sum_i X_i$, the location of the random walker after N steps. As shown in Chapter 2, although $\langle S_N \rangle = 0$, $\langle S_N^2 \rangle$ grows like N. This is typical of diffusive behavior.

---

[3] You can deal with finite memory by enlarging the space of interest.

Another example of a Markov process is percolitis (Chapter 4). One can take as a random variable the number of sick people on any particular day. The number sick the following day depends only on that. (I'm assuming there's no immunity.[4])

A convenient tool for study of Markov processes is the stochastic matrix, defined in the next subsection. For percolitis, in Footnote 30, Chapter 4, this is used to study extinction times, the expected duration of the "disease" even above the threshold for its persistence.

D.5   Stochastic dynamics

This topic mixes probability theory with time evolution. Say at time zero your system may be in one of several possible states, each with some probability. This can be described as a function of two variables, $p(x, t)$, where t, the time, starts at zero and x takes values in some set, finite or infinite. On the next—in this case the first—time step, there is a probability to go from the state x to another state, say x'. This is described by a matrix, call it R, with

$$R(x', x) = \Pr(x' \leftarrow x). \tag{D.24}$$

$\Pr(x' \leftarrow x)$ means the probability that the system goes from the state x to the state x' in one time step. Note that I'm taking R to be time independent and time itself to be discrete (*and* I am reading right to left). With this convention the probability distribution at time-$(t + \Delta t)$ is given in terms of that at time-t by

$$p(x, t + \Delta t) = \sum_y R(x, y) p(y, t) \tag{D.25}$$

(where I've changed the names of the variables). The size of the time step, Δt, is arbitrary and is usually taken to be either 1 or $1/N$, where x and y are in a space X and N is the size (cardinality) of X.

Since every state has to go *somewhere*, the probabilities for leaving x must add to 1, that is,

$$\sum_{x'} R(x', x) = 1 \quad \text{for every } x. \tag{D.26}$$

A square matrix with all non-negative entries and whose column sums are unity is known as *stochastic*. Eq. (D.26) can be looked upon as a left eigenvalue equation for

---

[4]If there is, say, 1 day of immunity, the space can be enlarged. But at some point the memory requirements become excessive.

the left eigenvector $A_0(x') \equiv 1$. But for every left eigenvector there's a right eigenvector with the same eigenvalue, so there exists p_0 such that

$$p_0(x) = \sum_y R(x,y)p_0(y) \,. \tag{D.27}$$

I quote a number of known results and definitions for stochastic matrices. In the following "R" is a stochastic matrix and p_0 a right eigenvector of eigenvalue 1.

- All eigenvalues of R fall on or within the unit circle in the complex plane (Frobenius–Perron theorem).
- (Definition) A stochastic matrix is said to be irreducible if any state can reach any other state (by repeated application of R). There are more abstract (equivalent) definitions: A reducible matrix is one for which you can rearrange rows and columns so that there is a non-trivial block on the lower left of all zeros. A matrix is irreducible if it is not reducible.
- If R is irreducible all entries of p_0 are strictly positive and p_0 is the only right eigenvector of eigenvalue 1.
- (Definition) Assuming p_0 to be unique, the current is $J(x,y) = R(x,y)p_0(y) - R(y,x)p_0(x)$ (no summation over doubled entries). If the current is zero (for all x and y) the system is said to satisfy detailed balance and is at equilibrium in the state p_0.
- If there are no eigenvalues *on* the unit circle, aside from 1, then the relaxation rate to the state p_0 is governed by the largest magnitude eigenvalue within the circle. This follows from a spectral expansion of R (including the possibility that R requires a Jordan form[5]).

In Chapter 5 I define a stochastic matrix (just before Exercise 14) that recovers the Boltzmann distribution for a given assignment of energies to the states. This is an equilibrium distribution and the current (just defined above) is zero. As you can imagine, not all stochastic matrices give rise to zero current and one obtains *non-equilibrium* steady states as the right eigenvector of eigenvalue 1, namely what we've been calling p_0.

There is a continuous time version of R. One can derive the form letting each time step be Δt and letting $\Delta t \to 0$. The result is

---

[5]... and if you don't know what a Jordan form is, you needn't be concerned. It doesn't come up in this book. For the record, it's a matrix J with degenerate eigenvalues and off-diagonal entries of the form (for example) $\begin{pmatrix} 0 & 1 & 0 \\ 0 & 0 & 1 \\ 0 & 0 & 0 \end{pmatrix}$. By transformations of the form SJS^{-1} it cannot be made fully diagonal. A raising operator is an example of such a matrix.

$$\frac{dp(x,t)}{dt} = \sum_y W(x,y)p(y,t) - p(x,t)\sum_y W(y,x). \tag{D.28}$$

In the physics literature Eq. (D.28) is often called the master equation. Under appropriate circumstances this can be related to a Fokker–Planck equation. See [226].

Exercise 34   Deduce the spectral properties of *W*.

Unfortunately I don't have any good references for this material. If you're willing to read the mathematical literature I can recommend [25, 65, 66, 109, 147]. A review paper on the observable representation is [198].

D.6   The notion of probability

There are philosophical discussions of whether "probability" is likelihood based on intuitive reasoning or whether it is based on trials, or maybe something else. As far as I can tell, these matters do not affect the calculations above, and I will not discuss this issue.

There is also the question of whether quantum mechanics is fundamentally probabilistic, concerning which there seems to be a consensus answer of "yes," except that both I and (*le-havdil*) Einstein stand in opposition.[6]

D.7   Stirling's approximation

The factorial *n*! can be approximated as

$$n! \sim \sqrt{2\pi n}\left(\frac{n}{e}\right)^n, \tag{D.29}$$

or less precisely as

$$\log n! \sim n(\log n - 1). \tag{D.30}$$

This can be derived from an asymptotic approximation for the Gamma function

$$\Gamma(z) = \int_0^\infty dt\, t^{z-1} e^{-t}. \tag{D.31}$$

Asymptotic approximations need not converge in the usual sense. Rather you fix the number of terms you're looking at and show that the remainder goes to zero faster

---

[6]Hebrew and Yiddish, meaning "as distinct from," literally, "to divide." In this case it means that I do not compare myself to Einstein, although we agree on this point.

than the last term in your series (in this case $n \to \infty$ or $\mathrm{Re}\, z \to \infty$). For example, if you used a more precise version of Stirling's approximation, such as ($x = \mathrm{Re}\, z$)

$$\Gamma(x) \sim \sqrt{2\pi x}\left(\frac{x}{e}\right)^x\left[1 + \frac{1}{12x}\right], \qquad (D.32)$$

then all you are guaranteed is that as $x \to \infty$ the error within the brackets would be smaller than order $1/x$.

To prove the integral form you first establish that $n! = \Gamma(n+1)$. This follows by induction. First, it's clear that both $\Gamma(1)$ and $\Gamma(2)$ are 1 (which incidentally justifies $0! = 1$). Then using integration by parts you show that

$$\Gamma(z) = t^{z-1}e^{-t}\big|_0^\infty + (z-1)\Gamma(z-1) = (z-1)\Gamma(z-1). \qquad (D.33)$$

For $z = n$, with n an integer greater than 1, this implies $\Gamma(n) = (n-1)\Gamma(n-1)$. Then by induction

$$n! = \Gamma(n+1) = \int_0^\infty dt\, t^n e^{-t} = \int_0^\infty dt\, \exp(n\log t - t). \qquad (D.34)$$

We are interested in this quantity for large n. It's clear that the integrand peaks somewhere (a "battle" between the growing t^n and shrinking e^{-t}), so that to find that peak we differentiate the logarithm of the integrand: $\phi(t) \equiv n\log t - t$ and $\phi' = n/t - 1$, implying that the maximum is at $t = n$. Higher derivatives of ϕ at $t = n$ are given by $\phi''(n) = -1/n$, with one higher power of n^{-1} for each additional derivative. We make two approximations: We drop derivatives higher than the second and we extend the left limit of integration to $-\infty$. They have the following consequences, respectively: higher derivatives would yield further terms in the asymptotic expansion (cf. the comment following Eq. (D.31)) and extending the range gives an error that is exponentially small in n (see, for example, [170]). The integral has thus become

$$n! = \int_{-\infty}^\infty dt\, \exp\left(\phi(n) + 0 + \frac{1}{2}(t-n)^2\phi''(n)\right) = \left(\frac{n}{e}\right)^n \int_{-\infty}^\infty du\, \exp\left(-\frac{u^2}{2n}\right), \qquad (D.35)$$

using $\phi'(n) = 0$. This leaves us with only a Gaussian integral to do, and making use of

$$\int_{-\infty}^\infty dx\, e^{-ax^2} = \sqrt{\frac{\pi}{a}}, \qquad (D.36)$$

we obtain the result Eq. (D.29). Eq. (D.36) is itself proved by squaring the integral and using polar coordinates in the resulting two-dimensional integration over the plane.[7]

---

[7]Proving Eq. (D.36) involves one of the nicest derivations that I've seen. I believe I first heard it from Bernard Epstein, my college instructor. Here is the derivation. Call the integral $J \equiv \int_{-\infty}^\infty dx\, e^{-ax^2}$. Then calling a second variable by a different name (y), you have $J^2 = \int_{-\infty}^\infty dx\, dy\, e^{-a(x^2+y^2)}$. But you can go over to polar coordinates in the plane and the angular integral is trivial (since the angle does not appear in the integrand). Thus $J^2 = 2\pi \int_0^\infty r\, dr\, e^{-ar^2}$, which immediately gives the result of Eq. (D.36).

D.8 Exercises in probability

The following are exercises drawing on an elementary knowledge of probability theory. They vary in their level of difficulty.

Exercise 35 How many different letter arrangements can be obtained from the letters of the word *statistically*, using all the letters?

Exercise 36 A deck of 52 cards is shuffled thoroughly. What is the probability that the four aces are all next to one another?

Exercise 37 If n balls are distributed randomly into k urns, what is the probability that the last urn contains j balls?

Exercise 38 If a five-letter word is formed at random (meaning that all sequences of five letters are equally likely), what is the probability that no letter occurs more than once?

Exercise 39 What is the coefficient of x^3y^4 in $(x+y)^7$?

Exercise 40 Two dice are rolled, and the sum of the face values is six. What is the probability that at least one of the dice came up a 3?

Exercise 41 A player throws darts at a target. On each trial, independently of the other trials, she hits the bull's-eye with probability 0.05. (So the trials are i.i.d.) How many times should she throw so that her probability of hitting the bull's-eye at least once is in excess of $1/2$?

Exercise 42 A cube whose faces are colored is split into 1000 small cubes ($10^3 = 1000$) by defining the small cubes through the intersection of three sets of nine equally spaced orthogonal planes, cutting the large cube. Only the original cube has colored faces. The (small) cubes thus obtained are mixed thoroughly. Find the probability that a cube drawn at random will have two colored faces.

Exercise 43 Monty Hall problem. Monty Hall was the host of the game show *Let's make a deal*. A contestant, call her Ms. A, was given a choice of three doors. Behind one is a car (which she can keep), and behind the other two, a goat (which—if she lives in a city—I wouldn't advise keeping). Ms. A makes a guess. But before the door is opened to reveal whether Ms. A has opened the right door, Mr. Hall (who knows where the car is) opens a door behind which a goat stands. Ms. A is then asked whether she wishes to change her choice. Should she?

Exercise 44 A lot contains m defective items and n good ones. From this lot s items are chosen at random and tested for quality. The first k of these are found not to be defective. What is the probability that the next item tested is good?

Exercise 45 A break occurs at a random point, C, on a telephone line $A\,B$ of length L. What is the probability that C is at a distance more than ℓ from the point A? See Fig. D.1.

Fig. D.1. A transmission line from A to B with a break at C.

Exercise 46 The random numbers x and y are drawn from uniform distributions on $[0, 1]$. (In other words, they are equally likely to be anywhere between 0 and 1.) What is the probability that they satisfy the following two conditions: their sum is equal to or less than one, and their product is equal to or less than $2/9$? (Hint: It is useful to consider the areas as subsets of the unit square.)

Exercise 47   Archer A has probability 0.8 of hitting a target, and Archer B has probability 0.7; what is the probability that at least one of them will hit the target? (Each shoots one arrow.)

Exercise 48   An urn contains three distinguishable balls, one red, one black and one white. You draw balls from it five times, one ball at a time, with replacement (i.e., after you draw a ball and check its color you put it back). What is the probability that the red and white balls will be drawn at least twice each?

Appendix E
The van der Waals gas

Every elementary text on thermal physics will have a discussion of the van der Waals gas. For example, see §12.9 of [15]. Instead of the ideal gas law, "$PV = Nk_BT$," the equation of state is $(P+aN^2/V^2)(V-Nb) = Nk_BT$, with a and b parameters of the gas. (P = pressure, V = volume, N = number of particles, k_B = Boltzmann's constant, and T = temperature.) This is still a mean field theory, but now there is a phase transition.

The quantity b represents a hard core repulsion between the particles, and is the volume occupied by a single particle (atom, molecule, whatever). Subtracting Nb from the volume can be considered a correction for that effect. The quantity a is a measure of the attractive force between gas particles and multiplies the number density squared. This is because the chance that two molecules are close enough to interact goes like the product of their two densities (in this case the densities are the same).

With this equation you get a critical point, a place where gas and liquid are the same. The "special values" mentioned in the main text are the location of this critical point, and in terms of the parameters just given are $V_c = 3Nb$, $k_BT_c = 8a/27b$ and $P_c = a/27b^2$.

Below the critical point and at fixed temperature you have a phase transition; there are two equilibrium volumes at the same pressure for a range of pressures (not all pressures). At which pressure are they at equilibrium? By a relatively simple argument (for the van der Waals model) you get the "Maxwell equal area" rule: it is that pressure such that a line drawn from the descending P vs. V curve to the next descending line is such that the areas of the two portions so delineated are equal. See Fig. E.1. (For proof, use the Gibbs–Duhem relation.)

In having a critical point, a gas obeying the van der Waals equation of state will resemble those studied by Guggenheim (Chapter 1). But the critical exponents, the ways in which correlation lengths, specific heats and other quantities approach infinity at the critical point, will be different. In particular the critical exponents for the van der Waals gas are those of mean field theory, the same as the corresponding quantities for the Curie–Weiss model of ferromagnetism.

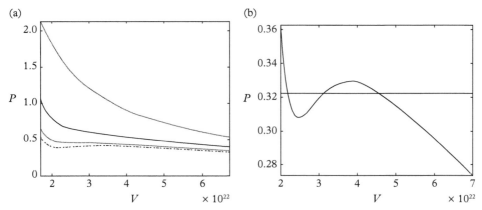

Fig. E.1 Graph of the van der Waals equation of state. (a) The lower three curves represent the function $P = \frac{Nk_B T}{V - Nb} - a\left(\frac{N}{V}\right)^2$ for three different temperatures (three values of T). The fourth (top most) curve is $P = Nk_B T/V$ for the highest temperature of the other three (the solid curve). The second curve from the bottom (dotted, (a)) is at the critical value. (b) The temperature is slightly below the critical temperature (with a different value of "b"). The area under the horizontal line (after the first intersection) at (about) $P = 0.322$ is (approximately) equal to the area above the line (prior to the last intersection), implying, using the Maxwell equal area rule, that there is a first-order phase transition at that temperature and pressure. (Units are $k_B = 1$.)

Appendix F
The logistic map

The logistic equation began in ecology as a differential equation

$$\frac{dN}{dt} = rN\left(1 - \frac{N}{K}\right),$$ (F.1)

where N is the population of some creature at time-t, dN/dt is its rate of change, r the reproduction rate and K the carrying capacity, meaning the number of creatures that a given region can support. This equation yields a well-behaved solution,

$$N(t) = \frac{K}{1 - \left(1 - \frac{K}{N_0}\right)e^{-rt}},$$ (F.2)

where N_0 is the initial population. $N(t)$ smoothly approaches K for any strictly positive N_0.

But then in the 1970s Robert May [150] tried a discrete version of Eq. (F.1) and found a surprise: for sufficiently large r a solution would bounce around even when the time was very large. The version usually used in physics confines variables (and changes their names) to $0 \leq x \leq 1$ and $0 \leq \lambda \leq 4$[1]

$$x' = \lambda x(1 - x).$$ (F.3)

This equation was further manipulated by Feigenbaum [58, 59], who found surprisingly universal results. But first I need to tell you what an attractor is. It's the limit of repeated application of the mapping Eq. (F.3). In other words if $x_{n+1} = \lambda x_n(1 - x_n)$, then the attractor is $\lim_{n \to \infty} x_n$, assuming that limit is a single point (hold on though— we'll have to generalize this definition).

- For $0 \leq \lambda \leq 1$ the only real, non-negative attractor is $x = 0$. As λ goes beyond 1, the root[2] 0 becomes unstable and the new attractor is $1 - 1/\lambda$. (Solve Eq. (F.3) with the left-hand side (x) the same as the right-hand side.)

---

[1]The fact that Eq. (F.1) appears to depend on two variables (r and K) is spurious and can be seen by substituting $u = N/K$. The discrete version replaces dN/dt by $N(t+1) - N(t)$.

[2]I call this "root" because an attractor that is a single point must be a root of the equation $x = \lambda x(1 - x)$, that is to say, the iteration brings you back to the same point.

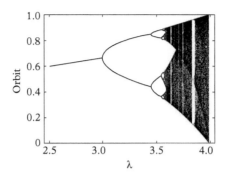

Fig. F.1 Attractors of the logistic map for $2.5 \le \lambda \le 4$.

- At $\lambda = 3$ the derivative of the function $F(x) \equiv \lambda x(1-x)$ becomes -1. This implies that the attractor for greater values of λ becomes unstable, *and* there is no attractor, that is, there ceases to be a point such that $\lim_{n \to \infty} x_n$ converges. This leads to a generalization of the concept of attractor: it becomes the *set of points* defined by $\lim_{n \to \infty} x_n$. Thus for (say) $\lambda = 3.1$ the attractor consists of the stable points that are fixed points of the mapping $x = F(F(x))$ (of which there are two). The phenomenon of the splitting of an attractor is called a bifurcation.

- For λ just a bit above 3 the attractor consists of two points.[3] Then as λ increases the attractor has four, then eight, and so forth until at $\lambda = 3.56994\ldots$ the attractor becomes infinite. There are two remarkable features involved in the approach to that point:

  * Let λ_n be the value of λ at which the nth bifurcation takes place. Then the

  $$\text{Feigenbaum number} = \lim_{n \to \infty} \frac{\lambda_n - \lambda_{n-1}}{\lambda_{n+1} - \lambda_n} = 4.669201\ldots \qquad \text{(F.4)}$$

  goes to the indicated limit.
  * The attractor at the limit point is *strange*. In particular, it has a fractal dimension that is less than 1.

- Beyond the critical value ($\lambda \approx 3.56994\ldots$) the attractor grows in its span although there are λ values for which you nearly get periodic points with periods not restricted to powers of 2. Finally at $\lambda = 4$ the mapping yields complete chaos with the range of x being the entire interval $(0,1)$ (for almost all initial conditions).

- See Fig. F.1 for an image of some of the attractors.

---

[3]There are four real solutions of $F(F(x)) = x$, but two of them (0 & $1 - 1/\lambda$) are unstable. It is the new solutions of that equation that are now the stable attractors.

So what? The amazing thing is that as you change parameters on certain differential equations the Poincaré section (to be defined in a moment) of the solution looks the same! Let me give an example. You can consider a nonlinear oscillator with damping and forcing, the Duffing equation,

$$\ddot{x} + 2\gamma\dot{x} + 1 \pm x^3 = f\cos(\omega t),\tag{F.5}$$

where the dot stands for d/dt, f, γ and ω are parameters and you want a solution for $x(t)$. (The form Eq. (F.5) is reduced: most significantly the x^3 had a coefficient, eliminated by a change of variables.) This equation makes appearances in many applications: electronic oscillators, biology, mechanical engineering, and really it's the next approximation to Hooke's law (linear dependence of force on displacement), with the option that at high forcing one can go to chaos. You can take a Poincaré section by taking the value of x each time the forcing term satisfies $\omega t = 2n\pi$ for integer n, a kind of stroboscopic view. Now plot this value of x versus ω for large f (the values used are $f = 25$, $\gamma = 0.1$ and the plus sign is taken in Eq. (F.5)). What happens is that at small ω there is but a single value of x in the Poincaré section, that is, at the times t when $\omega t/2\pi$ is an integer. As ω increases you get a bifurcation and two values of x occur, and it continues until at $\omega \approx 1.29$ you get an infinite cascade. Moreover, the values of $\lim_{n\to\infty}\frac{\omega_n - \omega_{n-1}}{\omega_{n+1} - \omega_n}$ are the Feigenbaum number! (There's the possibility of experimental error in the latter estimate, since the results are numerical.) I won't include a plot because it looks just like Fig. F.1, aside from rescaling.

There is a further step that some people take. One of the big questions is turbulence. It's because there is no analytic solution that gives reliable answers that airline manufacturers build million dollar wind tunnels. The sequence of bifurcations in the logistic equation is considered by some to be similar to what is found for the Navier–Stokes equations of fluid dynamics.

The connection to solutions of ordinary and partial differential equations (ODEs and PDEs) was behind the excitement that Feigenbaum's results led to. The simple recursion relation, Eq. (F.3), captured a universal feature of more complicated differential equations.

Appendix G
Lagrange multipliers

First we consider the simplest case, where there is only a single constraint.

G.1 Two variables, one constraint

You have $f(x, y)$, a function of two variables, but you also have a constraint, $g(x, y) = 0$. What you want is a stationary point of $f(x, y)$ on the set $\{(x, y) | g(x, y) = 0\}$. What you do is enlarge your question. Define a function w with

$$w(x, y) \equiv f(x, y) - \lambda g(x, y), \tag{G.1}$$

and with λ an undetermined number, to be known as the Lagrange multiplier that will also be fixed by this procedure. Require that w be stationary with respect to variation in x, y and λ. (As usual, we'll assume all functions have as many derivatives as needed for our proofs. Note too that w is now considered a function of x, y and λ.) Form the differential

$$\delta w(x, y, \lambda) = \delta f(x, y) - (\delta \lambda)g(x, y) - \lambda \delta g(x, y)$$
$$= \left(\frac{\partial f}{\partial x} - \lambda \frac{\partial g}{\partial x} \right) \delta x + \left(\frac{\partial f}{\partial y} - \lambda \frac{\partial g}{\partial y} \right) \delta y - g(x, y)\delta \lambda. \tag{G.2}$$

$\delta w = 0$ demands that the coefficients of δx, δy and $\delta \lambda$ all vanish. (Quick check: Counting variables, you now have three equations for three unknowns (values of x, y, and λ).) First you have the requirement that $g(x, y) = 0$. That's forced by having the coefficient of $\delta \lambda$ be zero. But if g vanishes, the variation of w must equal the variation in f.

Furthermore, for w to be stationary $\nabla f \equiv \frac{\partial f}{\partial x}\hat{x} + \frac{\partial f}{\partial y}\hat{y}$ must be parallel to $\nabla g \equiv \frac{\partial g}{\partial x}\hat{x} + \frac{\partial g}{\partial y}\hat{y}$, since there is a single factor λ connecting them. This is reasonable. Variation of g must be perpendicular to $g(x, y) = 0$ since g does not vary at all along the line $y = h(x)$ where h is defined as the function of x that makes g zero, that is, $g(x, h(x)) \equiv 0$. Moreover, ∇f must also be perpendicular to this line, since, if it were not, moving along this line would increase f (or decrease it—this is only a stationarity requirement). So they are parallel. But δw must vanish. That will be the requirement on λ. Thus λ is selected so that $\nabla f = \lambda \nabla g$.

To summarize, you solve

$$\left.\begin{array}{l} g(x,y) = 0 \text{ and} \\ \nabla w(x,y) = \nabla f(x,y) - \lambda \nabla g(x,y) = 0 \,, \end{array}\right\} \tag{G.3}$$

and find λ from the resulting equations.

The requirement $\nabla g(x,y) \| \nabla f(x,y)$ can also be found from an approach not requiring the introduction of an auxiliary function.

As above, you have a function of two variables, $f(x,y)$, and you want to find its extrema, subject to a constraint, $g(x,y) = 0$. In principle you can reduce the number of variables, using $g(x,y) = 0$ to write $y = h(x)$, so that $g(x, h(x)) = 0$. (This is the same h that was used earlier.) Then you'd want to have $f(x, h(x))$ stationary, so that $\delta f(x, h(x)) = 0$, or

$$\frac{\partial f}{\partial x} + \frac{\partial f}{\partial y}\bigg|_{y=h(x)} \frac{\partial h}{\partial x} = 0 \,. \tag{G.4}$$

Note that g must also have zero variation along the contours of its constraint (i.e., when $g(x,y) \equiv 0$). Thus $\partial g/\partial x + (\partial g/\partial y)\,(dy/dx) = 0$. But h and y are the same and therefore

$$\frac{dy}{dx} = -\frac{\frac{\partial f}{\partial x}}{\frac{\partial f}{\partial y}} = -\frac{\frac{\partial g}{\partial x}}{\frac{\partial g}{\partial y}} \tag{G.5}$$

or

$$\frac{\partial f}{\partial x} \cdot \frac{\partial g}{\partial y} = \frac{\partial g}{\partial x} \cdot \frac{\partial f}{\partial y} \,. \tag{G.6}$$

Think of the partial derivatives as two-dimensional gradients, for example $\nabla f = \frac{\partial f}{\partial x}\hat{x} + \frac{\partial f}{\partial y}\hat{y}$, then Eq. (G.6) implies (subject to various conditions, about which we will not concern ourselves) that there is a constant λ such that

$$\frac{\partial f}{\partial x} = \lambda \frac{\partial g}{\partial x} \quad \text{and} \quad \frac{\partial f}{\partial y} = \lambda \frac{\partial g}{\partial y} \,. \tag{G.7}$$

Not unexpectedly, a variable λ is introduced, and is the Lagrange multiplier.

G.2 Generalization, more than one constraint

Often one has more than two variables and more than one constraint. Call the variables $\{x_n\}$, with $n = 1, \ldots, N$, and $\{g_k\}$, with $k = 1, \ldots, K$. The generalization is that the function

$$w \equiv f(\{x_n\}) - \sum_k \lambda_k g_k(\{x_n\}) \tag{G.8}$$

is to be made stationary. In the example in Chapter 6 of the derivation of the Maxwell–Boltzmann distribution, the variable $\Pr(v)$ is parameterized by a continuous variable, v, and $\Pr(v)$ is the function f.

The generalization is straightforward, but note that there will be a separate "λ" for each constraint; in addition you have more than one constraint equation $(g_k(\{x_n\}) = 0$ for $k = 1, \ldots, N)$.

G.3   Example

Here is an example to illustrate the method. Let the constraint be $g(x, y) = x^2 + y^2 - 1 = 0$, that is, the point must lie on the unit circle. And let $f = x + y$. Then $w = f - \lambda g$, and (partial derivatives are indicated by subscripts) $f_x = 1$, $f_y = 1$, $g_x = 2x$, and $g_y = 2y$. And, bearing in mind the definition of ∇,

$$1 - 2\lambda x = 0, \qquad 1 - 2\lambda y = 0, \tag{G.9}$$

so that $\lambda = 1/2x = 1/2y$; in other words the stationary point is where the line $x = y$ crosses the circle. Imposing the constraint $g = 0$ then shows that the solution is $x = y = \pm\sqrt{2}$. Of course for this solution we still had to solve $g(x, y) = 0$ for y, but we didn't need the general functional form of this solution (as would be required by Eq. (G.4)). Note also that with the minus sign you actually get a minimum, so this method only picks out stationary points; they could be maxima or minima or in the multidimensional case, saddle points.

Finally in this case it *is* possible to work with Eq. (G.4) (since it is easy to invert $g = 0$, i.e., $y = \sqrt{1 - x^2}$) and one can verify the solution we've found.

Appendix H
Complexity in the observable representation

H.1 The observable representation

The *observable representation* (henceforth in this appendix: OR) is a tool for understanding complex systems. It's not uniquely adapted to them, but can be used to help in understanding the relationship between components of such systems or to allow classification of subgroups. In a way it's like the study of phase transitions. Certainly not all phase transitions are complex, but complex systems share properties with those phenomena. (The details involve stochastic matrices, in particular their eigenvectors. For more information on that subject see Appendix D.5.)

The OR is an embedding in a Euclidean space of the space on which a stochastic process is defined. That's a mouthful. Let me be more specific.

Let's start with a simple example: a random walk on a circle, with equal likelihood of going one step clockwise or counterclockwise. Let the sites be numbered 1 to N. If you're at site k you're equally likely to go to site $k+1$ or $k-1$, and the circular nature of the walk is expressed by having a counterclockwise step from site-N reach site-1, and a clockwise step from site-1 reach site-N. A way to describe this walk is by a matrix of transition probabilities: $R(x, y) = \Pr(x \leftarrow y) = 1/2$ if $|x - y| = 1$, where x and y are integers between 1 and N and points 1 and N are considered to be at a distance 1 from each other. (R is a stochastic matrix.)

For $N = 5$ the matrix looks like

$$R = \begin{pmatrix} 0 & 1/2 & 0 & 0 & 1/2 \\ 1/2 & 0 & 1/2 & 0 & 0 \\ 0 & 1/2 & 0 & 1/2 & 0 \\ 0 & 0 & 1/2 & 0 & 1/2 \\ 1/2 & 0 & 0 & 1/2 & 0 \end{pmatrix}. \tag{H.1}$$

This matrix can be analytically diagonalized. It is the sum of permutations. In any case, define

$$B = \begin{pmatrix} 0 & 1 & 0 & 0 & 0 \\ 0 & 0 & 1 & 0 & 0 \\ 0 & 0 & 0 & 1 & 0 \\ 0 & 0 & 0 & 0 & 1 \\ 1 & 0 & 0 & 0 & 0 \end{pmatrix}, \tag{H.2}$$

so that $R = \frac{1}{2}(B + B^\dagger)$, where $B^\dagger$ is the complex conjugate, transpose of B. Since B is real, this is also just the transpose, which we write B^t. In this particular case, B is also the inverse of B^t so that $BB^t = 1$. Of course this generalizes to any N and in MATLAB^{TM} notation B could be defined by

```
B=diag(ones(N-1,1),1); B(N,1)=1;
```

which means that there are all ones above the diagonal and a one in the lower left corner. The eigenvalues of this operator are $\exp(i\phi_\ell)$, with $\phi_\ell = 2\pi\ell/N$ and $\ell = 0, 1, \ldots, N-1$.[1] The eigenvectors are $u_k^{(\ell)} = \mathcal{N}\exp(ik\phi_\ell)$ for some normalization, $\mathcal{N}$. Since $R = \frac{1}{2}(B + B^t)$ and $[B, B^t] = 0$, the eigenvalues of R are $\cos\phi_\ell$. However, $\cos\phi$ and $\cos(2\pi - \phi)$ are the same, so except for $\ell = 0$ (and for even N, $\ell = N/2$) the eigenvalues are doubly degenerate.

Thus for $\ell = 1$ and $\ell = N - 1$ one gets the eigenvalue $\cos(2\pi/N)$, close to 1, but subsequent eigenvalues ($\cos(2\pi\ell/N)$ for ℓ near its boundaries) are also close to 1. We look at the two leading non-trivial left eigenvectors and plot one against the other. In using the word "non-trivial" I exclude the eigenvalue 1 ($\ell = 0$), and refer only to those with eigenvalues less than one. (Since R is symmetric its left and right eigenvectors are the same.) We take real eigenvectors, which in this case are $\cos(2\pi k/N)$ and $\sin(2\pi k/N)$ (with L_2 normalization they both get factors $\sqrt{2/N}$, but it doesn't matter here). Now comes the punch line: plot the eigenvectors against each other. In other words, for each point in X, namely the integers $k = 1$ through N, mark the point at $(\cos(2\pi k/N), \sin(2\pi k/N)) \in \mathbb{R}^2$ for each k. This is shown in Fig. H.1.

This is *not* a plot of the original space. It is a plot of the first non-trivial left eigenvalue against the second. The OR thus gives you an image of the space on which the walk takes place. This is true for a variety of spaces. If the Brownian motion is on a figure Y (the walk in on three lines and at their mutual intersection it can go to any one of them) then the OR gives a Y.

Even more remarkable are walks on non-planar graphs and on fractals. A non-planar graph is one that cannot be embedded in a plane. We [75] worked out a walk on the graph $K_{3,3}$ (which is non-planar) and found that when only two eigenvectors

---

[1]To diagonalize, think translation operator on the circle. Thus let a (putative) eigenvector be $u \equiv \exp(ik\phi)$, $k = 1, \ldots, N$ (up to normalization). Then $(Bu)_k = \exp(i\phi)u_k$ for $k = 1, \ldots, N-1$ and this will be an eigenvector if the Nth components also match, namely, $\exp(i(N+1)\phi) = \exp(i\phi)$. This implies $\exp(iN\phi) = 1$, so that ϕ takes the values $\phi = 2\pi\ell/N$ with $\ell = 1, \ldots, N$ (or 0 to $N-1$).

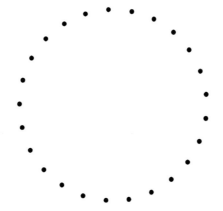

Fig. H.1 OR for a random walk on a circle. This is *not* a plot of the original space. Rather it is a plot of the first two non-trivial eigenvectors against one another, in other words (for this case) $(\cos(2\pi k/N), \sin(2\pi k/N)))$ for $k = 1,\dots,N$ (and $N = 25$).

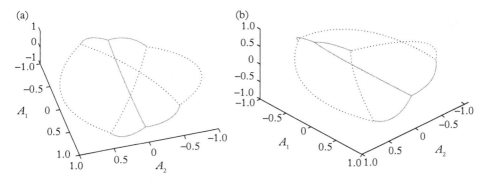

Fig. H.2 OR for a random walk on the non-planar graph, $K_{3,3}$. Two views are shown, hopefully to make clear that when rotated the lines of the figures do not overlap.

were used the graph invariably crossed itself. But with three—using MATLAB^{TM}'s ability to rotate figures—we could have a unique, uncrossed embedding. This is shown in Fig. H.2. Again, this is a plot of $(A_1(x), A_2(x), A_3(x))$ for $x \in$ the space in question (with 1, 2 and 3 the first eigenvalues *after* $A_0(x) \equiv 1$, eigenvalue 1).

Regarding fractals it was obviously not possible to produce a full fractal, with sizes going to zero, but there were as many levels as we had patience for the computer to run. Fig. H.3 shows first the OR for Brownian motion on a fractal, and next to it the images of the eigenfunctions. As usual, the figure on the left is not an image of the fractal, although it looks the same, but the observable representation, a plot of $(A_1(x), A_2(x))$ or $A_2(x)$ against $A_1(x)$ (for $x \in X =$ the space containing the fractal).

A second and somewhat different situation obtains if there are several, say m, eigenvalues near 1, whereas subsequent eigenvalues are much smaller; that is, $1 \equiv \lambda_0 >$

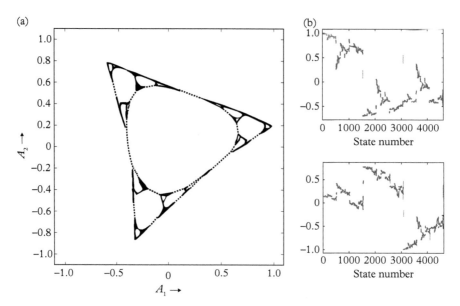

Fig. H.3 OR for a random walk on a fractal constructed of triangles. (a) The OR; (b) The eigenfunctions of the transition matrix.

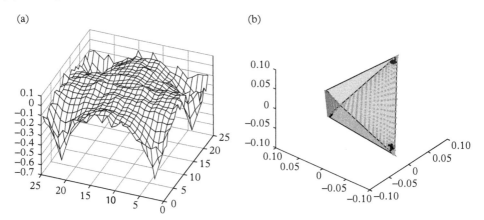

Fig. H.4 (a) The potential seen by the random walker. It is slightly randomized and smoothed, but is essentially one over a constant plus the distance to the nearest minimum $(1/(\text{const} + \text{distance}))$. The "temperature," measuring the probability of taking an uphill step (the higher the temperature, the more likely the step), is fairly low. (b) The OR for the transition matrix probabilities. As usual, this is a plot of the first three non-trivial eigenvectors $(A_1(x), A_2(x), A_3(x))$ for $x \in X = 625$ spatial positions. The number 3 is chosen for the number of eigenvectors because $\lambda_3 \gg \lambda_4$ and therefore $m = 3$.

$\lambda_1 \geq \lambda_2 \dots \geq \lambda_m \gg |\lambda_{m+1}|$, where the λ's are eigenvalues of R and for simplicity I've assumed all of the first m eigenvalues are real. (Note the "$\gg$" between λ_m and $|\lambda_{m+1}|$.) This is a situation where the OR is good for classifying things.

Consider a random walk in a potential with four minima, as in Fig. H.4a. The transition probability on this 25-by-25 grid is a random walk, but it is less likely to walk uphill than downhill, that is, the transition probability for increasing the potential is less than that for decreasing it. The transition matrix is 625-by-625 ($= 25^2 \times 25^2$) and can be diagonalized twice in a flash on a laptop (once for the stationary state (R) and once for left eigenvectors (R-transpose)). After $\lambda_0 = 1$ the next few eigenvalues are 0.9961, 0.9961, 0.9922, 0.9843, 0.9843, 0.9804, 0.9804. That doesn't seem like much of a dropoff, but looking at the 10th powers one gets 0.9614, 0.9614, 0.9244, 0.8538, 0.8533, 0.8202, 0.8202, and one can see a dropoff after the first two and especially after the third. In Fig. H.4b I plot the OR for the first three (non-trivial) left eigenvectors and it is clearly a tetrahedron. There are only four states with high probability (a large value of $p_0(x)$ for x a state, that is, a point in the 625-dimensional coordinate space) in the stationary state and these are shown with larger symbols, in black. Near them are states of slightly lower probability in smaller symbols (it may be difficult to see, but with MATLAB^{TM}'s figure rotation capability this is clear). And finally, smaller yet, is a multitude of states of lower probability.

There is further structure, namely to a good approximation for each $x \in X$, $P \equiv (A_1(x), A_2(x), A_3(x))$, can be written in the form $P = \mu_1 E_1 + \mu_2 E_2 + \mu_3 E_3$. The points $\{E_k\}$ are the extremal points and are images of particular points in X. This is a vector equation (i.e., P & $\{E_k\}$ are vectors in $\mathbb{R}^3$) and the $\{\mu_k\}$ are called the *barycentric coordinates* of P. These μ's add to one and are the *probabilities that a point starting from x ends in one or another extremum* [72].

This formalism was first developed [70, 71] for a definition of metastability.[2] But it turns out it is effective for sorting things out, for classifying. For example, there is a Karate club studied by Zachary [241]. There were about 60 members of this club (the place and names have not been revealed) and, of these, Zachary was able to collect information about personal relationships among 34 of them. By "personal relationship" I mean an affinity beyond a common interest in karate (admittedly a bit fuzzy). As in the usual course of human events the club split into two, based on an increase in fees that the instructor wished to charge: many agreed, the club president and others disagreed. In Fig. H.5 I have indicated their ORs using the affinities that Zachary found in the individuals' activities.

---

[2]It's amusing that according to "rigorous" statistical physics, where one insists on taking the "thermodynamic limit" (volume V, or # particles N, $\to \infty$) there is no such thing as metastability— even though you can find the density of supercooled water at $-2°$C in handbooks. The problem is that the time for formation of a critical droplet (a droplet with a 50% chance of growth) is also large. This would require a non-uniform limit (N and $\frac{\text{time-to-decay}}{\text{short-time}}$), excluded by assumption. In [71] the system may be large, but is finite, and a single parameter, the proximity of the eigenvalue just smaller than 1 to 1, fixes the phases, whether stable or metastable.

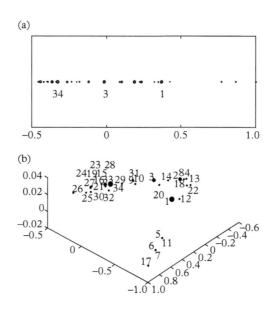

Fig. H.5 The points represent individuals, and their coordinates—34 of them. (a) is A_1 alone, with weighted sizes for the points. Clusters formed around #1 and #34 (the instructor and president) while #3 is in the middle. Various criteria were used to form clusters and our first criterion put #3 in the correct cluster, but, as is evident, it could have gone the other way. (b) shows what can happen when too many left eigenvectors are plotted. Symbols (solid circles) are still correlated with weight in the stationary state, which is highest for the two people who were the nuclei for the new groups (karate instructor and club president). As is clear, these two, numbers 1 and 34, have the largest weight and are centrally located within the cluster that went with them. As above, number 3, someone who was friends with many on both sides, is ambiguous. In the three-dimensional plot, some numbers have been moved for clarity and certain weights (some small ones) are covered by numbers.

The OR clusters give a good indication of how the group broke up. The only ambiguity was number 3 (a person), and indeed other analyses of the same data gave an incorrect prediction. Nevertheless this is not as clean as was the case for Fig. H.4. The spectrum does not drop off sharply; the first few non-trivial eigenvalues are $(0.868, 0.713, 0.613, 0.388, 0.351)$, and correspondingly the clusters, even using A_1 alone, must be interpreted.

Nevertheless, this example illustrates how a simple matrix can be used to separate a group into clusters. An adjacency matrix has similar properties. In Fig. H.6 I show both the OR and the spectrum for a particular matrix. (See the figure caption for details.) Not all adjacency matrices have tetrahedrons and even if there are four regions things can get complicated. But when it works, it can show the cluster structure effectively.

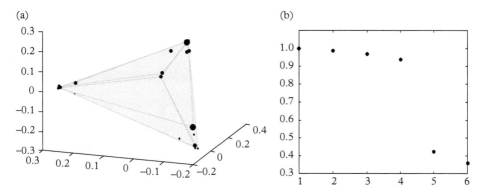

Fig. H.6 These figures are based on an adjacency matrix (not shown) with four tightly interconnected regions and with weaker connections between them. From the adjacency matrix a stochastic matrix is constructed, as described in the text. (a) shows the three-dimensional OR for this stochastic matrix and (b) illustrates the first few eigenvalues, starting with the value 1 (on the left), present for all stochastic matrices. Note the dropoff after the first three non-unit eigenvalues, implying an approximate tetrahedron for the OR.

Yet another application of the OR is to foraging animals. Here the extrema are the places where the animal finds food and if there are more than four food sources there can be problems representing them in 3-space. This system is not as complex as the stock market, but more complex than an ideal gas. I will not go into detail (see [236]), but what helps in this case is the fact that the actual food sources are located in two-(or sometimes three-) dimensional space, so that there is an additional mapping to the space in which the food is found.

There is a further application of the OR that I have not so far mentioned—it is part of a dream and perhaps doesn't belong in what might become a textbook. Nevertheless, I'll share my image with those who have worked their way through this appendix.

As has been repeatedly mentioned there is no accepted definition of *complexity*. Yes, there is Kolmogorov complexity and several institutes with that word in their title (probably the first of these was the Santa Fe Institute). But an abstract definition eludes us. This is where currents come in. Currents are not a single number; they have a topological structure and other "complex" properties. Attempts at more general definitions are given in [199] and a vague general theorem is proposed.

One example of a measure for currents is the following: Consider a loop expansion of a current matrix. First, by the "current matrix" I mean the positive part: $J_+(x,y) \equiv \frac{1}{2}(J_{xy} + |J_{xy}|)$. Next define "loops" to be current matrices that don't branch anywhere. (You could also call them "permutations.") By an expansion I mean that the (positive) current matrix is written $J_{+xy} = \sum_n \alpha_n J^n_{xy}$ where each J^n_{xy} is a loop. (It's easy to prove that this can be done.) For simple transition probability matrices there might be a single loop: For the matrix R,

$$R = \begin{pmatrix} 0.4629 & 0.1015 & 0.1597 \\ 0.5116 & 0.4053 & 0.4649 \\ 0.0255 & 0.4932 & 0.3754 \end{pmatrix}, \tag{H.3}$$

the ground state is $p_0 = (0.1918, 0.4472, 0.3610)$ and the (positive) current is

$$J_+ = \begin{pmatrix} 0 & 0 & 0.0528 \\ 0.0528 & 0 & 0 \\ 0 & 0.0528 & 0 \end{pmatrix}, \tag{H.4}$$

which means there is but a single loop. But for much larger matrices, the loops become ambiguous, and in fact the coefficients $\{\alpha_n\}$ form a simplex, with each α_n bounded from below by 0 and bounded from above by a constant that depends on J_+. Properties of this simplex become potential measures of complexity, for example a collection of maximum values of the $\{\alpha_n\}$.

To continue the speculation, I remark that entropy is the logarithm of the number of states available. This is not appropriate for complex systems. Entropy is extensive and measures the amount of missing information. But what about seeds? One grass seed can start an entire lawn (a ridiculous crop). Two grass seeds? A bit faster, but leading to the same result. If you have one pair of rabbits, that's enough to yield a lot of rabbits (ask Australians). A second pair will add a bit of diversity, but the basic message of "rabbits" is already present in a single pair. For this reason I would suggest a log-log, rather than a log.

Then there is the issue of survival. This brings to mind the quest for machines that have maximum *power*, rather than maximum efficiency, as in the Carnot cycle [46, 156, 240]. The idea—much extended—is that there are other criteria than efficiency. In the case of "survival," anything goes. It's like cryptography. The criterion is, does it work? There are no rules. In any case, a living creature—a surviving living creature—must produce chemicals that neutralize the weapons of those that seek to benefit from its benign activities, making sugar, making meat or engaging in reproduction.

And finally there is the ultimate dream. Every possible niche is filled. Each creature takes the work provided (ultimately derived from the Sun, in most cases) and uses it to maintain life, passing on to the next level an even smaller amount of profit to be derived from solar negentropy. The are limits. Compare Earth and Venus. Venus is even more ripe for life, but it doesn't happen.[3]

Above you'll find my thoughts on complexity, when I started writing this book. I've since come to an additional point of view. As remarked, there's more than one kind of complexity. The natural language name, and the names of various institutes throughout the globe, carry multiple meanings, sometimes at a single institute. I am a frequent visitor in Dresden, at the Max Planck Institute *for the Physics of Complex Systems*. Their notion of *complex* is different from that at the Santa Fe Institute and that will differ from the myriad of like named organizations.

None of these places (that I know of) is talking about Solomonoff–Kolmogorov–Chaitin complexity [1] (usually called Kolmogorov complexity—according to some an example of the Matthew effect[4]). This *complexity* is the length of the shortest program needed to define a string and is the complexity associated with that string. The programming language is fixed. (And the definition can be extended to other mathematical objects, but our purpose here is not to dwell on this kind of complexity.)

The other kinds of complexity have to do with interactions among many constituents, whether they be conscious or not. What I consider a byproduct of complexity is not on Lloyd's list [145], although he must deal with it elsewhere. Rather I start from the commonality of power laws and arrive at the equivalent feature, scaling. (See also [231].)

In Chapter 7 it was shown that for power laws, $p(x) = ax^{-b}$, there was scale invariance. That is, given the form indicated there was a function of a scaling variable γ such that there is a function $f(\gamma)$ for which $p(\gamma x) = f(\gamma)p(x)$. The distribution looked the same, up to a factor, if you look more—or less—closely. Moreover, the function was necessarily $f(\gamma) = \gamma^{-b}$ and $b = -\left.\frac{df}{d\gamma}\right|_{\gamma=1}$. Now for any natural phenomenon there are always errors and limits. This is the reason there are (valid) disputes about whether a given function is a power law or perhaps log-normal or some other function. Thus, in reality, $p(x) = ax^{-b} + \delta(x)$, with $x_0 \leq x \leq x_1$ and small values of $\delta(x)$. Now consider two additional points (in the interval), say x_2 and x_3, and suppose γx_2 and γx_3 are also in the interval. Then if there is a competing function the new δ's ($\bar{\delta} \equiv p(\gamma x) - f(\gamma)ax^{-b}$ and the same for the other function[5]) are also small—otherwise the decision about

---

[3]Maybe there is life on Venus. See *The New York Times* article, "On Venus, cloudy with a chance of microbial life" by Dennis Overbye, September 19, 2020.

[4]"...to everyone who has more will be given..." (Matthew 25:29) as suggested by Ming Li and Paul Vitányi in their 2008 book *An Introduction to Kolmogorov Complexity and its Applications* (doi:10.1007/978-0-387-49820-1_1).

[5]As an example, suppose the other function was a log-normal, then the other delta (call it δ') would be $\delta' = p(\gamma x) - f(\gamma)\frac{1}{x\sigma\sqrt{2\pi}}\exp\left(-\frac{(\log x - \log x_0)^2}{2\sigma^2}\right)$.

whether or not it's a power law would be simple. This means that it's not important whether the fitting function is a power law or other function. What's important is that—within the range—there is scaling, the b in the function $f(\gamma) = \gamma^{-b}$.

I don't think scaling is necessarily a measures of complexity, but I would say it's a requirement. You would want to know the range, (x_0, x_1), and exponent, what I've been calling b. Because of the dependence on data the numbers x_0 and x_1 are necessarily a bit fuzzy, but maybe that suits complexity. The closeness of b to unity probably also plays a role.

Given a definition in terms of scaling it's clear that language, finance, music, and so on are complex. How about the coast of Norway [57] or carbon resistors [51]? There are many factors involved (fjords, traps) and I suppose those are also complex. But even the concept of power law or scaling is fraught with ambiguity. As pointed out in [45] a sum of power laws is *not* a power law.

There is also the issue of renormalization—which does not seem to have attracted attention. That too depends on scale invariance (things should stay the same as you look less and less closely). Does this mean that words (à la Zipf) and cities belong in the same "universality class" since they seem to have similar power laws? And do the coasts of England, Norway and South Africa belong in different classes, because they have different exponents?

What scaling has to do with complexity baffles me. They seem to be different and yet many, many complex systems scale. Each complex system may have an explanation (see, for example, [231]) but it is the commonality, the fact that many complex systems scale, that troubles me. Is this a coincidence or is it a topic for research? I don't know.[6]

---

[6] ... fitting ending to this book.

Appendix I
A Quotation

Since so much of this book is concerned with curve fitting, I will append a quotation from our "sages:"

> In desperation I asked Fermi whether he was not impressed by the agreement between our calculated numbers and his measured numbers. He replied, "How many arbitrary parameters did you use for your calculations?" I thought for a moment about our cut-off procedures and said, "Four." He said, "I remember my friend Johnny von Neumann used to say, with four parameters I can fit an elephant, and with five I can make him wiggle his trunk." With that, the conversation was over.

from Dyson [53].

Appendix J
Solutions to exercises

Footnotes refer to the chapter from which the exercise is taken.

Exercise 1 from Chapter 3

Explain why more iterations are needed when the initial point is close to 1/2. (Hint: Go back to Eq. (3.5). Note that the longer times are not just proportional to the distance (in p) to be covered.)

Solution

Without loss of generality I can consider the fixed point $\bar{p} = 0$, for which Eq. (3.5) predicts $\delta p' = 3(\delta p)^2 - 2(\delta p)^3$. Close to the point $p = 1/2$ the first iteration only cuts the probability down by $3/4 - 1/4 = 1/2$, not as extreme as subsequently. Later in the iteration δp^2 is much smaller and the cubic term is completely negligible.

Exercise 2 from Chapter 3

Remember that $p \equiv p_L$. So in addition to Eq. (3.6) there will be a corresponding equation for p_R'. Check that $p_L' + p_R' = 1$.

Solution

The equation for $p_L' \equiv p'$ is $p_L' = q \cdot \frac{1}{2} + (1 - q) \cdot p_L^2 (3 - 2p_L)$ and the corresponding equation for p'_R is $p_R' = q \cdot \frac{1}{2} + (1 - q) \cdot p_R^2 (3 - 2p_R)$. Adding them gives 1.

Exercise 3 from Chapter 3

Check the stability of the root, $\bar{p} = \frac{1}{2}$. Hint: See Sec. 3.2. For $q \geq 1/3$ the fixed point 1/2 had better be stable—it's the only one. For $q < 1/3$ that fixed point should become unstable. (The real values given in Eq. (3.9) become stable.)

Solution

Let $\bar{p} = \frac{1}{2}$, $\delta = p - \bar{p}$ and $\delta' = p' - \bar{p}$. Using Eq. (3.6) (and some algebra) you find $\delta' = \delta \left(\frac{3}{2}\right)(1 - q) + O(\delta^3)$. For $q > \frac{1}{3}$ this gives—neglecting $O(\delta^3)$—stability. For $q < \frac{1}{3}$ the ratio of δ' to δ is greater than one, and the other roots occur. Exactly at $q = \frac{1}{3}$ the linear term in δ vanishes and we must look to the cubic terms, which are $-2(1-q) = -\frac{4}{3}$, so $\delta' = -\frac{4}{3}\delta^3$ ($|\delta| \leq \frac{1}{2}$ so δ' is definitely smaller.) Of course Sec. 3.4.1 shows this to be an artifact of mean field theory.

Exercise 4 from Chapter 3

Here's something that I don't know how to do. Maybe someone will solve this, and, who knows, maybe it has some practical value. The peak in Fig. 3.3 is not quite at $1/2$, and is closer to 0.495. Why? I've given a reason to move away from $1/3$, but that's just conversation. Can a precise value for the phase transition be found?

Solution

Good luck!

Exercise 5 from Chapter 3

Write a program to play the game (in MATLAB™ or in your favorite programming language). Check my results above.

Solution

No claims for great program writing, but the following will do the job. "L" is left, i.e., rubber band is on the left hand. Similarly for "R." The symbol % means the material to its right is commented out. Note that there are two different ways of changing parameters in a program. Usually I use the program setdefault, and then the change in parameter value can be made by using as input the structure "in." A second way is simply to change the program. The latter is the case below.

```
function rubber_band_going_wrong
% Rubber band game. This program is written in the matlab language.
% Keeps track of final configuration as a function of the initial conditions.
% L is (the rubber band is on) the left hand. R is right.

N=100;                   % Number of participants. Can be changed.
Nrepeat=1e3;             % Number of random initial conditions to try. Can be changed.
                         % Changes can be made by editing this file.

% Derived quantities for use in the program.
Nover2=ceil(N/2);        % Convenient to define. Nover2 = integer, even if N is odd.
                         % (This is due to the command "ceil," standing for "ceiling."
                         % Because of "ceil" the program rounds up (7/2 becomes 4).)
del=min(40,Nover2-1);    % Trials are limited to L between del and N-del, assuming that
                         % those closer to all-L or all-R always go to L or R,
                         % respectively. (See next definition, nLs.)
```

```
nLs=(Nover2-del):Nover2+del; % Initial number of L. I exclude states
                             % that are initially nearly all L or R.
                             % (That's "del.")
nnL=length(nLs);      % Number of initial conditions tested.


NL=zeros(1,N); NR=NL; % Predefining NR and NL speeds the calculation.
                      % NL will be the number that ends up with the rubber
                      % band on the left wrist (for each initial value of
                      % the probability that an individual begins on the left).
for k=1:nnL,
  nL=nLs(k);      % Initial # on left wrist.
  pL=nL/N;        % Initial probability of left.
  for kk=1:Nrepeat,
      p=pL;                       % Initial # left for all Nrepeat trials.
      while abs(p*(1-p))>1e-14; % Beginning of a trial.
                                % The abs(...) enforces condition that all
                                % are L or R. (p is fraction on left, L's)
                                % When the product of p and (1-p) is small
                                % enough (10^{-14} in TeX language), the
                                % iteration is halted.


      r=rand;                   % Here is where randomness plays a role.
      pp=3*p*(1-p);
      if r<p*pp,
          p=p+1/N;              % Increase #L if it's 2 out of 3 left.
      elseif (1-r)<(1-p)*pp,
          p=p-1/N;              % Decrease #L with prob 3(1-p)^2p.
                                % p evolves in time: can be different
                                % for each string of random numbers.
      % If neither criterion is satisfied (group already has 3 L's or R's),
      % and there will be no change.
      end
    end
```

```
      if abs(p)<1e-10,NR(nL)=NR(nL)+1;        % Checks whether final state
      elseif abs(1-p)<1e-10,NL(nL)=NL(nL)+1; % is L or R.
      else disp('something is wrong')
      end
   end
end


% Will assume that have gone far enough for NL and NR to be 0 or 1
% outside the range examined

% Clean things up at the beginning and end.
effectively_zero=Nrepeat/50;

if NL(nLs(1))<effectively_zero,NL(1:nLs(1)-1)=0;end
if NL(nLs(end))>N-effectively_zero,NL(nLs(end)+1:N)=Nrepeat;end

if NR(nLs(end))<effectively_zero,
   NR(nLs(end)+1:N)=0;
end
if NR(nLs(1))>N-effectively_zero,
   NR(1:nLs(1)-1)=Nrepeat;
end


% Plot results.
figure(100),plot(1:N,NL/Nrepeat,'o',1:N,NR/Nrepeat,'x'),
title(['N = ',num2str(N)]),
axset(1,nLs(1)-1/2,2,nLs(end)+1/2);
figure(101),plot((1:N)/N,NL/Nrepeat,'o',(1:N)/N,NR/Nrepeat,'x'),
title(['N = ',num2str(N)])


function axset(axname1,newvalue1,axname2,newvalue2,axname3,newvalue3,...
          axname4,newvalue4,axname5,newvalue5,axname6,newvalue6)
% axset.m  Sets "axis" in a figure to the value given after the label
% Syntax: axset(ax1,newval1,ax2,newval2,ax3,newval3,ax4,newval4,
%          % ax5,newval5,ax6,newval6);
% axn is a number from 1 to 6, newvaln is the new value for that "axis" entry
% axset takes an even number of arguments; maximum is 12, but can use 2, 4, etc.
ax=axis;     % The old values for the axes.
```

```
N=nargin/2;   % Half the arguments (N=2, for figure 100 above)
              % N should be an integer.
for k=1:N,
   kk=num2str(k);
   eval(['ax(axname',kk,')= newvalue',kk,';']);
end
axis(ax);
```

Exercise 6 from Chapter 4

Among some scientists the hallmark of a phase transition is a breakdown in analyticity.[1] This stems from the seminal work of Onsager [171] and Lee and Yang [137,
138] showing that there is a breakdown in analyticity at a phase transition. (I remark
that these ideas fail completely when it comes to understanding the metastable state.)
Now there are two functions, $s_0(r)$, the physical probability of finding someone with
percolitis, and something I'll call $\bar{s}_0(r)$, the solution(s) of the transcendental equation
$s = 1 - \exp(-rs)$. For example, letting $\bar{s}_0 = x + iy$, there's a complex root that to a
good approximation has $y = \frac{19\pi}{8r}$ and $x = 1 - y/\tan(ry)$ (which gets better as $r \to 1$).
For $s_0(r)$ I've already demonstrated that it can't be analytic. What about $\bar{s}_0(r)$?

Solution

We focus on the neighborhood of the phase transition and write $r = 1 + z$ (the notation
"z" will recall that z is to be thought of as complex). With a bit of algebra, the equation
for $\bar{s}_0(r)$ becomes

$$z = -1 - \frac{\log(1 - s)}{s} \tag{J.1}$$

(where I've dropped the 0 and bar (‾) in $\bar{s}_0$). As $s \to 0$, $\log(1-s)/s$ approaches -1 so the
s in the denominator is not a hallmark of a singularity. Note that the multivaluedness
of the logarithm replaces that of the exponential. It follows that

$$\frac{ds}{dz} = \frac{s^2}{\log(1 - s) - \frac{s}{1-s}}, \tag{J.2}$$

so that $s(z)$ has branch points where the logarithm of $1 - s$ does. In particular, there
is a branch point at $s = 1$, but not at $s = 0$. The function $\log(1 - s)$ is multivalued at 1

---

[1]Being *analytic* is a powerful condition on a function. (If you are first encountering this concept, I'd suggest skipping this exercise.) The definition sounds simple: let D be a domain (an
open, arcwise-connected set) of the complex plane and $z \in D$. Then $f(z)$ is analytic if its derivative,
$df/dz = \lim_{h \to 0}(f(z + h) - f(z))/h$, exists for h coming from any direction in D. This seemingly mild
restriction is enough to guarantee that there are infinitely many derivatives at z and that there exists
a power series for f (as well as many other properties).

and there is also a pole, but the function $s(z)$ (i.e., $\bar{s}_0(z)$) is analytic at $z = 0$ (which is $r = 1$).

Exercise 7 from Chapter 4
Check this. Estimate the error in Eq. (4.14) for large times.

Solution
The equation is

$$\delta(t+1) = \frac{2}{t + t_0 + 1} \stackrel{?}{=} \delta(t) - \delta(t)^2/2 = \frac{2}{t + t_0} - \frac{1}{2}\left(\frac{2}{t + t_0}\right)^2. \tag{J.3}$$

We look at the difference,

$$\text{Difference} = \frac{2}{t + t_0 + 1} - \frac{2}{t + t_0} + \frac{1}{2}\left(\frac{2}{(t + t_0)}\right)^2 = \frac{2}{(t + t_0)^2(t + t_0 + 1)} \tag{J.4}$$

(where the second equality involves some algebra). It is reasonable that this is the error: it is third order in $1/t$, and that's been neglected all along.

Exercise 8 from Chapter 4
Explain the small drop in the number of iterations slightly to the right of $r = 1$. (Nothing profound to be learned from this exercise—just a check to see whether you're paying attention.)

Solution
The iteration is to find the value of $s_0(r)$ for that r value and starts from an initial guess for s_0 of $1/2$. For the location of the dip (which is an r of about 1.39) the value of s_0 is already close to $1/2$.

Exercise 9 from Chapter 4
Show that, as r approaches 1 from above, the analogously defined τ also diverges (and in the same way).

Solution
For this value of r we define ϵ as $r = 1 + \epsilon$, with $\epsilon > 0$. The ratio $\frac{\delta(t+1)}{\delta(t)}$ was calculated to be (in the paragraph preceding Footnote 11) $\log(1 - \epsilon)$. This is set to equal $\exp(-1/\tau)$, giving (after some algebra) $\tau = 1/\epsilon$, and diverging as r approaches 1, from above.

Exercise 10 from Chapter 4
Immunity. For an epidemic, immunity should change the nature of our solution. Also, immunity is important in other applications of a percolitis-like formalism (as discussed in Sec. 4.9). In brain modeling, the Oz citizens become neurons, "infection"

corresponds to the firing of the neuron. Neurons in general exhibit a refractory period during which they do not re-excite. (An important difference though is that neurons can fire both excitatory signals and inhibitory signals.) Similarly, in the star formation process (the application to galactic morphology), a region where star formation has recently taken place is unlikely to allow a second episode of star formation for a considerable period—the gas is too hot.

We therefore develop a theoretical description in which there is d-day immunity to percolitis. First consider 1 day. Show that the mean field expression is

$$1 - s(t+1) = \exp(-rs(t)) + s(t) - s(t)\exp(-rs(t)). \tag{J.5}$$

Hint: As in Sec. 4.5, being sick, transmitting disease and similar variables can be considered "yes" or "no," "on" or "off." These are modeled as Boolean random variables and take (only) the values 0 and 1. For two such variables, A and B, the **or** operator (symbolized by a vertical line, "|") satisfies $A|B = A + B - AB$.

Find the critical value r_c such that the disease fades away for $r < r_c$.

Solution

The equation to be justified is

$$1 - s(t+1) = \exp(-rs(t)) + s(t) - s(t)\exp(-rs(t)). \tag{J.6}$$

On the left is the probability of being healthy on day-$(t+1)$. This can happen if no one transmits the disease—that's $\exp(-rs(t))$ (in the limit of N large). The next term on the right is $s(t)$, that is, if you were sick the previous day. But there's over-counting: if no one transmitted *and* you were sick the previous day. So you need to subtract the joint probability. That's the third term on the right.

The critical value is unchanged (it's still 1) since when s is small the probability of being sick on the previous day is negligible. In equations, after some algebra, the equation becomes $\exp(-rs) = 1 - \frac{s}{1-s}$. But for small s this reduces to the previous one (for non-immunity), which is $\exp(-rs) = 1 - s$, having the critical value 1.

Exercise 11 from Chapter 4

Consider another kind of immunity, one that operates through the transmission probability. Instead of a constant probability r (more precisely, r/N), one could take a variable rate that depends on the sickness level on the previous time step, for example $r[1 - s(t)]$. This model could apply where the transmission depends on some disease resource that is depleted when the illness is widespread. In the galaxy model (discussed in Sec. 4.9) this is a natural assignment, since star formation (what corresponds to being ill with percolitis) depends on having enough cool gas around, something that is less likely if there was a recent episode of star formation. Take the model then to have the iteration scheme

$$s(t+1) = 1 - e^{-rs(t)[1-s(t)]}.$$ \hfill (J.7)

(r is a constant.) You can easily check that the threshold, $r = 1$, is unchanged (why?). But this scheme has another threshold, a value of r for which the system *does not settle down*. It oscillates between two levels indefinitely (very much like the logistic map[2]). Can you find that next threshold analytically or numerically? (Answer: it's about 8.02659.[3]) As for the logistic map (see Appendix F) there is also a transition to chaos at r in the upper teens (later bifurcations: 14.361, 16.61, 17.15, 17.27, with chaos at about 17.3). There is evidence that such oscillations exist in galaxies.

Solution

The numbers (answers) were given in the main text. But here I will give a MATLAB™ program that provides them. No guarantees are tendered concerning the quality of the programming. The symbol % means the material to its right is commented out.

```
function [rs,S]=immune2
% Immunity with the rule s(t+1) = 1 - exp(-rs(t)(1-s(t)))
Niter = 1e8;            % Number of iterations (will depend on how close
                        % to a bifurcation you are). Change if necessary.
M=50;                   % Number of values to plot
s=zeros(1,Niter+1);     % Predefining for faster processing.
rs=8.0265:.00001:8.0267; % Set of r values to look at. Looking closely
                        % at r in this neighborhood.
nr=length(rs);          % Cardinality of the set of r's.
S=zeros(nr,M);          % More predefining. S is the final M values of s.
for k=1:nr,             % Begin a loop, in which each r value is used.
    r=rs(k);
    s(1)=1/2;           % Initial s value
    for t=1:Niter,
        s(t+1)=1-exp(-r*s(t)*(1-s(t))); % Finally! The iteration.
```

[2]$x(t+1) = \lambda x(t)(1-x(t))$ is the physics form of the logistic map (see Appendix F). The curves $x(1-x)$ and $1 - \exp(-rs(1-s))$ have the same properties: for $0 \le (x \text{ or } s) \le 1$ there is a 1-to-1 mapping $x = X(s,r)$ and $\lambda = \Lambda(s,r)$ (one bump in the middle, with a monotonic decrease on both sides). Note that this does not mean the mapping can be written down as a simple expression.

[3]If you're really feeling ambitious you can check the usual criterion for the first bifurcation. Define $F(s) = 1 - \exp(-rs(1-s))$. Then at the first bifurcation you should have $F'(s) = -1$, where the prime means $\partial/\partial s$.

```
    end
    S(k,:)=s(end-M+1:end);% Keep only the last M values.
end
figure(100)
%J=find(a<.47);a(J)=nan; % This can be useful when searching
                         % for a range of r values
plot(rs,S,'.')           % Plotting the final M values vs. r
axis('tight')
rs=rs';
Sbar=mean(mean(S));      % Mean value of S
rr=mean(rs);             % Mean value of r
rr*(1-Sbar)*(1-2*Sbar);  % This is the derivative at the mean value of r
```

Exercise 12 from Chapter 4

As an example of the effect of fluctuations and a finite size effect, consider the issue of extinction for $r > 1$. For $N \to \infty$ the number of people sick follows Eq. (4.4). However, for any finite system it can happen—and *will* happen with finite (even if small) probability—that *no one* gets sick on a given time step. If r is close to 1 ($= r_{\text{critical}}$), this can be important and terminate the "disease." Do numerical simulations of the finite-N process (not using Eq. (4.4)) to determine the actual lifetimes as a function of N. I suggest using $r = 1.1$ and 1.5 and using rather small values of N.

and

Exercise 13 from Chapter 4

For $r > 1$ and out of the critical region, estimate the necessarily finite time before extinction of the disease. The idea is that, with N people, it will eventually happen that—despite its being unlikely—all attempts at transmission fail.

See Footnote 30 (this chapter) for a discussion of extinction.

Solution

Instead of doing the indicated simulations I refer to the suggestions of Footnote 30 and the method of Sec. 4.7.

Exercise 14 from Chapter 5

Show that with these rules the Boltzmann distribution (Eq. (5.2) or Eq. (5.9)) is recovered.

Solution

The proof given here is slightly more general than that given in the main text (following the posing of this exercise).

Recall the definition, $R_{ij} = \Pr(i \leftarrow j)$, so that for some (sufficiently small) constant c and $i \neq j$ we have

$$R_{ij} = \begin{cases} c\exp(-\beta\Delta E), & \text{where } \Delta E = E_i - E_j > 0, \\ c, & \text{if } E_i - E_j \leq 0. \end{cases} \tag{J.8}$$

(It's the c that makes this proof more general.) Now suppose k has spin down and ℓ is spin up, and $E_\ell > E_k$, so the energy of state ℓ is higher (less desirable) than that of state k (and the transition $\ell \to k$ has the rate $c\exp(-\beta\Delta E)$ with $\Delta E > 0$). Suppose that ℓ and k have probabilities p_ℓ and p_k and under the transition rules let their new probabilities be p'_ℓ and p'_k. Then

$$p'_\ell = p_\ell(1 - c) + p_k\, c\exp(-\beta\Delta E), \tag{J.9}$$

$$p'_k = p_\ell\, c + p_k(1 - c\exp(-\beta\Delta E)). \tag{J.10}$$

Let r be the ratio of p_ℓ to p_k (with a corresponding definition of r'). Then dividing the top equation by the bottom and dividing the right-hand side by p_k we get

$$r' = \frac{r(1 - c) + c\exp(-\beta\Delta E)}{rc + 1 - c\exp(-\beta\Delta E)}. \tag{J.11}$$

For a stationary solution we require that $r' = r$, which translates to

$$cr^2 + r - cr\exp(-\beta\Delta E) = r - cr + c\exp(-\beta\Delta E). \tag{J.12}$$

Subtract r from both sides and divide by c, giving

$$r^2 + r(1 - \exp(-\beta\Delta E)) = \exp(-\beta\Delta E). \tag{J.13}$$

This quadratic equation has two solutions, -1 and $\exp(-\beta\Delta E)$. The -1 is not of interest since both probabilities are positive. This implies that

$$p_\ell \exp(\beta E_\ell) = p_k \exp(\beta E_k), \tag{J.14}$$

or that, for all j, $p_j \exp(\beta E_j)$ is a constant (call it $1/Z$ since this constant is the partition function). This gives the Boltzmann distribution, $p_j = (1/Z)\exp(-\beta E_j)$. The constant c can be anything as small as or smaller than 1 over the maximum of the column sums of R.

Note that this proof has been concerned with the ratio of probabilities for $i \neq j$. The correction to the diagonal (the MATLAB™ instructions immediately preceding this exercise in Chapter 5) are therefore irrelevant and deal with transitions $i \to i$ and $j \to j$.

Exercise 15 from Chapter 5

Propose other stochastic matrix choices for $R(\tilde{\sigma}_1, \tilde{\sigma}_0)$. Allow multiple spin flips as well. In all cases you need to satisfy what I prove below for the matrix just defined: it should give the Boltzmann distribution as its stationary state.

Solution

Allow changes in two (or more) spins and take the same rules. The constant c (cf. Eq. (J.8)) may have to be changed. (Obviously these are not the only possible changes.)

Exercise 16 from Chapter 5

Do the algebra.

Solution

It all depends on recognizing that $M = A - (N - A)$. This follows because the total magnetization is the number of spins up (A) minus the number down $(N - A)$. Since $M = 2A - N$, $\mu = \frac{M}{N} = \frac{2A}{N} - 1 = 2\alpha - 1$. Therefore, $\alpha = \frac{\mu+1}{2}$.

Exercise 17 from Chapter 5

Use Eq. (5.24) to show that, for T below (but near) T_c, $d^2 f/d\mu^2 < 0$ for non-zero μ_{spon}.

Hint: For the solution, see Footnote 18.

Solution

See the indicated footnote.

Exercise 18 from Chapter 5

Calculate $d\mu/dh$ for h near 0 and $T = 1$. This is the magnetic susceptibility and is analogous to the specific heat. Determine its behavior as $h \to 0$.

Solution

Work from the relation $\mu = \tanh(\beta\mu + h)$, which (for $\beta = T = 1$) becomes $\mu = \tanh(\mu + h)$; using $\tanh x = x - \frac{1}{3}x^3 + \ldots$, we obtain $\mu = (3h)^{1/3} + \text{smaller terms}$. Then $\frac{d\mu}{dh} = \frac{1}{(3h)^{2/3}}$.

Exercise 19 from Chapter 5

I'll call this an exercise, but actually I don't know the answer, so maybe I should call it a project. Here's the question: We've found that per-particle fluctuations go like $N^{-1/4}$. The reason is that correlations among the underlying variables have become significant. You don't really have N independent participants, because the dynamics has them doing things in a correlated way. How many independent motions are there? I would guess there are $\sqrt{N}$; I say that because if this number of degrees of freedom acted independently, they would give rise to per-particle fluctuations of order $1/\sqrt{\sqrt{N}}$. So the questions is, what are these $\sqrt{N}$ degrees of freedom? Or maybe show something a bit easier, that—effectively—there *are* that many independent degrees of freedom, even if you can't explicitly identify them.

Solution

You can't Fourier transform a spatial variable in mean field theory since there's no space. But there may be other possibilities.

Exercise 20 from Chapter 6

Suppose that a binary signal—a "0" or a "1"—has probability p of being transmitted correctly and probability q of having an error $(q + p = 1)$.
What is the missing information for each transmitted signal?
Now use redundancy to enhance reliability: For each bit (0 or 1) in your bit stream you send three signals. At the receiving end they have the same reliability as in the one-signal case. But, you now take a vote. If two or three are ones, you take the original signal to have been 1; otherwise, you take it to have been 0.
What is the probability that your answer is correct?
Show that for $1/2 < p < 1$ there is improved reliability.
What is the missing information? For small q, what is the leading term in q? in q^2? If the leading term in q^2 is 0, find the next term.
What level of redundancy (i.e., how many bits for each initial bit), along with a voting rule, is necessary to ensure one part in 8 million accuracy (so a megabyte file is likely to be copied correctly)?
For numerical answers assume $q = \exp(-6) \approx 1/400$.

Solution

For the first part, I'll ignore the overall multiplicative constant, "k" (which could be $(\log 2)^{-1}$ or k_B, etc.). The missing information (as usual) is given by "$-\sum p \log p$" and is therefore

$$I = -(1 - q)\log(1 - q) - q\log q \approx -q\log q, \tag{J.15}$$

The leading term for small q.
The probability of a correct answer is

$$p_{\text{correct}} \equiv p' = (1 - q)^3 + 3(1 - q)^2 q. \tag{J.16}$$

To show that this is more reliable you need to show that for $1/2 < p < 1$ this is greater than p, that is, $p' = p^3 + 3p^2(1 - p) > p$. Since we are not interested in $p = 0$ (and can therefore divide by p) this amounts to $p^2 + 3p(1 - p) - 1 > 0$. This can be written $(p + 1)(p - 1) + 3p(1 - p) > 0$. Again, since we are not interested in $p = 1$, we can divide this by $(1 - p)$, yielding $2p > 1$. This is satisfied (recall $p > 1/2$).

The missing information requires $q' \equiv 1 - p' = 3q^2 - 2q^3$, so that

$$I_{2 \text{ out of } 3} = -(1 - q') \log(1 - q') - q' \log q' \approx -6q^2 \log q. \tag{J.17}$$

I call this $O(q^2)$ although strictly speaking it is $O(q^2 \log q)$ and the $\log q$ is much larger (for small q).

To get one part in 8 million accuracy, you need an integer k such that $n \equiv 2k + 1$ (with n the number of votes) and $\sum_{\ell=0}^{k} \binom{2k+1}{\ell} p^{n-\ell} (1-p)^{\ell} > 1 - \frac{1}{8 \times 10^6}$. This is equivalent to finding $q = 1 - p$ such that $\sum_{\ell=0}^{k} \binom{2k+1}{\ell} q^{n-\ell} (1-q)^{\ell} < \frac{1}{8 \times 10^6}$. The first n for which this is true is $n = 7$.

Exercise 21 from Chapter 6

There's an additional constraint on the probabilities, namely that they all be non-negative. Why didn't we have to enforce it? (And it would be difficult to enforce, being non-holonomic.)

Solution

Let the desired function be the logarithm of the probabilities, instead of the probabilities themselves. (There might be something more profound at work. But this suffices for the present.)

Exercise 22 from Chapter 6

Check that $\frac{1}{2} m \langle v^2 \rangle = \frac{3}{2} k_B T$. Remember that <u>*v* is three-dimensional</u> and use the appropriate integration weight.

Solution

$Z(T) = \int_0^\infty dv \, 4\pi v^2 \exp\left(-\frac{mv^2/2}{k_B T}\right)$ and use $\sqrt{\frac{\pi}{a}} = \int_{-\infty}^{\infty} dv \, \exp(-av^2)$ and derivatives of the previous formula with respect to $-a$. Note that $\langle v^2 \rangle$ involves a *fourth* power of v.

Exercise 23 from Chapter 7

For the directed network with $p = 0$, show that the dropoff indeed has power 2.

Solution

The only change from the derivation that appears for the non-directed network is that for each new node there is but a single new connection. Following that derivation, "$\sqrt{\frac{t}{t_u}}$" is replaced by $\frac{t}{t_u}$, and the eventual power is lowered by 1.

Exercise 24 from Chapter 8

Show that $\text{Tr } A^n = \sum a_k^n$, where $\{a_k\}$ are the eigenvalues of A.

Solution

Use $A = U A_D U^{-1}$ (where A_D is the diagonal matrix of eigenvalues of A) and $\text{Tr } AB = \text{Tr } BA$ for A and B matrices. The matrix under consideration, L, is diagonalizable, but even for matrices requiring a Jordan form it remains true. (Occasionally, one comes across a situation where a Jordan form is required. I was surprised a few years ago when I encountered such matrices [200].) This implies, by the way, that the determinant is the product of eigenvalues.

Exercise 25 from Chapter 8

Find the eigenvalues of L (given in Eq. (8.6)).

Solution

Recall that the eigenvalues, λ, must satisfy

$$0 = \det(L - \lambda \mathbf{1}) = \det \begin{pmatrix} e^{K+H} - \lambda & e^{-K} \\ e^{-K} & e^{K-H} - \lambda \end{pmatrix} \tag{J.18}$$

(where $\mathbf{1}$ is the identity matrix) and, therefore,

$$\lambda^2 - 2\lambda e^K \cosh H + 2\cosh(2K) = 0. \tag{J.19}$$

It follows (using $1 = \cosh^2 u - \sinh^2 u$) that $\lambda_\pm = e^K \left[\cosh H \pm \sqrt{\sinh^2 H - e^{-4K}} \right]$.

Exercise 26 from Chapter 8

For these *normalized* eigenvectors, check that $\langle n|n' \rangle = \delta_{nn'} = $ Kronecker delta $= 1$ when $n = n'$ and 0 otherwise ($n, n' \in \{1, 2\}$ in this case).

Solution

Recall that $\phi_1 = \frac{1}{\sqrt{2}} \begin{pmatrix} 1 \\ 1 \end{pmatrix}$ and $\phi_2 = \frac{1}{\sqrt{2}} \begin{pmatrix} 1 \\ -1 \end{pmatrix}$. We need to check whether 1 (the scalar) is $\phi_1^\dagger \phi_1 = \frac{1}{2} \begin{pmatrix} 1 & 1 \end{pmatrix} \begin{pmatrix} 1 \\ 1 \end{pmatrix}$, which, following the rules of matrix multiplication, it is. Ditto for ϕ_2. We also need to check $\phi_1^\dagger \phi_2 = \frac{1}{2} \begin{pmatrix} 1 & 1 \end{pmatrix} \begin{pmatrix} 1 \\ -1 \end{pmatrix} = 0$, which, again following the rules for matrix multiplication, it is. Ditto for $\phi_2^\dagger \phi_1$.

Exercise 27 from Chapter 8

Check the spectral expansion. See that you recover L, as in Eq. (8.18).

Solution

We need to check that L is equal to $\lambda_+ \phi_+ \phi_+^\dagger + \lambda_- \phi_- \phi_-^\dagger$. Bearing in mind that $\phi_+ \phi_+^\dagger = \begin{pmatrix} 1 & 0 \\ 0 & 1 \end{pmatrix}$, $\phi_- \phi_-^\dagger = \begin{pmatrix} 1 & 0 \\ 0 & -1 \end{pmatrix}$ and $\lambda_\pm = 2 \left\{ \begin{smallmatrix} \cosh \\ \sinh \end{smallmatrix} \right\} K$, we see that they are equal.

Exercise 28 from Chapter 8

We earlier found that the correlation length is given by $\xi = -1/\log(\tanh \mathcal{J})$. Now we find that with $\mathcal{J}'$, a particular function of $\mathcal{J}$ (given by Eq. (8.33)), $\xi' = \frac{1}{2}\xi$. Are these consistent?

Solution

Yes, they are consistent. The proof involves a lot of algebra. This implies that the relation Eq. (8.33) was implicit in the correlation length relation. Or, to put it differently, the complicated calculation leading to Eq. (8.33) can be replaced by the obvious $\xi' = \frac{1}{2}\xi$ and the correlation length deduced.

Exercise 29 from Chapter 8

Can you find a different example of "indifference to details?" I've given something simple in which taking one decimation goes from $\beta \cdot$ Hamiltonian $= -\mathcal{J}_a \sum_j \sigma_{2j}\sigma_{2j+1} - \mathcal{J}_b \sum_j \sigma_{2j+1}\sigma_{2j+2}$ to $-\mathcal{J}' \sum_j \sigma_j\sigma_{j+1}$. Perhaps if the interaction involved a next nearest neighbor or a distribution of coupling constants (with a mean and standard deviation) the scheme would work.

Solution

I don't think it can be made exact, but there might be an approximate procedure. (On the other hand, I'd welcome a good example.)

Exercise 30 from Chapter 9

Find the dependence of p_0 on the level of zealotry.

Solution

See Sec. 9.2.

Exercise 31 from Chapter 9

This is not a complete proof. Show that the 80 minute trip is an attractor. If one driver takes the fast route there is a saving, but others will follow and finally all lose.

Solution

Suppose you went from 2000–2000 to 2001–1999, which is to say one driver switches from ACD to ABCD, using the newly constructed BC connection. This person will take $\frac{2001}{100} + \frac{2001}{100} \approx 40$ minutes. So others follow. Finally when 500 switch (so 2500 use the new road) it takes those who go ABCD (on the new road) 65 minutes, but the other 1500 are still taking ACD, which means they require $45 + \frac{2000+2000}{100} \approx 85$ minutes, so they surely switch. For further details, see the Wikipedia article on Braess's paradox (accessed April 2020).

Exercise 32 from Appendix C

Show that the cardinality of the Cantor set is the same as that of the line of all real numbers.

Solution

To say that two sets have the same cardinality means that there is a one-to-one mapping between them. In binary notation ("base-2") any point between zero and one on the real line can be written

$$x = \sum_{n=1}^{\infty} \frac{a_n}{2^n}, \tag{J.20}$$

with $a_n \in \{0, 1\}$. The Cantor set removes successive middle thirds, leaving only points of the form (in base-3 notation)

$$y = \sum_{n=1}^{\infty} \frac{b_n}{3^n}, \tag{J.21}$$

with $b_n \in \{0, 2\}$. In other words, the process that produced the Cantor set removed all points such that there was any term in $\sum_{n=1}^{\infty} \frac{c_n}{3^n}$ (with $c_n \in \{0, 1, 2\}$) for which $c_n = 1$. The one-to-one mapping of $x \leftrightarrow y$ is $a_n \leftrightarrow b_n$ (since the set of possible b_n only has two entries). By the way, it is also known that the cardinality of any non-trivial interval of the real line is equal to the cardinality of the entire line.

Exercise 33 from Appendix D

Show that $\langle X \rangle \langle Y \rangle = \langle XY \rangle$ implies that $\Pr(X{=}x \,\&\, Y{=}y) = \Pr(x)\Pr(y)$, the converse of Eq. (D.4).

Solution

By definition $\langle X \rangle \langle Y \rangle = \sum_{x,y} \Pr(x)x\Pr(y)y$ and by assumption this is equal to $\langle XY \rangle = \sum_{x,y} \Pr(X{=}x \,\&\, Y{=}y)\,xy$. Comparing the two sums, we have $\Pr(X{=}x \,\&\, Y{=}y) = \Pr(x)\Pr(y)$.

Exercise 34 from Appendix D

Deduce the spectral properties of W.

Solution

W is $\lim_{\Delta t \to 0} \frac{(R-1)}{\Delta t}$, where by "1" I mean the matrix form of the Kronecker delta (1's on the diagonal, 0 elsewhere). The matrix $(R-1)$ simply has the spectrum of a unit circle, but displaced by -1, so that its center is at -1 and its maximum eigenvalue is zero. Now divide it by Δt with $\Delta t \to 0$. The spectrum of W can then lie anywhere on the left side of the complex plane, that is, if λ is in W's spectrum, $\mathrm{Re}\,\lambda \le 0$. (I have assumed all relevant limits exist.)

Exercise 35 from Appendix D
How many different letter arrangements can be obtained from the letters of the word *statistically*, using all the letters?

Solution
There are 13 letters. The letter t is repeated thrice; the letters s, a, l, and i are each repeated twice. c and y each appear once. If all letters were distinct there would be 13! sequences. But that would be over-counting since the order of, for example, the t's is irrelevant. The actual number is thus the multinomial coefficient

$$\frac{13!}{3!\,(2!)^4} = 64,864,800\,.$$

Exercise 36 from Appendix D
A deck of 52 cards is shuffled thoroughly. What is the probability that the four aces are all next to one another?

Solution
There are 52! ways of ordering the deck. Now consider the case where the four aces are together. Suppose you go through the deck card by card. The first ace can come in any one of 49 positions, from 1st to 49th. Among the aces there are 4! ways of their being ordered. The other 48 cards have 48! orderings. The probability is the ratio of orderings satisfying the condition to all possible orderings, specifically

$$p = \frac{49 \cdot 4! \cdot 48!}{52!} = \frac{1}{5525}\,.$$

Exercise 37 from Appendix D
If n balls are distributed randomly into k urns, what is the probability that the last urn contains j balls?

Solution
There is probability $1/k$ of putting the ball in the last urn and probability $(1-1/k)$ of putting it in one of the others. So for each particular sequence that puts exactly k balls in the last urn there is probability $(1/k)^j\,(1-(1/k))^{n-j}$ of its occurrence. But there are many such sequences, specifically n-choose-j. Thus the desired probability is

$$\binom{n}{j}(1/k)^j[1-(1/k)]^{n-j}\,.$$

Exercise 38 from Appendix D
If a five-letter word is formed at random (meaning that all sequences of five letters are equally likely), what is the probability that no letter occurs more than once?

Solution

Pick the five letters (returning your pick to the alphabet soup immediately, i.e., "with replacement"). The first can be anything. The second has probability 25/26 of being different from the first. The next letter must be different from the first two; and so on. Thus the probability is

$$\frac{25}{26} \cdot \frac{24}{26} \cdot \frac{23}{26} \cdot \frac{22}{26} = \frac{303,600}{456,976} \approx 0.6644 \,.$$

Exercise 39 from Appendix D

What is the coefficient of $x^3 y^4$ in $(x+y)^7$?

Solution

In the product $(x+y) \cdot (x+y) \cdot (x+y) \cdot (x+y) \cdot (x+y) \cdot (x+y) \cdot (x+y)$ you will get $x^3 y^4$ by taking x from three of the factors in parenthesis, y from the other four. There are $\binom{7}{3}$ ways of doing this.

Exercise 40 from Appendix D

Two dice are rolled, and the sum of the face values is six. What is the probability that at least one of the dice came up a 3?

Solution

There are five ways to get 6, $(5,1)$, $(4,2)$, $(3,3)$, $(2,4)$, $(1,5)$, where I am able to distinguish which die gave which number. As it happens, if one comes up 3 the other must also (to get 6), but that does not affect the count. The probability is thus 1/5.

Exercise 41 from Appendix D

A player throws darts at a target. On each trial, independently of the other trials, she hits the bull's-eye with probability 0.05. (So the trials are i.i.d.) How many times should she throw so that her probability of hitting the bull's-eye at least once is in excess of 1/2?

Solution

The probability of (complete) failure after n attempts is $p_f = (1 - 0.05)^n$. The probability of at least one success is thus $1 - p_f$. We want this to be 1/2, so we set $1/2 = 1 - (1 - 0.05)^n$ and solve for n. The answer is 13.51, so at least 14 throws are needed to have the probability of success exceed 0.5. Note that the assumption of independence may be incorrect if, for example, the player gets nervous after 10 failures. Or, alternatively, the player may learn from mistakes.

Exercise 42 from Appendix D

A cube whose faces are colored is split into 1000 small cubes ($10^3 = 1000$) by defining the small cubes through the intersection of three sets of nine equally spaced orthogonal planes, cutting the large cube. Only the original cube has colored faces. The (small) cubes thus obtained are mixed thoroughly. Find the probability that a cube drawn at random will have two colored faces.

Solution

The number of small cubes is 1000. A cube has 12 edges so that on each edge there are eight small cubes that have two colored faces (the ones on the very end have three colored faces). Therefore the number with two colored faces is $12 \cdot 8 = 96$, so that $p = 96/1000 = 0.096$.

Exercise 43 from Appendix D

Monty Hall problem. Monty Hall was the host of the game show *Let's make a deal*. A contestant, call her Ms. A, was given a choice of three doors. Behind one is a car (which she can keep), and behind the other two, a goat (which—if she lives in a city— I wouldn't advise keeping). Ms. A makes a guess. But before the door is opened to reveal whether Ms. A has opened the right door, Mr. Hall (who knows where the car is) opens a door behind which a goat stands. Ms. A is then asked whether she wishes to change her choice. Should she?

Solution

Suppose Ms. A has chosen door #1. There is a 1 chance in 3 that it has the car. Below are the results of switching or not switching for all possible (random and initially secret) configurations of car and goats. If Ms. A switches (from door #1) she has two chances out of three to win. So she should switch! If you are troubled by this result, check Wikipedia (`https://en.wikipedia.org/wiki/Monty_Hall_problem`; accessed March 2020).

Behind door #1	Behind door #2	Behind door #3	Stick to #1	Switch
Car	Goat	Goat	Car	Goat
Goat	Car	Goat	Goat	Car
Goat	Goat	Car	Goat	Car

Exercise 44 from Appendix D

A lot contains m defective items and n good ones. From this lot s items are chosen at random and tested for quality. The first k of these are found not to be defective. What is the probability that the next item tested is good?

A ——————————— C   B

Fig. J.1 A transmission line from A to B with a break at C.

Solution

Bear in mind that you know the number of defective items, which is not the usual case for this kind of testing, where you'd like to find out the fraction defective. For this situation it's as if your total population is now $n-k$ of which m are defective. Therefore the probability that the next one is good is $p = (n-k)/(n-k+m)$.

Exercise 45 from Appendix D

A break occurs at a random point, C, on a telephone line AB of length L. What is the probability that C is at a distance more than ℓ from the point A? See Fig. J.1.

Solution

$p = 1 - \frac{\ell}{L}$.

Exercise 46 from Appendix D

The random numbers x and y are drawn from uniform distributions on $[0, 1]$. (In other words, they are equally likely to be anywhere between 0 and 1.) What is the probability that they satisfy the following two conditions: their sum is equal to or less than one, and their product is equal to or less than $2/9$? (Hint: It is useful to consider the areas as subsets of the unit square.)

Solution

The two conditions are equivalent to $x + y \leq 1$ and $xy \leq 2/9$. The pair of numbers can be anywhere in the unit square. Consider the situation if they only had to satisfy $x + y \leq 1$. Then the region of the square bounded by that condition would be acceptable. This is the region bounded by $x + y = 1$ and would be the triangle on the lower left. Since the area of this triangle is $1/2$, and the area of the square is 1, the probability would be $1/2$. But what about the second condition? We *also* want $xy \leq 2/9$. The boundary defined by this condition is $xy = 2/9$, which is a hyperbola that intersects the line $x + y = 1$ twice. To find the meeting points impose both conditions: for x this implies $x(1-x) = 2/9$ whose solutions are $1/2 \pm 1/6$. See Fig. J.2.

The desired area is the intersection of the two conditions. This means it is $1/2$ minus the area of the small region between the curves, that is,

$$\text{Probability} = \frac{\text{Area}}{1} = \frac{1}{2} - \int_{1/3}^{2/3} dx \left(x - \frac{2}{9x} \right) = \frac{1}{2} - \frac{x^2}{2} \Big|_{1/3}^{2/3} + \frac{2}{9} \log x \Big|_{1/3}^{2/3} = \frac{1}{3} + \frac{2}{9} \log 2.$$

$$(J.22)$$

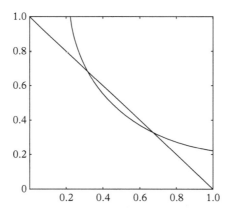

Fig. J.2 The two curves are $x + y = 1$ and $xy = 2/9$. They intersect at $1/2 \pm 1/6$ or $1/3$ and $2/3$.

Exercise 47 from Appendix D

Archer A has probability 0.8 of hitting a target, and Archer B has probability 0.7; what is the probability that at least one of them will hit the target? (Each shoots one arrow.)

Solution

The probabilities that they miss are 0.2 and 0.3, respectively. So the probability that both miss is the product. Thus the probability that at least one succeeds is $p = 1 - (0.2)(0.3) = 0.94$.

Exercise 48 from Appendix D

An urn contains three distinguishable balls, one red, one black and one white. You draw balls from it five times, one ball at a time, with replacement (i.e., after you draw a ball and check its color you put it back). What is the probability that the red and white balls will be drawn at least twice each?

Solution

The total number of ways to pick the balls is 3^5. There are many ways to satisfy the given conditions. First you could pick two red, two white and one black. Since the order in which you pick them is irrelevant, there would be $\left(\frac{5!}{2!\,2!\,1!}\right)$ ways of doing this, which is 30. But you also might have gotten three red and two white, which can be done in $\left(\frac{5!}{3!\,2!\,0!}\right)$ ways, which is 10. Similarly you could have gotten three white and two red, which is also 10. Thus the total probability is $p = \frac{30+10+10}{243} = \frac{50}{243} \approx 0.2058$.

References

[1] There are several early references. Priority goes to Ray Solomonoff, with his "A Preliminary Report on a General Theory of Inductive Inference," Feb. 1960, Report V-131. As far as I can tell, next came Andrey Kolmogorov with "On Tables of Random Numbers," Sankhya Ser. A. 25:369–375 in 1963. Finally, and independently, there was Gregory J. Chaitin with "On the Simplicity and Speed of Programs for Computing Infinite Sets of Natural Numbers," J. ACM 16:407–422 in 1969.

[2] Acebrón, Juan A.; Bonilla, L. L.; Vicentez, Conrad J. Pérez; Ritort, Félix; and Spigler, Renato (2005). The Kuramoto model. *Rev. Mod. Phys.*, **77**, 137.

[3] Advani, Madhu; Lahiri, Subhaneil; and Ganguli, Surya (2013). Statistical mechanics of complex neural systems and high dimensional data. *J. Stat. Mech.*, **2013**(03), P03014.

[4] Aguilar, J.; Monaenkova, D.; Linevich, V.; Savoie, W.; Dutta, B.; Kuan, H.-S.; Betterton, M. D.; Goodisman, M. A. D.; and Goldman, D. I. (2018). Collective clog control: Optimizing traffic flow in confined biological and robophysical excavation. *Science*, **361**, 672–677.

[5] Albert, Réka and Barabási, Albert-László (2002). Statistical mechanics of complex networks. *Rev. Mod. Phys.*, **74**, 47–97.

[6] Alpher, R. A. and Herman, R. (1949). Remarks on the evolution of the expanding universe. *Phys. Rev.*, **75**, 1089–1095. The estimate of present temperature ("of the order of $5\,\mathrm{K}$") follows Eq. 12d in that paper. For astrophysics this is remarkable agreement.

[7] Amit, Daniel J.; Gutfreund, Hanoch; and Sompolinsky, H. (1985). Spin-glass models of neural networks. *Phys. Rev. A*, **32**, 1007–1018.

[8] Amit, Daniel J.; Gutfreund, Hanoch; and Sompolinsky, H. (1985). Storing infinite numbers of patterns in a spin-glass model of neural networks. *Phys. Rev. Lett.*, **55**, 1530–1533.

[9] Anderson, P. W. (1972). More is different. *Science*, **177**, 393–396.

[10] Aoki, I. (1982). A simulation study on the schooling mechanism in fish. *Bull. Japan. Soc. Sci. Fisheries*, **48**, 1081–1088.

[11] Arshad, Sidra; Hu, Shougeng; and Ashraf, Badar Nadeem (2018). Zipf's law and city size distribution: A survey of the literature and future research agenda. *Physica A*, **492**, 75–92.

[12] Attanasi, Alessandro; Cavagna, Andrea; Castello, Lorenzo Del; Giardina, Irene; Grigera, Tomas S.; Jelic, Asja; Melillo, Stefania; Parisi, Leonardo; Pohl, Oliver; Shen, Edward; and Viale, Massimiliano (2014). Information transfer and behavioural inertia in starling flocks. *Nature Phys.*, **10**, 691–696.

[13] Auerbach, Felix (1913). Das Gesetz der Bevölkerungskonzentration ("The Law of Population Concentration"). *G. Petermanns Mitteilungen*, **59**, 73–76.

[14] Avron, J. E.; Roepstorff, G.; and Schulman, L. S. (1981). Ground state degeneracy and ferromagnetism in a spin glass. *J. Stat. Phys.*, **26**, 25–36.

[15] Baierlein, R. (1999). *Thermal Physics*. Cambridge Univ. Press, New York.

[16] Bain, Read (1959). Review of Zipf's book, [243]. *Social Forces*, **28**(3), 340–341.

[17] Bak, Per and Tang, Chao (1989). Earthquakes as a self-organized critical phenomenon. *J. Geophys. Res.*, **94**, 15635–15637.

[18] Bak, Per; Tang, Chao; and Wiesenfeld, Kurt (1988). Self-organized criticality. *Phys. Rev. A*, **38**, 364–374.

[19] Bak, Per; Tang, Chao; and Wiesenfeld, Kurt (1988). Self-organized criticality: an explanation of $1/f$ noise. *Phys. Rev. Lett.*, **59**, 381–375.

[20] Barger, Vernon D. and Olssen, Martin G. (1995). *Classical Mechanics* (2nd edn). McGraw-Hill, New York.

[21] Barmpalias, George; Elwes, Richard; and Lewis-Pye, Andrew (2016). Unperturbed Schelling segregation in two or three dimensions. *J. Stat. Phys.*, **164**, 1460–1487.

[22] Beggs, John M. (2008). The criticality hypothesis: how local cortical networks might optimize information processing. *Phil. Trans. R. Soc. A*, **366**, 329–343.

[23] Beggs, John M. and Plenz, Dietmar (2003). Neuronal avalanches in neocortical circuits. *J. Neurosci.*, **23**, 11167–11177.

[24] Bennett, Matthew; Schatz, Michael F.; Rockwood, Heidi; and Wiesenfeld, Kurt (2002). Huygens's clocks. *Proc. R. Soc. Lond. A*, **458**, 563–579.

[25] Berman, A.; Neumann, M.; and Stern, R. J. (1989). *Nonnegative Matrices in Dynamic Systems*. Wiley, New York.

[26] Bialek, William; Cavagna, Andrea; Giardina, Irene; Mora, Thierry; Silvestri, Edmondo; Viale, Massimiliano; and Walczak, Aleksandra M. (2012). Statistical mechanics for natural flocks of birds. *Proc. Natl. Acad. Sci. USA*, **109**, 4786–4791. See also the "Supporting information" that accompanies the article.

[27] Biham, Ofer; Middleton, A. Alan; and Levine, Dov (1992). Self-organization and a dynamical transition in traffic-flow models. *Phys. Rev. A*, **46**, R6124–R6127.

[28] Bloom, Harold (1991). *The Book of J.* Vintage, New York. Translated from the Hebrew by David Rosenberg.

[29] Boghosian, Bruce (November 2019). Is inequality inevitable? (a.k.a. The inescapable casino). *Sci. Am.*, **321**, 70–77.

[30] Bonner, W. B. (2000). Local dynamics and the expansion of the universe. *Gen. Relativ. Gravit.*, **32**, 1005–1007.

[31] Braess, Dietrich; Nagurney, Anna; and Wakolbinger, Tina (2005). On a paradox of traffic planning. *Trans. Sci.*, **39**, 446–450. This is a translation of the original, Über ein Paradoxon aus der Verkehrsplanung, by D. Braess, *Unternehmensforschung* **12**, 258–268 (1968).

[32] Bricker, J.; Dettling, L.J.; Henriques, A.; Hsu, J.W.; Moore, K.B.; Sabelhaus, J.; Thompson, J.; and Windle, R. (2014). Changes in U.S. family finances from 2010 to 2013: Evidence from the survey of consumer finances. *Fed. Reserve Bull.*, **100**, 1–41.

[33] Buck, J. B. (1938). Synchronous rhythmic flashing of fireflies. *Q. Rev. Biol.*, **13**, 301–314.

[34] Buck, J. B. (1988). Synchronous rhythmic flashing of fireflies II. *Q. Rev. Biol.*, **63**, 265–289. See references therein.

[35] Burridge, R. and Knopoff, L. (1967). Model and theoretical seismology. *Bull. Seismol. Soc. Am.*, **57**, 341–371.

[36] Calude, Andreea S. and Pagel, Mark (2011). How do we use language? Shared patterns in the frequency of word use across 17 world languages. *Phil. Trans. R. Soc. B*, **366**, 1101–1107.

[37] Carlson, J. M. and Doyle, John (2002). Complexity and robustness. *Proc. Natl Acad. Sci. USA*, **99**(Suppl. 1), 2538–2545.

[38] Carter, Bruce and Mancini, Ron (2009). *Op Amps for Everyone.* (3rd edn), pp. 366–370. Texas Instruments, Newnes; Elsevier, New York. The relevant formula appears as a Laplace transform of what is given here.

[39] Cavagna, Andrea; Castello, Lorenzo Del; Giardina, Irene; Grigera, Tomas S.; Jelic, Asja; Melillo, Stefania; Mora, Thierry; Parisi, Leonardo; Silvestri, Edmondo; Viale, Massimiliano; and Walczak, Aleksandra M. (2015). Flocking and turning: a new model for self-organized collective motion. *J. Stat. Phys.*, **158**, 601–627.

[40] Chowdhury, Debashish; Santen, Ludgar; and Schadschneider, Andreas (2000). Statistical physics of vehicular traffic and some related systems. *Phys. Rep.*, **329**, 199–329. See especially Sec. 8 and Appendix C.

[41] Clifford, P. and Sudbury, A. (1973). A model for spatial conflict. *Biometrika*, **60**, 581–588.

[42] Coleman, Ph; Pietronero, L; and Sanders, R. H. (1988, 7). Absence of any characteristic correlation length in the CfA galaxy catalog. *Astron. Astrophys.*, **200**(1–2), L32–L34.

[43] Cosmai, L.; Fanizza, G.; Sylos Labini, F.; Pietronero, L.; and Tedesco, L. (2019). Fractal universe and cosmic acceleration in a Lemaitre-Tolman-Bondi scenario. *Class. Quantum Grav.*, **36**, 045007.

[44] Cover, Thomas A. and Thomas, Joy A. (2006). *Elements of Information Theory* (2nd edn). Wiley, New York. Proof of the properties quoted is on p. 28, Thm. 2.6.3 of the 2nd edition.

[45] Cristelli, Matthieu; Batty, Michael; and Pietronero, Luciano (2012). There is more than a power law in Zipf. *Sci. Rep.*, **2**, 812.

[46] Curzon, F. L. and Ahlborn, B. (1975). Efficiency of a Carnot engine at maximum power output. *Am. J. Phys.*, **43**, 22–24.

[47] Darroch, J. N. and Ratcliff, D. (1972). Generalized iterative scaling for log-linear models. *Ann. Math. Stat.*, **43**, 1470–1480.

[48] Davis, C. (2004). Physicists and traffic flow, *APS News*, April 8.

[49] Devitt-Lee, Adrian; Wang, Hongyan; Li, Jie; and Boghosian, Bruce (2018). A nonstandard description of wealth concentration in large-scale economies. *SIAM J. Appl. Math.*, **78**, 996–1008.

[50] Drossel, B. and Schwabl, F. (1992). Self-organized critical forest-fire model. *Phys. Rev. Lett.*, **69**, 1629–1632.

[51] Dutta, P. and Horn, P. M. (1981). Low-frequency fluctuations in solids: $1/f$ noise. *Rev. Mod. Phys.*, **53**, 497–516.

[52] Dutton, Aaron A. (2009). On the origin of exponential galaxy discs. *Mon. Not. R. Astron. Soc.*, **396**, 121–140.

[53] Dyson, Freeman (Jan. 22, 2004). A meeting with Enrico Fermi. *Mature*, **427**, 297

[54] Elmegreen, Bruce G. and Struck, Curtis (2013). Exponential galaxy disks from stellar scattering. *Astrophys. J. Lett.*, **775**, L35.

[55] Englman, R.; Gur, Y.; and Jaeger, Z. (1983). Fluid flow through a crack network in rocks. *J. Appl. Mech.*, **50**, 707–711.

[56] Estoup, Jean-Baptiste (1916). *Gammes sténographiques* (4th edn). Institut Sténographique de France, Paris.

[57] Feder, Jens (1988). *Fractals*. Plenum, New York. See Fig. 2.7.

[58] Feigenbaum, Mitchell J. (1978). Quantitative universality for a class of nonlinear transformations. *J. Stat. Phys.*, **19**, 25–52.

[59] Feigenbaum, Mitchell J. (1979). The universal metric properties of nonlinear transformations. *J. Stat. Phys.*, **21**, 669–706.

[60] Fernández-Gracia, Juan; Suchecki, Krzysztof; Ramasco, José J.; San Miguel, Maxi; and Equíluz, Victor M. (2014). Is the Voter Model a model for voters? *Phys. Rev. Lett.*, **112**, 158701. See also the Physics "Focus" note about this paper.

[61] Finkelstein, David (1958). Past-future asymmetry of the gravitational field of a point particle. *Phys. Rev.*, **110**, 965–967.

[62] Fischetti, Mark (2001). Drowning in New Orleans. *Sci. Am.*, September 5.

[63] Gabaix, Xavier (1999). Zipf's law for cities: An explantion. *Q. J. Econ.*, August, 739–767.

[64] Gabrielli, A.; Sylos Labini, F.; Joyce, M.; and Pietronero, L. (2004). *Statistical Physics for Cosmic Structures.* Springer Verlag, New York.

[65] Gantmacher, F. R. (1959). *Matrix Theory*, vol. I. Chelsea, New York. Translated from the Russian, Teoriya Matrits, by K. A. Hirsch.

[66] Gantmacher, F. R. (1959). *Matrix Theory*, vol. II. Chelsea, New York. Translated from the Russian, Teoriya Matrits, by K. A. Hirsch.

[67] Gardner, Martin. He had a series of columns in the "Mathematical games" section of *Scientific American* in the early 1970s (several issues) dealing with the Game of Life, as "played" by John Conway, a British–American mathematician. Initially Conway's game was a way to make a self-reproducing cellular automaton, but the ideas were extended in many ways, including galactic morphology.

[68] Gaveau, B. and Schulman, L. S. (1991). Mean-field self-organized criticality. *J. Phys. A*, **24**, L475–L480.

[69] Gaveau, B. and Schulman, L. S. (1994). Fluctuations in mean field self organized criticality. *J. Stat. Phys.*, **74**, 607–630.

[70] Gaveau, B. and Schulman, L. S. (1996). Master equation based formulation of non-equilibrium statistical mechanics. *J. Math. Phys.*, **37**, 3897–3932.

[71] Gaveau, B. and Schulman, L. S. (1998). Theory of non-equilibrium first order phase transitions for stochastic dynamics. *J. Math. Phys.*, **39**, 1517–1533.

[72] Gaveau, B. and Schulman, L. S. (2006). Multiple phases in stochastic dynamics: Geometry and probabilities. *Phys. Rev. E*, **73**, 036124.

[73] Gaveau, B. and Schulman, L. S. (2011). Violation of the zeroth law of thermodynamics for a non-ergodic interaction. *J. Stat. Phys.*, **145**, 1458–1471.

[74] Gaveau, B. and Schulman, L. S. (2015). Is ergodicity a reasonable hypothesis for macroscopic systems? *Eur. Phys. J. Spec. Top.*, **224**, 891–904.

[75] Gaveau, B.; Schulman, L. S.; and Schulman, L. J. (2006). Imaging geometry through dynamics: the observable representation. *J. Phys. A*, **39**, 10307–10321.

[76] Gerola, H.; Seiden, P. E.; and Schulman, L. S. (1980). Theory of dwarf galaxies. *Astrophys. J.*, **242**, 517–527.

[77] Gerstein, Mark (1997). A structural census of genomes: Comparing bacterial, eukaryotic, and archaeal genomes in terms of protein structure. *J. Mol. Biol.*, **274**, 562–576.

[78] Ghiradella, Helen and Schmidt, John T. (2004). Fireflies at one hundred plus: A new look at flash control. *Integr. Comp. Biol.*, **44**, 203–213.

[79] Glass, Leon and Mackey, Michael C. (1988). *From Clocks to Chaos: The Rhythms of Life*. Princeton Univ. Press, Princeton.

[80] Gold, Thomas (1967). *The Nature of Time*. Cornell Univ. Press, Ithaca, New York.

[81] Golfinopoulis, T. Critical opalescence. See `http://web.mit.edu/8.334/www/grades/projects/projects08/TheodoreGolfinopoulos/text3a.html` (accessed April 2020).

[82] Gotelli, Nicholas J. (1998). *A Primer of Ecology* (2nd edn). Sinauer, Sunderland, MA.

[83] Gould, James L. (1982). *Ethology*, ch.15. McLeod, Toronto.

[84] Granovsky, Boris L. and Madras, Neal (1995). The noisy voter model. *Stoch. Process. Their Appl.*, **55**, 23–43.

[85] Grassberger, Peter (2002). Critical behaviour of the Drossel-Schwabl forest fire model. *New J. Phys.*, **4**, 17.

[86] Gravish, Nick; Gold, Gregory; Zangwill, Andrew; Goodisman, Michael A. D.; and Goldman, Daniel I. (2015). Glass-like dynamics in confined and congested ant traffic. *Soft Matter*, **11**, 6552–6561.

[87] Gravish, Nick; Monaenkova, Daria; Goodisman, Michael A. D.; and Goldman, Daniel I. (2013). Climbing, falling and jamming during ant locomotion in confined environments. *Proc. Natl Acad. Sci. USA.*, **110**, 9746–9751.

[88] Greenwood, Veronique (2014). The thermodynamic theory of ecology. *Quanta*. September 3.

[89] Grégoire, Guillaume and Chaté, Hugues (2004). Onset of collective and cohesive motion. *Phys. Rev. Lett.*, **92**, 025702.

[90] Grégoire, Guillaume; Chaté, Hugues; and Tu, Yuhai (2003). Moving and staying together without a leader. *Physica D*, **181**, 157–170.

[91] Gregory, Stephan A. and Thompson, Laird A. (1982). Superclusters and voids in the distribution of galaxies. *Sci. Am.*, March, 88–96.

[92] Guggenheim, E. A. (1945). The principle of corresponding states. *J. Chem. Phys.*, **13**, 253–261.

[93] Gunter, Jen (2019). The myth of period syncing. *The New York Times*, June 9. Based on research by "Clue" reported in FIGO. See `https://www.figo.org/news/period-syncing-myth-debunked-0015541` (accessed October 24, 2019).

[94] Gutenberg, B. and Richter, C. F. (1956). Magnitude and energy of earthquakes. *Annali di Geofisica*, **9**, 1–15.

[95] Hahn, Gerald; Ponce-Alvarez, Adrian; Monier, Cyril; Benvenuti, Giacomo; Kumar, Arvind; Chavane, Frédéric; Deco, Gustavo; and Frégnac, Yves (2017). Spontaneous cortical activity is transiently poised close to criticality. *PLoS Comput. Biol.*, **13**(5), 1–29.

[96] Haldane, J. B. S. (1949). *What is Life? The Layman's View of Nature*, p. 248. Alcuin Press, London. Like many other of the quotations in this book, the reference may be apocryphal. I read (online, at the quoteinvestigator.com) a learned discussion—with citations—on the origin of this phrase. Apparently, according to Haldane, the Creator was also fond of stars.

[97] Hall, M.; Christensen, K.; di Collobiano, S. A.; and Jensen, H. J. (2002). Time-dependent extinction rate and species abundance in a Tangled-Nature model of biological evolution. *Phys. Rev. E*, **66**, 011904.

[98] Harmon, Dion; de Aguiar, Marcus A. M.; Chinellato, David D.; Braha, Dan; Epstein, Irving R.; and Bar-Yam, Yaneer (2011). Predicting economic market crises using measures of collective panic. Technical Report 2010-08-01, New England Complex Systems Institute, Cambridge, MA (https://arxiv.org/pdf/1102.2620.pdf).

[99] Harrison, Paul M. and Gerstein, Mark (2001). Studying genomes through the aeons: Protein families, pseudogenes and proteome evolution. *J. Mol. Biol.*, **318**, 673–681.

[100] Harte, John (2011). *Maximum Entropy and Ecology: A Theory of Abundance, Distribution, and Energetics*. Oxford Univ. Press, Oxford.

[101] Harte, John (2018). Maximum entropy and theory construction: A reply to Favretti. *Entropy*, **20**, 285.

[102] Harte, John; Conlisk, Erin; Ostling, Annette; Green, Jessica L.; and Smith, Adam B. (2005). A theory of spatial structure in ecological communities at multiple spatial scales. *Ecol. Monogr.*, **75**, 179–197.

[103] Harte, John; Smith, Adam B.; and Storch, David (2009). Biodiversity scales from plots to biomes with a universal species–area curve. *Ecol. Lett.*, **12**, 789–797.

[104] Harte, John; Zillio, T.; Conlisk, E.; and Smith, A. B. (2008). Maximum entropy and the state-variable approach to macroecology. *Ecology*, **89**, 2700–2711.

[105] Helbing, Dirk (2014). Pedestrian, crowd, and evacuation dynamics. See `http://webarchiv.ethz.ch/soms/teaching/CrowdstoCrisesHS14/pedestrians.pdf` (accessed August 2020).

[106] Helbing, Dirk; Farkas, Illés; and Vicsek, Tamás (2000). Simulating dynamical features of escape panic. *Nature*, **407**, 487–490.

[107] Hethcote, H. W. (2000). The mathematics of infectious deseases. *SIAM Rev.*, **42**, 599–653.

[108] Hoogendoorn, Serge (2013). Presentation at a Monash University conference on traffic. See `http://users.monash.edu.au/~mpetn/files/talks/Hoogendoorn.pdf` (accessed March 25, 2021).

[109] Horn, R. A. and Johnson, C. R. (1985). *Matrix Analysis*. Cambridge Univ. Press, New York.

[110] Hubbell, Stephen P. (2001). *The Unified Neutral Theory of Biodiversity and Biogeography*. Princeton Univ. Press, Princeton.

[111] Huntley, D. J. (2006). An explanation of the power-law decay of luminescence. *J. Phys. Condens. Matter*, **18**, 1359–1365.

[112] Huth, Andreas and Wissel, Christian (1992). The simulation of the movement of fish schools. *J. Theor. Biol.*, **156**, 365–385.

[113] Ispolatov, S.; Krapivsky, P. L.; and Redner, S. (1998). Wealth distributions in asset exchange models. *Euro. Phys. J. B*, **2**, 267–276.

[114] Jarzynski, C. (1997). Equilibrium free-energy differences from nonequilibrium measurements: A master-equation approach. *Phys. Rev. E*, **56**, 5018.

[115] Jarzynski, C. (1997). Nonequilibrium equality for free energy differences. *Phys. Rev. Lett.*, **78**, 2690.

[116] Jaynes, Edwin T. (1957). Information theory and statistical mechanics. *Phys. Rev.*, **106**, 620–630.

[117] Jaynes, Edwin T. (1957). Information theory and statistical mechanics II. *Phys. Rev.*, **108**, 171–190.

[118] Jiminez, Sergio (2015). Material taken from his work, appearing in the Wikipedia article on Zipf's law https://en.wikipedia.org/wiki/Zipf%27s_law, accessed April 2020. This is a plot of the rank versus frequency for the first 10 million words in 30 Wikipedias (dumps from October 2015) on a log-log scale.

[119] Joyce, M.; Sylos Labini, F.; Gabrielli, A.; Montuori, M.; and Pietronero, L. (2005). Basic properties of galaxy clustering in the light of recent results from the Sloan Digital Sky Survey. *Astron. Astrophys.*, **443**, 11–16.

[120] Kadanoff, Leo P. (2009). More is the same; Phase transitions and mean field theories. *J. Stat. Phys.*, **137**, 777–797.

[121] Kadanoff, Leo P. (2009). Phases of matter and phase transitions; from mean field theory to critical phenomena. See https://jfi.uchicago.edu/~leop/Rejected Papers/ExtraV1.2.pdf (accessed June 2020). If you're copying this reference, be sure the letters and symbols are the same as you see—pdf printing as well as some fonts do not copy well.

[122] Kaizoji, T. and Sornette, D. (2008). *Market Bubbles and Crashes.* (https://arxiv.org/abs/0812.2449) This is an extended version of a review for the *Encyclopedia of Quantitative Finance.*

[123] Katz, Amnon (1967). *Principles of Statistical Mechanics: The Information Theory Approach.* Freeman, San Francisco. Unfortunately this book may be difficult to get hold of, but the presentation of information theory is excellent.

[124] Kennickell, A. B. (2017). Modeling wealth with multiple observations of income: Redesign of the sample for the 2001 Survey of Consumer Finances. *Stat. J. IAOS*, **33**, 51–58.

[125] Kerner, Boris S. (1999). The physics of traffic. *Physics World*, **12**(8), 25–30.

[126] Kittel, C. (1996). *Introduction to Solid State Physics* (7th edn). Wiley, New York.

[127] Kluckhorn, Clyde (1950). Review of Zipf's book, [243]. *Am. Anthropol.*, **52**, 268–270.

[128] Koonin, Eugene V. (2011). Are there laws of genome evolution? *PLoS Comput. Biol.*, **7**, e1002173.

[129] Koonin, Eugene V.; Wolf, Yuri I.; and Karev, Georgy E. (2006). *Power Laws, Scale-Free Networks and Genome Biology.* Springer, New York.

[130] Krug, J. and Spohn, H. (1988). Universality classes for deterministic surface growth. *Phys. Rev. A*, **38**, 4271–4282.

[131] Krugman, Paul R. (1996). Confronting the mystery of urban hierarchy. *J. Jpn Int. Econ.*, **10**, 399–418.

[132] Krugman, Paul R. (1996). *The Self-Organizing Economy*, ch. 3. Blackwell, Cambridge, MA.

[133] Kuramoto, Y. (1975). Self-entrainment of a population of coupled non-linear oscillators. In *International Symposium on Mathematical Problems in Theoretical Physics* (ed. H. Araki), p. 420. Lecture Notes in Physics, vol 39. Springer, Berlin.

[134] Sylos Labini, F.; Montuori, M.; and Pietronero, L. (1998). Scale invariance of galaxy clustering. *Phys. Rep.*, **293**, 61–226.

[135] Landau, D. P. and Alben, R. (1973). Monte Carlo calculations as an aid in teaching statistical mechanics. *Am. J. Phys.*, **41**, 394.

[136] Lee, E. D.; Broedersz, C. P.; and Bialek, W. (2015). Statistical mechanics of the US Supreme Court. *J. Stat. Phys.*, **160**, 275–301.

[137] Lee, T. D. and Yang, C. N. (1952). Statistical theory of equations of state and phase transitions. I. Theory of condensation. *Phys. Rev.*, **87**, 404–409.

[138] Lee, T. D. and Yang, C. N. (1952). Statistical theory of equations of state and phase transitions. II. Lattice gas and Ising model. *Phys. Rev.*, **87**, 410–419.

[139] Lestrade, Sander (2017). Unzipping Zipf's law. *PLoS ONE*, **12**, e0181987.

[140] Levitin, Daniel J.; Chordia, Parag; and Menon, Vinod (2012). Musical rhythm spectra from Bach to Joplin obey a $1/f$ power law. *Proc. Natl Acad. Sci. USA*, **109**, 3716–3720.

[141] Lewis, Sara M. and Cratsley, Christopher K. (2008). Flash signal evolution, mate choice, and predation in fireflies. *Annu. Rev. Entomol.*, **53**, 293–321.

[142] Li, Jie; Boghosian, Bruce M.; and Li, Chengli (2019). The affine wealth model: An agent-based model of asset exchange that allows for negative-wealth agents and its empirical validation. *Physica A*, **516**, 423–442.

[143] Lin, Henry W. and Loeb, Abraham (2016). Zipf's law from scale-free geometry. *Phys. Rev. E*, **93**, 032306.

[144] Lipton, Alexander and Pentland, Alex (2018). Breaking the bank. *Sci. Am.*, **318**(1), 28–31.

[145] Lloyd, Seth (2001). Measures of complexity: a non-exhaustive list. *IEEE Mag.*, **21**, 7–8.

[146] Mandelbrot, Benoit (1962). On the theory of word frequencies and on related Markovian models of discourse. In *Structure of Language and its Mathematical Aspects. Proceedings of Symposia in Applied Mathematics. Vol. XII* (ed. R. Jakobson), pp. 190–219. American Mathematical Society, Providence, RI.

[147] Marcus, M. and Minc, H. (1992). *A Survey of Matrix Theory and Matrix Inequalities*. Dover, New York. Originally published in 1964 by Prindle, Weber & Schmidt, Boston.

[148] Maris, Humphrey J. and Kadanoff, Leo P. (1978). Teaching the renormalization group. *Am. J. Phys.*, **46**, 652–657.

[149] Mattuck, R. D. (1992). *A Guide to Feynman Diagrams in the Many-Body Problem* (2nd edn). Dover, New York.

[150] May, R. M. (1976). Simple mathematical models with very complicated dynamics. *Nature*, **261**(5560), 459–467.

[151] Metropolis, Nicholas; Rosenbluth, Arianna W.; Rosenbluth, Marshall N.; Teller, Augusta H.; and Teller, Edward (1953). Equation of state calculations by fast computing machines. *J. Chem. Phys.*, **21**, 1087.

[152] Mirollo, R. E. and Strogatz, S. H. (1990). Synchronization of pulse-coupled biological oscillators. *SIAM J. Appl. Math.*, **50**, 1645–1662.

[153] Mlodinow, Leonard (2008). *The Drunkard's Walk: How Randomness Rules Our Lives*. Pantheon, New York.

[154] Montemurro, Marcelo A. (2001). Beyond the Zipf-Mandelbrot law in quantitative linguistics. *Physica A*, **300**, 567–578.

[155] Montgomery, James (2020). Classroom notes (Soc(iology) 376) by Montgomery (University of Wisconsin, Madison). Ultimately this is intended as a book, but he has given me permission to use this material.

[156] Moreau, M.; Gaveau, B.; and Schulman, L. S. (2012). Efficiency of a thermodynamic motor at maximum power. *Phys. Rev. E*, **85**, 021129.

[157] Mountain, R. D. and Thirumalai, D. (1989). Measures of effective ergodic convergence in liquids. *J. Phys. Chem.*, **93**, 6975–6979.

[158] Moussaid, Mehdi; Helbing, Dirk; and Theraulaz, Guy (2011). How simple rules determine pedestrian behavior and crowd disasters. *Proc. Natl Acad. Sci. USA*, **108**, 6884–6888.

[159] Muldoon, Ryan; Smith, Tony; and Weisberg, Michael (2012). Segregation that no one seeks. *Philos. Sci.*, **79**, 38–62.

[160] Nagel, Kai and Schreckenberg, Michael (1992). A cellular automaton model for freeway traffic. *J. Phys. I EDP Sciences*, **2**, 2221–2229.

[161] Newburgh, Ronald; Peidle, Joseph; and Rueckner, Wolfgang (2006). Einstein, Perrin, and the reality of atoms: 1905 revisited. *Am. J. Phys.*, **74**, 478–481.

[162] Newman, C. M. and Schulman, L. S. (1986). One dimensional $1/|j-i|^s$ percolation models: The existence of a transition for $s \leq 2$. *Commun. Math. Phys.*, **104**, 547–571.

[163] Newman, M. E. J. (2005). Power laws, Pareto distributions and Zipf's law. *Contemp. Phys.*, **46**, 323–351.

[164] Nichols, Donald A. and Reynolds, Clark W. (1971). *Principles of Economics*, p. 465. Holt, Rinehard and Winston, New York.

[165] Nielsen, Søren Nors; Müller, Felix; Marques, Joao Carlos; Bastianoni, Simone; and Jørgensen, Sven Erik (2020). Thermodynamics in ecology: An introductory review. *Entropy*, **22**, 820.

[166] Nikl, M.; Boháček, P.; Mihóková, E.; Kobayashi, M.; Ishii, M.; Usuki, Y.; and Babin, V. (2000). Excitonic emission of scheelite tungstates AWO4 (A= Pb, Ca, Ba, Sr). *J. Lumin.*, **87**, 1136–1139.

[167] Nikl, M.; Mihóková, E.; Mareš, J. A.; Vedda, A.; and Martini, M. (2000). Traps and timing characteristics of LuAG: Ce^{3+} scintillator. *Physica Status Solidi A Appl. Res.*, **181**(1), R10–R12.

[168] Nikl, M. and Yoshikawa, A. (2015). Recent R&D trends in inorganic single-crystal scintillator materials for radiation detection. *Adv. Opt. Mat.*, **3**(4), 463–481.

[169] Oliveira, H. M. and Melo, L. V. (2015). Huygens synchronization of two clocks. *Sci. Rep.*, **5**, 11548.

[170] Olver, F. W. J. (1974). *Asymptotics and Special Functions*. Academic Press, New York.

[171] Onsager, Lars (1944). Crystal statistics. I. A two-dimensional model with an order-disorder transition. *Phys. Rev.*, **65**, 117–149.

[172] Paczuski, Maya; Maslov, Sergei; and Bak, Per (1996). Avalanche dynamics in evolution, growth, and depinning models. *Phys. Rev. E*, **53**, 414–443.

[173] Parker, Belinda S.; Rautela, Jai; and Hertzog, Paul J. (2016). Antitumour actions of interferons: implications for cancer therapy. *Nat. Rev. Cancer*, **16**, 131–144.

[174] Perlmutter, S.; Aldering, G.; Goldhaber, G.; Knop, R. A.; Nugent, P.; Castro, P. G.; Deustua, S.; Fabbro, S.; Goobar, A.; Groom, D. E.; Hook, I. M.; Kim, A. G.; Kim, M. Y.; Lee, J. C.; Nunes, N. J.; Pain, R.; Pennypacker, C. R.; Quimby, R.; Lidman, C.; Ellis, R. S.; Irwin, M.; McMahon, R. G.; P. Ruiz-Lapuente; Walton, N.; Schaefer, B.; Boyle, B. J. and A. V Filippenko and Matheson, T.; Fruchter, A. S.; Panagia, N.; H. J. M. Newberg and Couch, W. J. (1999). Measurements of ω and λ from 42 high-redshift supernovae. *Astrophys. J.*, **517**, 565–586.

[175] Peskin, Charles S. (1975). *Mathematical Aspects of Heart Physiology*. Courant Institute of Mathematical Sciences, New York. Lecture notes; see especially pp. 268–278.

[176] Piantadosi, Steven T. (2014). Zipf's word frequency law in natural language: a critical review and future directions. *Psychon. Bull. Rev.*, **21**, 1112–1130.

[177] Pietronero, L. and Vespignani, A. (1995). Fractals, self-organized-criticality and the fixed scale transformation. *Chaos Solitons Fractals*, **6**, 471–480.

[178] Pollakis, Alexandros; Wetzel, Lucas; Jörg, David J; Rave, Wolfgang; Fettweis, Gerhard; and Jülicher, Frank (2014). Synchronization in networks of mutually delay-coupled phase-locked loops. *New J. Phys.*, **16**, 113009.

[179] Pomeroy, Harold and Heppner, Frank (1992). Structure of turning in airborne rock dove (Columba livia) flocks. *Auk*, **109**, 256–267.

[180] Portugal, Steven J.; Hubel, Tatjana Y.; Fritz, Johannes; Heese, Stefanie; Trobe, Daniela; Voelkl, Bernhard; Hailes, Stephen; Wilson, Alan M.; and Usherwood, James R. (2014). Upwash exploitation and downwash avoidance by flap phasing in ibis formation flight. *Nature*, **505**, 399–404.

[181] Powers, David M. W. (1998). Applications and explanations of Zipf's law. In *New Methods in Language Processing and Computational Natural Language Learning* (ed. D. M. W. Powers), pp. 151–160. Association for Computational Linguistics, Stroudsburg, PA.

[182] Proverbs (1927). 6.6. The translation is from the Jewish Publication Society.

[183] Qian, Jiang; Luscombe, Nicholas M.; and Gerstein, Mark (2001). Protein family and fold occurrence in genomes: Power-law behaviour and evolutionary model. *J. Mol. Biol.*, **313**, 673–681.

[184] Redner, S. (1998). How popular is your paper? An empirical study of the citation distribution. *Euro. Phys. J. B*, **4**, 131–134.

[185] Reed, W. J. and Hughes, B. D. (2002). From gene families and genera to incomes and internet file sizes: Why power laws are so common in nature. *Phys. Rev. E*, **66**, 067103. As the title suggests, these authors deal with other power laws as well.

[186] Reynolds, Craig (1987). Flocks, herds and schools: A distributed behavioral model. *Comput. Graph.*, **21**, 25–34. SIGGRAPH '87.

[187] Richardson, Lewis F. (1961). The problem of contiguity: An appendix to statistics of deadly quarrels. *Soc. Gen. Syst. Res.*, **6**, 139–187.

[188] Roudi, Yasser; Tyrcha, Joanna; and Hertz, John (2009). Ising model for neural data: Model quality and approximate methods for extracting functional connectivity. *Phys. Rev. E*, **79**, 051915.

[189] Samorodnitsky, Gennady and Taqqu, Murad S. (1994). *Stable Non-Gaussian Random Processes*. Chapman & Hall, New York.

[190] Schelling, Thomas C. (1971). Dynamic models of segregation. *J. Math. Sociol.*, **1**, 143–186.

[191] Schneidman, Elad; Berry II, Michael J.; Segev, Ronen; and Bialek, William (2006). Weak pairwise correlations imply strongly correlated network states in a neural population. *Nature*, **440**, 1007–1012.

[192] Schneidman, Elad; Still, Susanne; Berry II, Michael J.; and Bialek, William (2003). Network information and connected correlations. *Phys. Rev. Lett.*, **91**, 238701.

[193] Schrödinger, Erwin (1944). *What is life?* Cambridge Univ. Press, Cambridge.

[194] Schulman, L. S. (1981). *Techniques and Applications of Path Integration*. Wiley, New York. Wiley Classics 1996; Dover 2005, with supplements.

[195] Schulman, L. S. (1997). *Time's Arrows and Quantum Measurement*. Cambridge Univ. Press, New York. See p. 1; also p. 14.

[196] Schulman, L. S. (2009). Source of the observed thermodynamic arrow. *J. Phys.: Conf. Series*, **174**, 012022. Proceedings of the conference, DICE2008, Castiglion-cello, Italy.

[197] Schulman, L. S. (2017). Bacterial resistance to antibiotics: a model evolutionary study. *J. Theor. Biol.*, **417**, 61–67.

[198] Schulman, L. S. (2019). The observable representation. *Entropy*, **21**, 310.

[199] Schulman, L. S. and Gaveau, B. (2003). Complex systems under stochastic dynamics. *Att. Fond. G. Ronchi*, **58**, 805–818. (https://arxiv.org/abs/cond-mat/0312711)

[200] Schulman, L. S.; Luck, J. M.; and Mehta, A. (2012). Spectral properties of zero temperature dynamics in a model of a compacting granular column. *J. Stat. Phys.*, **146**, 924–954.

[201] Schulman, L. S. and Seiden, P. E. (1978). Statistical mechanics of a dynamical system based on Conway's Game of Life. *J. Stat. Phys.*, **19**, 293–314.

[202] Schulman, L. S. and Seiden, P. E. (1982). Percolation analysis of stochastic models of galactic evolution. *J. Stat. Phys.*, **27**, 83–118.

[203] Schulman, L. S. and Seiden, P. E. (1986). Hierarchical structure in the distribution of galaxies. *Astrophys. J.*, **311**, 1–5.

[204] Schulman, L. S. and Seiden, P. E. (1986). Percolation and galaxies. *Science*, **233**, 425–431.

[205] Seabrook, J. (2002). The slow lane. *The New Yorker*, September 2.

[206] Seiden, P. E. and Schulman, L. S. (1990). Percolation model of galactic structure. *Adv. Phys.*, **39**, 1–54.

[207] Seiden, P. E.; Schulman, L. S.; and Elmegreen, B. G. (1984). A galactic disk is not a true exponential. *Astrophys. J.*, **282**, 95–100.

[208] Sellis, Diamantis and Almirantis, Yannis (2009). Power-laws in the genomic distribution of coding segments in several organisms: An evolutionary trace of segmental duplications, possible paleopolyploidy and gene loss. *Gene*, **447**, 18–28.

[209] Shannon, Claude E. (1948). A mathematical theory of communication. *Bell System Tech. J.*, **27**, 379–423.

[210] Shew, Woodrow L. and Plenz, Dietmar (2013). The functional benefits of criticality in the cortex. *Neuroscientist*, **19**(1), 88–100.

[211] Silk, Joseph; Szalay, Alexander S.; and Zel'dovich, Yakov B. (1983). The large-scale structure of the universe. *Sci. Am.*, **249**, 72–80.

[212] Simon, H. A. (1955). On a class of skew distribution functions. *Biometrika*, **42**, 425–440.

[213] Sorbaro, Martino; Herrmann, J. Michael; and Hennig, Matthias (2018). Statistical models of neural activity, criticality, and Zipf's law. (https://arxiv.org/abs/1812.09123v1 [q–bio.NC], accessed August 2021). Originally prepared as a book chapter for the volume *The Functional Role of Critical Dynamics in Neural Systems*, edited by Udo Ernst, Nergis Tomen and Michael Herrmann.

[214] Sornette, Didier. There is a newsletter published by the "Chair of Entrepreneurial Risks" (which is Sornette's chair in ETH, Zurich). See `http://www.er.ethz.ch` (accessed March 2020), for criteria related to bubbles.

[215] Sornette, Didier (2003). *Why Stock Markets Crash: Critical Events in Complex Financial Systems*. Princeton University Press, Princeton.

[216] Sornette, Didier and Cauwels, Peter. (2014). Financial bubbles: mechanisms and diagnostics. Notenstein white paper series, 2014. (https://arxiv.org/abs/1404.2140 [q-fin.RM], accessed August 2021).

[217] Spaeth, H. J.; Epstein, L.; Ruger, T. W.; Whittington, K.; Segal, J. A.; and Martin, A. D. (2011). The Supreme Court database. See `http://scdb.wustl.edu/index.php` (accessed September 2020).

[218] Strogatz, S. H. (1994). *Nonlinear Dynamics and Chaos*, pp. 103–106. Perseus Books, Reading, MA.

[219] Stumpf, Michael P. H. and Porter, Mason A. (2012). Critical truths about power laws. *Science*, **335**, 665–666.

[220] Suzuki, Ryuji; Buck, John R.; and Tyack, Peter L. (2004). The use of Zipf's law in animal communication analysis. *Anim. Behav.*, **69**, F9–F17.

[221] Swadesh, M. (1952). Lexicostatistic dating of prehistoric ethnic contacts. *Proc. Am. Phil. Soc.*, **96**, 452–463.

[222] Tadic, Bosiljka; Dankulov, Marija Mitrovic; and Melnik, Roderick (2017). Mechanisms of self-organized criticality in social processes of knowledge creation. *Phys. Rev. E*, **96**, 032307.

[223] Toner, John and Tu, Yuhai (1995). Long-range order in a two-dimensional dynamical XY model: How birds fly together. *Phys. Rev. Lett.*, **75**, 4326–4329.

[224] Traub, R. and Miles, R. (1991). *Neuronal Networks of the Hippocampus*. Cambridge Univ. Press, New York.

[225] Ueda, Hiroki R.; Hayashi, Satoko; Matsuyama, Shinichi; Yomo, Tetsuya; Hashimoto, Seiichi; Kay, Steve A.; Hogenesch, John B.; and Iino, Masamitsu

(2004). Universality and flexibility in gene expression from bacteria to human. *Proc. Natl Acad. Sci. USA*, **101**, 3765–3769.

[226] van Kampen, N. G. (1981). *Stochastic Processes in Physics and Chemistry.* North-Holland (Elsevier), Amsterdam.

[227] Vicsek, Tamás; Czirók, Andras; Ben-Jacob, Eshel; Cohen, Inon; and Shochet, Ofer (1995). Novel type of phase transition in a system of self-driven particles. *Phys. Rev. Lett.*, **75**, 1226–1229.

[228] Vicsek, Tamás and Zafeiris, Anna (2012). Collective motion. *Phys. Rep.*, **517**, 71–140.

[229] Voss, Richard F. and Clarke, John (1975). $1/f$ noise in music and speech. *Nature*, **258**, 317–318.

[230] Voss, Richard F. and Clarke, John (1978). $1/f$ noise in music: Music from $1/f$ noise. *J. Acoust. Soc. Am.*, **63**, 258.

[231] West, Geoffrey (2017). *Scale: The Universal Laws of Growth, Innovation, Sustainability, and the Pace of Life in Organisms, Cities, Economies, and Companies.* Penguin Press, London.

[232] Wetzel, Lucas; Jörg, David J.; Pollakis, Alexandros; Rave, Wolfgang; Fettweis, Gerhard; and Jülicher, Frank (2017). Self-organized synchronization of digital phase-locked loops with delayed coupling in theory and experiment. *PLoS ONE*, **12**, e0171590.

[233] Wilber, Mark Q.; Kitzes, Justin; and Harte, John (2015). Scale collapse and the emergence of the power law species-area relationship. *Global Ecol. Biogeogr.*, **24**, 883–895.

[234] Winfree, A. T. (1967). Biological rhythms and the behavior of populations of coupled oscillators. *J. Theoret. Biol.*, **16**, 15–42.

[235] Winslow, Nathan (1992). Introduction to self-organized criticality & earthquakes. Available as course notes from the University of Chieti, Pescare, Italy, at `ftp://ftp.rm.ingv.it/pro/terrasol/materiale_consultazione/SOCEQ.pdf`. If you have trouble reaching that site, look at `http://www2.econ.iastate.edu/classes/econ308/tesfatsion/SandpileCA.Winslow97.htm` (accessed September 2020).

[236] Wosniack, M. E.; Santos, M. C.; Raposo, E. P.; Viswanathan, G. M.; da Luz, M. G. E.; and Schulman, L. S. (2019). Identifying dynamical structures in the physical space of stochastic processes. *Euro. Phys. Lett.*, **125**, 20004.

[237] Young, H. Peyton (1998). *Individual Strategy and Social Structure.* Princeton Univ. Press, Princeton.

[238] Yu, Shuiyuan; Xu, Chunshan; and Liu, Haitao (2018). Zipf's law in 50 languages: its structural pattern, linguistic interpretation, and cognitive motivation. (https://arxiv.org/abs/1807.01855 [cs.CL], accessed August 2021).

[239] Yule, G. U. (1925). II.—A mathematical theory of evolution, based on the conclusions of Dr. J. C. Willis, F.R.S. *Phil. Trans. R. Soc. Lond. B*, **213**, 402–410.

[240] Yvon, J. (1955). in Proceedings of the International Conference on Peaceful Uses of Atomic Energy, Geneva (unpublished), p. 387.

[241] Zachary, W. W. (1977). An information flow model for conflict and fission in small groups. *J. Anthropol. Res.*, **33**, 452–473.

[242] Zhang, Junfu (2004). A dynamic model of residential segregation. *J. Math. Sociol.*, **28**, 147–170.

[243] Zipf, G. (1936). *The Psychobiology of Language*. Routledge, London.

[244] Zipf, George Kingsley (1949). *Human Behavior and the Principle of Least Effort: An Introduction to Human Ecology*. Addison-Wesley, Cambridge, MA.

Index

1/f noise, 77, 238

Accelerated expansion, 185
Acebrón, JA, 164
Advani, M, 68
Agent based simulation, 13, 18, 106, 112,
        113, 122–124, 131, 135, 137, 138,
        146, 150, 151, 156, 188
  Integrate and fire, 145
Aguilar, J, 148
Ahlborn, B, 236
Alben, R, 54
Albert, R, 82
Albert-Barabási algorithm, 82
Aldering, G, 186
Almirantis, Y, 77, 152
Alpher, RA, 185
Amit, DJ, 68, 177
Analytic calculation, 12, 15, 26, 106, 130,
        164, 189, 229
Analyticity, 23, 24, 114, 244, 245
Anderson, PW, 1
Antennate, 148
Antibiotics, 153, 155
  Resistance, 4, 156
Ants, 147–151, 160
  Idleness, 148
Aoki, I, 157
Appendix A Notation, 196
Approximation
  e^u cutoff, 39
  Asymptotic, 217
  Stirling, 56, 114, 216
Arrow of time, 55, 185, 200
Arshad, S, 121
Ashraf, BN, 121
Asymptotics, 81, 82, 216, 217
Attanasi, A, 159
Auerbach, F, 76, 115, 119
Avalanche, 86, 178
Avron, JE, 201

Babin, V, 180
Baierlein, R, 201, 221
Bain, R, 79
Bak, P, 34, 86, 88, 178
Ballistic motion, 159
Bar-Yam, Y, 112, 129
Barabási, A-L, 82

Barger, VD, 188
Barmpalias, G, 124
Barycentric coordinates, 233
Batty, M, 238
Beggs, J, 178
ben Abraham, L, 104
Ben-Jacob, E, 157, 158
Bennett, M, 146
Benvenuti, G, 176
Berman, A, 216
Berry II, MJ, 68, 71
Betterton, MD, 148
Bialek, W, 68, 71, 162
Biham, O, 135, 136
Biochemistry, 141
Biological sciences, 140
Biorobotics, 147–151
Black body radiation, 8, 185
Bloom, H, 156
Boghosian, B, 107–109
Boháček, P, 180
Bohr, N, 184
Boltzmann
  Constant, 6, 53, 197, 200, 221
  Distribution, 51, 54, 93, 201, 202, 215, 248
  H-theorem, 198, 200
Boltzmann, L, 53, 198
Bonilla, LL, 164
Bonner, WB, 186
Boolean variable, 29, 32
Boyle, BJ, 186
Braess's paradox, 133, 254
Braess, D, 133
Braha, D, 112, 129
Bricker, J, 109
Broedersz, CP, 71
Brown, GE, 1
Brownian motion, 198, 229–231
Buck, JB, 141
Buck, JR, 79
Burridge, R, 88

Calude, AS, 117
Canonical ensemble, 201
Cantor set, 204, 205, 255
Cardinality, 255
Carlson,JM, 4
Carter, B, 170